An Introduction to Human F
A Beta Vers

MW00934541

John D. Lee
University of Wisconsin-Madison

Christopher D. Wickens
Colorado State University

Yili Liu
University of Michigan

Linda Ng Boyle
University of Washington

January 25, 2017

Acknowledgments

We thank all of the following for their helpful comments on earlier drafts of this book.

Kristi Bauerly
Lauren Chiang
Erin Chiou
John Campbell
Johan Engström
Donald Fisher
Mahtab Ghazizadeh
Lorelie Grepo
Briana Krueger
Ja Young Lee
Xiaoxia Lu
Erika Miller
Tyler Parbs
Barb Sweet
Vindhya Venkatraman
Trent Victor
Matt Wilson

The page layout and formatting were implemented in LaTex based on a style developed by J. Peatross and M. Ware: http:// optics.byu.edu /CLSFile.aspx

Preface

This book is a beta version of the third edition of "*An Introduction to Human Factors Engineering*." It contains the first 11 of 18 chapters of that book. This volume is not a final version, but a prototype that we hope to improve with feedback. We are particularly interested to receive input from our primary audience— undergraduate students who are seeing the discipline of human factors engineering for the first time. Based on the input we gather we will revise and publish the full version. Comments can be sent to: jdlee@engr.wisc.edu

Contents

I   Cognitive Considerations   79

Chapter 1

Introduction

At the end of this chapter you will be able to...

1. prioritize the human factors design goals for high-risk systems, the workplace, and consumer products

2. describe how the understand, create, and evaluate elements of the human factors design cycle address the cognitive, physical, and organizational aspects of design

3. discuss the order of application of the six general human factors interventions that is typically most effective

4. explain the scope of human factors engineering in terms of application domains, interventions, and related disciplines

5. explain why technology that doesn't work for people doesn't work

6. explain why intuition is insufficient in designing for people

Figure 1.1 Moving pig iron at a factory similar to Bethlehem Steel. (Photo of Coltness Iron Co., Scotland, 1910)

As a new manager at Bethlehem Steel, Fred Taylor was confronted with a problem: how to move pig iron out of storage faster. Taylor studied workers moving pig iron in detail, recording the time required for each motion and the amount of effort people could expend over the day. This detailed study made it possible to select the one in eight workers suited to the task and to specify the rest periods needed for the workers' muscles to recover. Enforcing these rest cycles improved capacity well beyond the rest periods that workers had chosen for themselves, as in Figure 1.1. By identifying the one best way to move pig iron, Taylor increased the amount of iron a worker could move from an average of 12.5 tons per day to 47 tons per day [1].

A B-25 C fully loaded with bombs and fuel was taking off from a 2500-foot strip with trees at both ends. The pilots recounted: "We crossed the end of the runway at an altitude of 2 feet and were pulling up over the trees shortly ahead when I gave the wheels up signal. The airplane mushed and continued to brush the tree tops at a constant 125 mph speed with T.O. [Take Off] power." The co-pilot had pulled up the flaps instead of the wheels, almost causing a deadly crash. Paul Fitts and Richard Jones reviewed 460 such mishaps to identify design changes to prevent future mishaps, saving thousands of lives [2].

In 2007, Apple debuted a revolutionary design when it released the iPhone to compete with an already crowded market dominated by Blackberry. How did Blackberry go from having almost half of the smartphone market in 2010 to less than 1% in 2015? "By all rights the product should have failed, but it did not," said David Yach, [Research in Motion] RIM's chief technology officer. To Mr. Yach and other senior RIM executives, Apple changed the competitive landscape by shifting the raison d'être of smartphones from something that was functional to a product that was beautiful. 'I learned that beauty matters....RIM was caught incredulous that people wanted to buy this thing,' Mr. Yach says.[3]" Blackberry ceased production in 2016, and Apple transformed the smartphone market with an intense focus on designing a beautiful product that provided an integrated experience.

1.1   What is Human Factors Engineering?

Human Factors Engineering makes technology work for people.

Human factors engineering aims to make technology work for people. This aim is very broad as shown by the three vignettes at the beginning of the chapter. These three vignettes also highlight the history of the field, with early developments focused on improving workplace productivity and efficiency. Taylor, the father of *"Scientific Management"*, introduced time studies and the scientific study of work in the early 1900s. He focused on increasing productivity, but not safety or satisfaction. During the Second World War more pilots died because of human error, as in the second vignette, than died in combat. Fitts and Jones studied these errors and identified the importance of designing for safe operations, which accelerated the growth of human factors engineering so that the field has become an important part of the Department of Defense and NASA. Blackberry's downfall underscores the importance of user satisfaction. The Apple design team demonstrated the importance of making products aesthetically pleasing and usable. The advent of the graphical user interface and the internet have made computers a part of billions of lives and the value of creating products that satisfy and delight has become central to the success of many companies. The historical development of human factors engineering shows the importance of considering people in design and how neglecting human involvement invites disaster.

At 9:30 AM, the morning of July 5, 2006, a patient suffering from a strep infection arrived at St Mary's hospital in Madison, Wisconsin to deliver her baby. The patient was concerned about pain during delivery and so the nurse retrieved a bag of epidural pain medication from the dispensary down the hall. Epidural pain medication must be delivered to the space between the spine and the spinal cord, and is deadly if delivered into the bloodstream. As the nurse returned to the patient's room, another nurse handed her a bag of intravenous penicillin to treat the patient's strep infection. She connected the bag to the infusion pump, which began the flow of medication into the patient. Within minutes, the patient fell into cardiovascular collapse, and despite nearly an hour and a half of resuscitation efforts, the patient died. The nurse had confused similar-looking bags, and was able to connect the intravenous tubing to the epidural bag that allowed the epidural medication to flow through the infusion pump. Factors that contributed to the confusion included production pressure to prepare the room in advance of the anesthesiologist arriving to deliver the epidural medication, 16 hours of work the previous day, and distractions from family members. She was arrested and charged with a felony for her error [4].

While the first three vignettes highlight the positive contributions of human factors engineering, this fourth vignette highlights what can happen when human factors engineering is neglected. These four vignettes illustrate the importance of human factors engineering. All systems include people and meeting their needs is the end goal of engineers and designers—*if a system doesn't work for people, it doesn't work*—human factors engineering makes technology work for people [5]. These vignettes illustrate the benefits of tailoring technology to fit the capabilities and needs of people and how neglecting these considerations can lead to problems. When things go wrong, as in the tragedy at St Mary's hospital, people often call for a diagnosis and solution. Understanding these situations— rather than attributing the cause to human error and blaming the nurse—represents an important contribution of human factors to system design. Human factors engineering can also identify unrecognized needs that can help avoid such mishaps and even delight customers. *Human factors engineering* is a discipline that considers the cognitive, physical, and organizational influences on human behavior to improve human interaction with products and processes.

Human error is a symptom of a poor design.

1.2　Goals and Process of Human Factors Engineering

Human factors engineering improves people's lives by making technology work well for them. Most broadly, human factors engineering aims to improve human interaction with systems by enhancing:

- Safety: Reducing the risk of injury and death

- Performance: Increasing productivity, quality, and efficiency

- Satisfaction: Increasing acceptance, comfort, and well-being

In considering these goals, safety is always a critical concern, but the relative emphasis of each goal depends on the particular area of application: high-risk systems, such as a B787 cockpit; workplace design, such as a manufacturing assembly line; consumer products, such as a iPhone. Design of high-risk systems must focus on *safety*. In contrast, design of workplaces focuses more on *performance*, and design of consumer products focuses more on *satisfaction*. Figure 1.2 shows the relative emphasis of each of these goals for each application area.

There are clearly tradeoffs among these goals: it is not always possible to maximize both safety and performance. For example, performance is an all-encompassing term that may involve increasing the speed of production. Increasing speed may cause people to rush through assembly and inspections, which can lead to more operator errors and undermine safety. As another example, some companies may cut corners on time-consuming safety procedures to meet productivity goals.

Fortunately, good human factors designs can avoid these tradeoffs. Human factors interventions often can satisfy safety, performance, and satisfaction simultaneously. For example, one company that improved its workstation design reduced workers compensation losses from $400,000 to $94,000 [6]. Workers were able

Figure 1.2 The goals of human factors engineering and application areas. The lengths of the lines indicate the relative emphasis of each human factors goal.

to work more (increasing performance), while greatly reducing the risk of injury (increasing safety), and increasing their engagement with work (satisfaction).

The three goals of human factors are accomplished through the human factors design cycle, shown in Figure 1.3. The design cycle begins with understanding the people and system they interact with, proceeds with creating a solution, and completes with evaluating how well the solution achieves the human factors goals. The outcome of this evaluation becomes an input to the cycle because it typically leads to a deeper understanding of what people need and identifies additional opportunities for improvement. Because designs are imperfect and people adapt to designs in unanticipated ways, the design process is iterative, repeating until a satisfactory design emerges. The design cycle repeats many times as a product or process is developed and continues even after the first version is released.

Understanding the people and system includes understanding the opportunities for a new product or process, or problems with existing systems. Most fundamentally, this understanding identifies a need that a design can satisfy. Understanding what people need must be coupled with an understanding of the cognitive, physical, and organizational issues involved. As an example, a human factors engineer would combine an analysis of the events that led up to the tragedy that occurred at St Mary's with an understanding of principles of communication (Chapter 6), decision making (Chapter 7), display design (Chapter 8), and performance degradation under stress (Chapter 15) to provide a more complete understanding of the causes of the St Mary's tragedy and offer recommendations for improvement.

This book contains four main sections. focuses on methods for understanding people's needs and evaluating whether those needs are met. Figure 1.4 shows understanding at the start of the process and evaluation at the end. Chapter 2 describes specific methods for understanding people's needs, such as observation and task analysis. Chapter 3 discusses specific methods for evaluating systems, such as heuristic evaluation and usability testing. The center of Figure 1.4 shows six human factors design interventions can create a design solution [7]. Using these interventions depends on a core knowledge of the nature of the physical body (its size, shape, and strength), the mind (its information-processing characteristics and limitations), and the social interactions as part of teams or larger organizations. The least effective of these design interventions are selection and training: design should fit technology to person rather than fit the person to the technology. In fact, design should strive to accommodate all people.

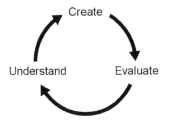

Figure 1.3 Understand, create, evaluate design cycle.

ᵗᵣ∧Consider training and selection after other interventions.

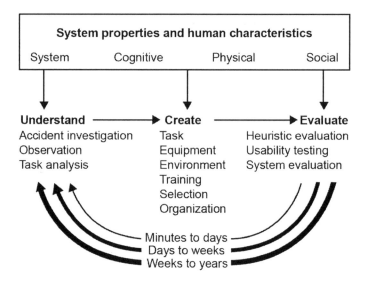

Figure 1.4 The human factors design cycle informed by human cognitive, physical and organizational characteristics and system properties. The process of understanding, creating and evaluating is repeated across days, weeks and years of a system's lifecycle.

Task design focuses more on changing what operators do than on changing the devices they use. A workstation for an assembly-line worker might be redesigned to eliminate manual lifting. Task design may involve assigning part or all of tasks to other workers or to automated components. For example, a robot might be designed to lift the component.

Equipment design changes the physical equipment that people work with. Apple's design of the iPhone hardware and software demonstrates how important a focus on equipment design can be to a product's success. The tubing that allowed the epidural bag to be connected to the intravenous pump in the St Mary's tragedy could be redesigned to prevent medication errors.

Environmental design changes the physical environment where the tasks are carried out. This can include improved lighting, temperature control, and reduced noise. Noise attenuating headsets can improve communication in a noisy cockpit.

Training enhances the knowledge and skills of people by preparing them for the job environment. This includes teaching and practicing the necessary physical or mental skills. Training is most applicable when there will be many repetitions of a task or a long involvement with the job. Periodic training is also important for those tasks that are rare, but where performance is critical, such as fire drills and emergency first aid.

Selection changes the makeup of the team or organization by picking people that are best suited to the job. Just as jobs differ in many ways, people differ from each other along almost every physical and mental dimension. Performance can be improved by selecting operators who have the best set of characteristics for the job. In our opening vignette, Taylor carefully selected the one person in eight that could move hundreds of 92-pound pieces of pig iron each day.

Team and organization design changes how groups of people communicate and relate to each other, and provides a broad view that includes the organizational climate within which the work is performed. This might, for example, represent a change in management structure to allow workers more participation in implementing safety programs. Workers are healthier, happier, and more productive if they can control their work, which directly contradicts Taylor's approach of identifying the one best way of doing a task. We return to this discussion in Chapter 18.

These six human factors design interventions show that design goes well beyond the interface and the objects that people might see and touch. Design includes redefining tasks, interaction, and overall environment. The opening vignette described how the RIM executives belatedly discovered the importance of beauty, but what was not mentioned was the additional value of Apple's *interaction design*, which made it possible for people to "touch" information on the iPhone screen; Apple's organizational design also enabled thousands of people outside Apple to develop apps for the iPhone.

Historically, the role of human factors engineering has often focused on the evaluation, such as usability testing, which is performed after the design is complete. This role is consistent with our discussion of fixing problems, such as those associated with St. Mary's hospital. The role human factors engineering plays is just as relevant to designing systems that are effective and that avoid problems. Thus, the role of human factors in the design cycle can just as easily enter at the point of understanding people's needs rather than simply evaluating system design. Human factors engineers should be problem solvers as the design develops, not just problem finders and design fixers after the design is complete.

Considering human factors early in the design process can save considerable money and possibly human suffering. For example, early attention to equipment design could have prevented the medication error at St Mary's and many other similar errors that occur daily at other hospitals. The percentage cost to an organization of incorporating human factors in design grows from 1 percent of the total product cost when human factors is addressed at the earliest stages to more than 12 percent when human factors is addressed only in response to problems, after a product is in the manufacturing stage[8]. Ideally, the understand, create, and evaluate cycle would focus on early and rapid iterations shown in the center of Figure 1.4, and proceed to usability testing and overall system evaluation only after considerable attention to people's needs and capabilities. In Chapter 2 we talk in detail about the role of human factors in the design process.

1.3 Scope of Human Factors Engineering

Although the field of human factors engineering originally grew out of a fairly narrow concern for human interaction with physical

Human Factors desig cludes the design o face, interaction, e× and organization.

Intervention	High-risk Aircraft cockpits, nuclear power plants, cars	Workplace Manufacturing lines, office workstations, cars	Consumer products Websites, games, smartphones, cars
Task Chapters 2, 10, 11	●	●	●
Equipment Chapters 2, 12, 13	●	●	●
Environment Chapters 15, 18	●	●	○
Training Chapter 17	●	◑	○
Selection Chapter 17	●	◑	○
Organization Chapter 18	●	○	○

Table 1.1 This matrix of human factors interventions and application areas where their application is most central (●) moderately central (◑) and less central (○).

The priority of human goals and interventions depends on the application area.

devices (usually military or industrial), its scope has broadened greatly during the last few decades. Human factors engineering is not just concerned about making work more productive and safer, but also with improving the routines of daily life, such as cooking, and making the most out of leisure time.

The range of human factors applications leads to a huge range of career options. Options include working for software and computer companies in positions described as usability engineers or user experience designers. Human factors engineers also work to create safer workplaces in almost every large company by ensuring that offices are configured in an ergonomic fashion and manufacturing processes are safe. Human factors engineers also work for government agencies using human factors research to write regulations and guide industries towards safer and more efficient practices, such as in design of medical devices, cars, roads, tax forms, and aircraft. Human factors engineers also work in consulting and research companies conducting studies to understand how technology affects human behavior.

Table 1.1 shows one way of understanding the roles of human factors professionals. Across the top of the matrix are major types of systems that human factors engineers aim to improve. Major categories include high-risk environments, the workplace, and consumer products. Consumer products include watches, cameras, games, and smartphones. The workplace includes manufacturing plants, customer service, assembly lines, and office work. High-risk environments include nuclear power plants, chemical processes, and aircraft cockpits. Some products and processes cut across multiple categories, such as cars. Cars are consumer products and so satisfaction is a critical design goal; they are also central for the work of many people, such as taxi drivers, and so performance is an important consideration, and cars are certainly part of a high-risk environment, where mishaps can cause severe injuries and deaths.

As we discussed earlier, these application areas often imply different priorities for human factors goals. These application areas also imply different priorities for human factors interventions. For example, selection and training are an important part of high-risk environments like aviation, but not so with most consumer products. The rows of Table 1.1 indicate the human factors approaches and the cells contain symbols that indicate the relative emphasis of the intervention for each category. With consumer products, the focus tends to be the device and task, but with the workplace, environmental design (e.g., lighting and temperature) is more relevant. With high-risk systems, such as aviation, training, selection, and team design are critical; however, good task and equipment design can minimize or eliminate the need for training and selection, and so these human factors approaches should only be considered after the others. Table 1.1 also highlights the prominence of equipment and task design—all categories of human factors product and process design should minimize the need for training and selection through careful equipment and task design.

A second way of looking at the scope of human factors is to consider the relationship of the discipline with other related domains of science and engineering, as shown in Figure 1.5. Items within the figure are placed close to other items to which they are related. The core discipline of human factors is shown at the center of the circle, and immediately surrounding it are various subdomains of study within human factors. Moving from the top to the bottom of this figure implies a shift of emphasis from the individual to team and organization. Moving from the left to the right implies a shift of emphasis from cognitive considerations to physical considerations. The six closely related human factors disciplines are shown as circles within the broad umbrella of human factors. Finally, outside of the circle are other disciplines that are likely to overlap with some aspects of human factors, particularly as members of a design team.

Fields closely related to human factors engineering include engineering psychology, ergonomics, human-systems integration, macro ergonomics, cognitive engineering, and human-computer interaction. Historically, the study of ergonomics has focused on the aspect of human factors related to the workplace, particularly physical work: lifting, reaching, stress, and fatigue. This discipline is closely related to aspects of human physiology, hence its closeness to the study of anatomy, physiology and biomedical engineering. Ergonomics has also been the preferred label in Europe to describe all aspects of human factors. However, in practice the domains of human factors and ergonomics have been sufficiently blended on both sides of the Atlantic so that the distinction is often not maintained.

Engineering psychology is a discipline within psychology, whereas the study of human factors is a discipline within engineering. The distinction is clear: The ultimate goal of the study of human factors

Lillian Gilbreth (1850-1946) Designer of the modern kitchen and pioneering industrial engineer. "The idea that housework is work now seems like a commonplace. We contract it out to housekeepers, laundromats, cleaning services, takeout places. We divvy it up: You cooked dinner, I'll do the dishes. We count it as a second shift, as well as primary employment. But it wasn't until the early part of the 20th century that first a literature, and then a science, developed about the best way to cook and clean. The results of this research shape the way we treat housework today, and created a template for the kitchen that remains conceptually unchanged from the 1920s [9]. Photograph by Theodor Horydczak/Library of Congress, Prints & Photographs Division, Theodor Horydczak Collection, [reproduction number, LC-H814-T-2474-020 (interpositive)]

is system design, accounting for those factors, psychological and physical, that are properties of the human component. In contrast, the ultimate goal of engineering psychology is to understand the human mind as relates to design [10].

Cognitive engineering, also closely related to human factors, but focuses on the cognitive considerations, particularly in the context of safety of complex systems, such as nuclear power plant [11, 12]. It focuses on how organizations and individuals manage such systems with the aid of sophisticated displays, decision aids, and automation, which is the focus of Chapters 7 and 11.

Macro ergonomics, like cognitive engineering, takes complex systems as its focus. Macro ergonomics addresses the need to consider not just the details of particular devices or processes, but the need to consider the overall work system. Macro ergonomics takes a broad systems perspective to design not just devices, but teams and organizations, which is the focus of Chapters 16 through 18.

Human-systems integration takes an even broader view and considers how designs must consider how people interact with all systems, to the point of forecasting availability of qualified staff based on demographic trends and training requirements.

Human-computer interaction (HCI), is often linked to the broader field of user experience and tends to focus more on software and less on the physical and organizational environment. Computers already touch many aspects of life, and the internet of things and wearable computers make it likely that we will soon

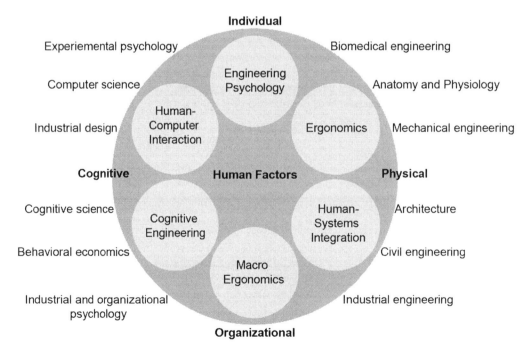

Figure 1.5 The domains of human factors.

be in nearly continuous interaction with computers. As a consequence, human-computer interaction and user experience increasingly overlap with other areas of human factors engineering. For example, as computers have been transformed from desktop machines to devices that are held in your hand or worn on your wrist, the physical aspects of reach and touch have become critically important.

1.4 Scientific Base of Human Factors Engineering

Unlike other system components, engineers and designers do not require specialized training to have some intuition for the human component of a design. Unfortunately, this intuition is often based on common sense and life experiences, and is not always a good indicator of how systems should be designed. Intuitions fail because people are not aware of how their minds and bodies operate: expectations change what people see, attention makes people blind to events that happen right in front of them, and default settings often make decisions for people. Intuition also fails to guide design because designers often differ substantially from the end users: they have different needs, priorities, and preferences. Designers might also have deep familiarity with technology, such as a computer mouse, that leads to *learned intuition* that might not be shared by those unfamiliar with computer technology, such as an 85-year-old who has never used a computer. Because preference for particular design features sometimes fails to support the best performance, even if designers can sense people's preferences, it might not lead to the best design. The scientific base of human factors engineering addresses the limits of intuition and provides a solid basis for design.

Intuition is often a poor guide for design.

The scientific base of human factors engineering also makes it possible to link human characteristics to engineering specifications. How bright lighting needs to be for efficient reading, how loud alarms need to be to capture attention, and how much a person can safely lift among many other design considerations have been quantified in units (e.g., lumens, dB, kilograms) that can guide design in a way that intuition cannot. This quantification is one instance of the more general ability of the human factors science base to support design [13]. In the problem understanding phase, investigators wish to *generalize* across classes of problems that may have common elements. As an example, the problems of communications between an air traffic control center and the aircraft may have the same elements as the communications problems between workers on a noisy factory floor or between doctors and nurses in an emergency room, thus enabling similar solutions to be applied to all three cases. Such generalization is more effective when it is based on a deep understanding of the physical and mental components of the human operator. It also is important to be able to *predict* that solutions designed to create good human factors will actually succeed when put into practice.

1.5   Overview of the Book

The forthcoming chapters are divided into four sections. In Chapters 2 and 3 we describe for design and evaluation methods. The second section addresses cognitive characteristics of people and their implications for design: visual and auditory and other sensory systems (Chapters 4 and 5), cognition (Chapter 6), macro cognition and decision making (Chapter 7), followed by the application of this information to display (Chapter 8), and control design (Chapter 9), as well as to human computer interaction (Chapter 10) and automation and increasingly autonomous systems (Chapter 11). The third section addresses many of the physical characteristics and their implications for design: workspace layout (Chapter 12), materials handling (Chapter 13), physiology (Chapter 14), and stress (Chapter 15). The final section focuses on organizational characteristics and their implications for design: safety (Chapter 16), training and selection (Chapter 17), and group and organizational performance (Chapter 18).

Human factors design includes cognitive, physical, and organizational considerations.

Additional Resources

Several journals address human factors issues that may be of interest to readers. These journals provide more depth on the theory and applications introduced in this book. Some recommendations include: *Ergonomics, Human Factors, Ergonomics in Design, Computer Human Interaction (CHI)* conference proceedings, and *Accident Analysis and Prevention.*

Several books cover similar material as this book: Sanders and McCormick [14] and Proctor and Van Zandt [15] offer comprehensive coverage of human factors. Norman [16] examines human factors manifestations in the kinds of consumer systems that most of us encounter every day.

1. Sanders, M. S., McCormick, E. J., & Sanders, S. (1993). *Human Factors in Engineering and Design.* New York: McGraw-Hill.

2. Proctor, R., and Van Zandt, T. (2008). *Human Factors in Simple and Complex Systems.* Boca Raton, FL: Taylor & Francis.

3. Norman, D. (2013). *The Design of Everyday Things: Revised and Expanded Edition.* Basic Books.

At a more advanced level, Wickens, Hollands, Banbury, and Parasuraman [10] provide coverage of engineering psychology, foregoing treatment of those human components that are not related to psychology (e.g., visibility, reach, and strength). In complementary fashion, Wilson and Corlett [17] and Chaffin, Andersson, and Martin [18] focus on the physical aspects of human factors. Finally, a comprehensive treatment of nearly all aspects of human factors can be found in the *Handbook of Human Factors and Ergonomics*

[19], and issues of system integration can be found the *Handbook of Human-Systems Integration*[20].

1. Wickens, C. D., Hollands, J. G., Banbury, S., & Parasuraman, R. (2016). *Engineering psychology and human performance* (Fourth). New York: Routledge, Taylor & Francis Group.

2. Wilson, J., & Sharples, S. (2015). *Evaluation of Human Work*. Boca Raton, FL: Taylor and Francis.

3. Chaffin, D. B., Andersson, J., G. B., & Martin, B. J. (2006). *Occupational Biomechanics* (Fourth). New York: John Wiley & Sons.

4. Salvendy, G. (2013). *Handbook of Human Factors and Ergonomics*. New York: John Wiley and Sons.

5. Boehm-Davis, D. A., Durso, F. T., & and Lee, J. D. (2015). *APA Handbook of Human System Integration*. APA press.

Questions

Questions for 1.1 What is Human Factors Engineering?

P1.1   What three general influences on human behavior are considered by human factors engineering in guiding design?

P1.2   What is "human error" a symptom of?

Questions for 1.2 Goals and Process of Human Factors Engineering

P1.3   What are the three goals of human factors engineering and what is their relative importance?

P1.4   How might the three goals of human factors engineering conflict with each other?

P1.5   How can potential conflicts in the goals of human factors engineering be resolved?

P1.6   What are three application areas that influence the priority of human factors engineering goals?

P1.7   How do the three goals of human factors engineering depend on the application area (e.g., high-risk, the workplace, and consumer products)?

P1.8   What are the three components of the human factors engineering design cycle?

P1.9   What is the most fundamental outcome of the understand element of the human factors design cycle?

P1.10   What human factors activities comprise the create element of the human factors design cycle?

P1.11   What characteristics of people and systems make the evaluation element of the human factors design cycle essential?

P1.12   Why is it best to design with humans in mind from the start?

P1.13   Why are human factors interventions related to the task or equipment considered to be of greater importance than organization, training or selection?

P1.14   Why should training and selection be considered only after other human factors design interventions?

Questions for 1.3 Scope of Human Factors Engineering

P1.15   Explain why some human factors interventions, such as training and selection, are most relevant to the workplace and high-risk application areas?

P1.16   How do the two dimensions used to describe the scope of human factors relate to the organization of this textbook?

Questions for 1.4 Scientific Base of Human Factors Engineering

P1.17   Why is intuition insufficient to guide design?

P1.18   Why is it important to remember that preference does not always equal performance?

P1.19   Beyond helping engineers go beyond the limits of their intuition, what general feature of the scientific base of human factors engineering is important for communicating with a design team?

Chapter 2

Design Methods

At the end of this chapter you will be able to...

1. identify appropriate design process for high-risk systems, the work place, and consumer products

2. apply human-centered design using the understand, create, and evaluate iterative cycle

3. identify the role of human factors in system design processes

4. identify design opportunities using focus groups, observations, and accident investigation

5. define design requirements using task analysis

6. create prototypes using iterative design and refinement

Thomas Edison was a great inventor but a poor businessman. Consider the phonograph. Edison invented it, he had better technology than his competitors, but he built a technology-centered device that failed to consider his customers' needs, and his phonograph business failed. One of Edison's failings was to neglect the practical advantages of the disc over the cylinder in terms of ease of use, storage, and shipping. Edison scoffed at the scratchy sound of the disc compared to the superior sound of his cylinders. Edison thought phonographs could lead to a paperless office in which dictated letters could be recorded and the cylinders mailed without the need for transcription. The real use of the phonograph, discovered by a variety of other manufacturers, was for prerecorded music. Once again, he failed to understand the real desires of his customers. Edison decided that big-name, expensive artists did not sound that different from the lesser-known professionals. He is probably correct. Edison thought he could save considerable money at no sacrifice to quality by recording those lesser-known artists. He was right; he saved a lot of money. The problem was, the public wanted to hear the well-known artists, not the unknown ones. Edison bet on his technology-centered analysis: He lost. The moral of this story is to know your customer. Being first, being best, and even being right do not matter; what matters is understanding what your customers want and need. Many technology-oriented companies are in a similar muddle. They develop technology-driven products without understanding customer needs and desires. (Adapted from Norman [16]).

The goal of a human factors specialist is to make systems successful by enhancing safety, performance, and satisfaction. This is achieved by applying human factors principles, methods, and data to the design of products or systems. The concept of "design" is very broad and can include activities such as:

- Creating new products, systems, and experiences

- Improving existing products to address human factors problems

- Ensuring safe environments in the workplace and in the home

- Implementing safety-related activities, such as hazard analyses, industrial safety programs, and safety-related training

- Developing performance support materials, such as checklists and instruction manuals

- Developing methods to train and assess groups and teams

- Guiding team and organizational design

In this chapter, we review some of the methods that human factors specialists use to support design, with particular emphasis on the early stages of design. Human factors methods and principles are applied in all product design phases: front-end analysis, prototyping, technical design, and final test and evaluation.

Although interface design may be the most visible design element, human factors specialists go beyond interface to design the tasks, interaction, overall experience, and even the organization of people and technology. Cooper [21] argues that focusing solely on interface design is ineffective and calls it "painting the corpse." Making a pretty, 3-D graphical interface cannot save a system that does not consider the job or organization it supports. A general recognition of the need to go beyond user interface (UI) design is reflected in the increasing emphasis on user experience (UX) design, which extends beyond the interface to include all aspects of a users' interaction with a system [22]. The material in this chapter provides an overview of the human factors process needed to address these broad considerations, and later chapters provide some of the basic content information necessary to carry out those processes. Later chapters also provide specialized processes needed to address considerations beyond user experience design, such as organizational design.

Human factors considerations go beyond the interface.

2.1 Human Factors in Design and Evaluation

Many products and systems are designed without adequate consideration of human factors. Designers tend to focus on the technology and its features without fully considering the use of the product from the human point of view. In a book that every engineer should read, Norman [16] writes:

> Why do we put up with the frustrations of everyday objects, with objects that we can't figure out how to use, with those neat plastic-wrapped packages that seem impossible to open, with doors that trap people, with washing machines and dryers that have become too confusing to use, with audio-stereo-television-video-cassette-recorders that claim in their advertisements to do everything, but that make it almost impossible to do anything?

Even when designers attempt to consider human factors, they often complete the product design first and only then hand off the blueprint or prototype to a human factors expert to evaluate. This expert is then placed in the unenviable position of having to come back with criticisms of a design that took several months to develop. It is not hard to understand why the design team would be less than thrilled to receive the results of a human factors analysis. Designers clearly believe in the design, and so are often reluctant to accept human factors recommendations. Bringing human factors analysis at the end of the design process places everyone involved at odds with one another. Because of the initial investment and the designer's

Considering human factors at the start of the design smooths the design process.

resistance to change, the result is often a product that is not particularly successful in supporting human safety, performance, and satisfaction. Effectively integrating human factors considerations depends on understanding the system design process.

2.1.1   Human Factors in the System Design Process

Systematic design processes specify a sequence of steps for product analysis, design, and production. Even though there are many different design processes, they generally include stages that reflect *understanding* the users needs (predesign or front-end analysis activities), *creating* a product or system (prototypes, pre-production models), *evaluating* how well the design meets user's needs; all of which is an iterative process that continually cycles back to understanding the user's needs. *Product lifecycle models*, are design processes that include product implementation, utilization and maintenance, and dismantling or disposal. Design processes differ to the degree they are defined by sequential steps or by iteration, flexibility, and adaption to uncertainty.

Figure 2.1 shows three common design processes, the first is the *Vee process*, which is often used in the design of large, high-risk systems, such as the design of a new aircraft, where sequential development is possible and verification, validation, and documentation are critical. The Vee shape starts with a broad system description and design requirements, which are decomposed into detailed requirements. For the dashboard of a car, these detailed requirements might include information elements, such as speed and level of the gas tank. Designs of these components are then integrated and verified by comparing them to the original system requirements. In the Vee process, the general specifications are well-defined at the start and emphasis is given to documenting a successful implementation of those specifications.

A second design model is the *Plan-Do-Check-Act cycle*, which is commonly used to enhance workplace efficiency and production quality [23]. The cycle begins with the target improvement. The create Plan describes objectives and specifies the targeted improvement, Plan is implemented in the Do stage where a product, prototype or process is created. The Check stage involves assessing the intervention defined by the Do stage to understand what effect it had. Act completes the cycle by implementing the intervention or developing a new Plan based on the outcomes. This cycle reflects the scientific management approach of Taylor in that each plan represents a hypothesis of how the system or product might be improved.

A third design model is the *Scrum approach*, which is more typical of consumer software products, such as smartphone and web applications, where an iterative and incremental approach is needed to resolve uncertainty in design requirements. The Scrum approach focuses on creating products and using those products to discover requirements [24]. Early prototypes reveal design opportunities that are visible only after the technology has been implemented. Central to the scrum approach is delivering system components quickly and accommodating requirements discovered during development. "Sprints," which are short duration efforts, typically 24 hours to 30 days, focus effort on quickly producing new iterations of the product. The scrum approach is well-suited to situations that demand high degree of innovation, such as those where technology changes rapidly and potential applications emerge abruptly. This flexibililty is why such techniiques are sometimes termed agile design The scrum approach relies on close interaction between co-located workers to develop solutions in an ad-hoc manner and therefore, the approach tends to place less emphasis on standardized work processes, documentation, and testing.

Vee processes on methodical implementation, PDCA guides incremental improvement, Scrum processes focuses on fast iteration.

As noted in the introduction, cars are increasingly becoming highly computerized consumer products. Consequently, one might think a Scrum approach might be appropriate for designing a car given the rapidly changing technology and the associated need for

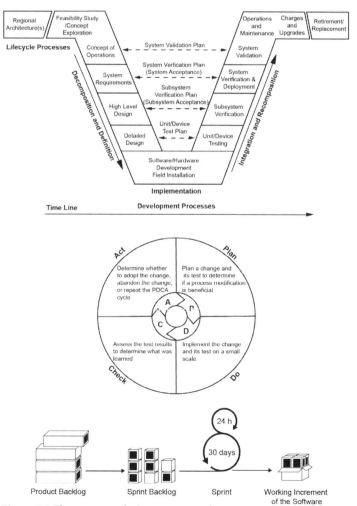

Figure 2.1 Three system design processes that correspond roughly to design of high-risk systems, the work-place, and consumer products.

innovation to stay ahead of competitors. Rapid technology change also makes it difficult to specify detailed requirements in advance. At the same time, cars have elements of high-risk systems that intensify the demands to verify and validate critical safety features, making the "Vee" model more appropriate. Such design situations demonstrate the need for a hybrid approach that combines elements of the Vee, Plan-Do-Check-Act, and Scrum.

Effectively integrating human factors considerations depends on matching the methods to the demands and opportunities of the particular design process. For example, with a short development timeline there may be no opportunities for time consuming human factors methods. Some of the methods described in this chapter, such as a comprehensive task analysis, provide an accurate description, but require weeks to months to complete. Such comprehensive methods best fit the Vee model. Other methods that provide a less accurate description, such as an informal observations or an internet-based survey, might be completed in days. Such rapid methods best fit the Scrum model. Human factors methods trade accuracy for speed. Understanding how to make this speed-accuracy tradeoff is critical for inserting human factors considerations into design.

↯ Select human factors methods that fit the demands of the design process.

2.1.2 Human-Centered Design

The overriding principle in human factors engineering is to center the design process around the user, thus making it a *human-centered design* process [25]. Other phrases that denote a similar meaning are "know the user" and "honor thy user." For a human factors specialist, system or product design revolves around the user: Design must meet people's needs, be useful, and be compatible with their capabilities [26].

We put this principle into practice by determining user needs and involving users in all stages of the design process. That is, the human factors specialist will study the users' job or tasks, elicit their needs and preferences, ask for their insights and design ideas, and collect data on their response to design solutions. User-centered design does not mean that the user designs the product. The goal of the human factors specialist is to find a system design that supports the user's needs rather than designing a system to which users must adapt, as shown in Table 2.1.

A holistic perspective or systems thinking is an important part of user-centered design. Rather than considering elements of the design as independent, unrelated parts, systems thinking focuses on the interaction and relationships of parts—a focus on the whole rather than just the parts. Such holistic thinking can identify important benefits for integrating elements of a device, such as making it possible to place a call from a smartphone by touching (rather than dialing) a number on a webpage. Systems thinking can also avoid unintended consequences. For example, shortening the shifts of healthcare workers to reduce fatigue and improve healthcare quality can have the unintended consequence of increasing the need to handoff a patient from one healthcare worker to another. More

Early focus on users, their environment, and their tasks

Empirical measurement, using questionnaires, usability studies, and usage studies focusing on quantitative performance data

Iterative design, using prototypes, where rapid changes can be made to the design

Participatory design, where users are part of the design team, providing input, but not defining the final design

Multidisciplinary team, which might include specialists in human factors, industrial design, computer science, mechanical engineering, and electrical engineering to name a few

Holistic perspective, where designers consider the complete user experience

Table 2.1 Elements of Human-Centered Design.

handoffs can undermine healthcare quality [27].

Human-centered design process can be simplified into three major phases: understand users, create a prototype, and evaluate the prototype [28]. *Understanding* users involves careful observations of people and the tasks they perform to the point of establishing empathy for their situation. *Creating* a prototype involves designers combining this understanding with a knowledge of human characteristics, interface guidelines, and principles of human behavior, which we will discuss in later chapters, to produce initial design concepts. Soon after these initial concepts are developed, designers evaluate these prototypes. *Evaluating* can include heuristic evaluations and usability tests with low-fidelity mock-ups or prototypes. Usability tests are particularly useful because they often help designers better understand the users and their needs. This enhanced understanding provides input to the next cycle of creating an improved prototype.

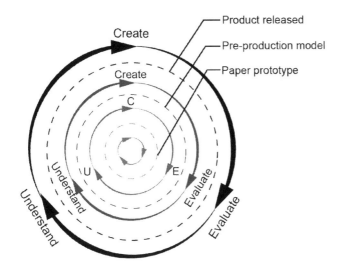

Figure 2.2 The Understand, Create, and Evaluate cycle describes design as an iterative cycle that repeats many times at multiple time scales.

Figure 2.2 shows the cycles of the design process, gravitating from inside to out, as the prototype evolves into a final product. The cycles vary in how long they take to complete, with the outer cycles taking months or years and inner cycles taking minutes. In the extreme, one might complete a cycle during an interview with a user where the designer creates a simple paper prototype of a possible solution, and the user provides immediate feedback. Taking hours rather than seconds, a *heuristic evaluation*, where the design principles and guidelines are applied to the prototype, can quickly assess how design might violate human capabilities. Usability tests typically take days or weeks but provide a more detailed and precise understanding of how people respond to a design. The inner elements of the design cycle provide rapid, but approximate information about how a particular design might succeed in meeting people's needs, and the outer elements of the cycle are more time consuming, but more precise. This speed-accuracy tradeoff means that the time and resources needed to understand, create, and evaluate should be matched to the system being developed. Rapidly changing markets place a premium on fast and approximate methods.

Usability tests are conducted multiple times as the interface design goes through modifications. Each repetition of the testing and modification cycle can produce significant improvements, and many iterations of design should be expected. At the beginning, it is not necessary to worry about the details of screen design or making the screens look elegant. Rather, the emphasis should be on identifying useful functions, and how the user responds to those functions. This iterative process has been shown to be incredibly valuable in refining software, hardware, and even work process designs. Although each usability test typically includes only five people (see Chapter 3 for more detail), as many as 60 cycles of testing can provide benefits that outweigh the costs [29]. At a minimum, three to five iterations should be considered and one can expect improvements of 25-40% for each iteration [30].

When the system design nears completion, it may be placed in an operational environment and a comprehensive test and evaluation may be performed (see Chapter 3). This evaluation can be considered the final step of product development. It can also be considered as the first step in developing a better understanding of the user for the next version of the product. The outermost cycle in Figure 2.2 indicates that even after the product is released, the cycle continues with data being collected to understand how people use the system. For many consumer products, early beta versions of products are released for this purpose, but this is not the case for high-risk systems. For high-risk systems post-release surveillance to detect design flaws is important. In the automotive industry, post-release surveillance occasionally results in recalls to fix design flaws that were not detected during the design process. The remainder of this chapter describes critical elements of each of these three phases—Understand, Create, and Evaluate–with a focus on understanding the user.

> Iterative design is central to understanding and meeting people's needs.

2.2   Understanding the User, Context, and Tasks

> "If you want to improve a piece of software all you have to do is watch people using it and see when they grimace, and then you can fix that." David Kelley [31]

The purpose of front-end analysis is to understand the users, their needs, and the demands of the work situation. Front-end analyses differ substantially in the time required to complete and need to be matched to the development timelines and priorities of a project. Not all activities are carried out in detail for every project, but in general, the designer will need to answer the following questions before design solutions are created:

1. Who are the users? (This includes not only users in the traditional sense, but also the people who will install, maintain,

monitor, repair, and dispose of the system.)

2. Why do users need the product and what are their preferences?

3. What are the environmental conditions under which the system or product will be used?

4. What is the physical and organizational context of the users' activity?

5. What major functions must be fulfilled by a person, team, or machine?

6. When must tasks occur, in what order, and how long do they take?

These questions are answered with various analyses that begin by collecting data, often by observing and talking with people. These data are then analyzed to support design decisions. We use the term task analysis to describe this process, which can vary substantially in its level of detail. In general, the more complex and critical the system, such as air traffic control, the more detailed the task analysis. It is not unusual to spend several months performing this analysis for a complex product or system.

Although direct observation is the primary technique for collecting information about tasks and activities, it is not always the most effective. Accident prevention is a major goal of the human factors profession, especially as humans are increasingly called upon to operate large and complex systems, and of course we rarely directly observe accidents. Accidents and critical incidents can be systematically analyzed to determine the underlying causes. In Chapter 1 we discussed how Fitts and Jones interviewed pilots after crashes and near crashes to identify opportunities to improve aircraft design.

Accident analysis has pointed to several cases where poor system design has resulted in human error. As an example, in 1987 Air Florida Flight 90 and crashed into the 14th Street Bridge on the Potomac River shortly after taking off from Washington National Airport, killing 74 of the 79 passengers and crew. Careful analysis of the events leading up to the crash identified training and decision errors that led the pilots to take off even though snow and ice had accumulated on the wings. Accidents usually result from several coinciding breakdowns, and so identifying human error is the first and not the last step in understanding accidents.

Practicing the "Five Whys" by tracing back the causes of an event by asking "why" at least five times can be useful in going beyond human error as cause of an accident. For the Air Florida Flight 90, this might mean asking why the aircraft had inadequate lift? (ice accumulated on the wings). This leads to the question: why did the aircraft take off with ice on its wings? (aircraft accumulated ice as it waited in taxi line for 49 minutes before takeoff), which then leads to the questions: why did the pilots decided not to return for deicing? (production pressure and lack of experience),

The "Five Whys" help identify the multiple causes of accidents.

why did the pilots did not notice the severity of the icing problem as they began taxied for takeoff? (captain failed to address concerns of first officer). These series of questions typically show multiple unsafe elements associated with training, procedures, controls and displays, that should be considered before rather than after an accident. This requires a proactive approach to system safety analysis rather than a reactive one such as accident analysis or accident reconstruction. This safety analysis is one particular example of understanding users, their operating environment, and the tasks they must perform and is addressed in greater detail in Chapter 16, where we discuss system safety.

In contrast to accident analysis, which typically focuses on system safety, *time-motion studies* developed by Taylor (Chapter 1), focuses on improving performance of manual work. Taylor's detailed observations dramatically improved steelworker productivity by precisely recording the movements and timing of actions of workers on assembly lines. These time motion studies continue to be an important technique to improve efficiency and avoid injury associated with manual materials handling, which we discuss in more detail in Chapter 14 on biomechanics. Human factors engineers can incorporate knowledge gained in time motion studies to understand user needs, the context that the work is to be conducted, and the sequences of tasks.

Understanding users' needs for computer systems and consumer products often benefits from an approach termed contextual inquiry [32]. Contextual inquiry provides an understanding of users' needs by going to users' workplace or wherever the system would be used, and adopting master-apprentice relationship. The interviewer acts as an apprentice learning from the master regarding how to perform a particular activity. As an apprentice, the interviewer asks the master to verify his or her understanding by commenting on task descriptions and prototypes, as simple as a series of sketches to show screens of a potential application.

Understanding the user tasks is essential, whether the goal is to prevent accidents, make assembly lines more efficient, or create delightful products. This understanding goes beyond simply documenting activity, but involves establishing a deep empathy for the user. Without some form of front-end analysis, designers and engineers creating products often find it hard to create systems that serve peoples' needs effectively. Depending on how the data are collected and analyzed, the methods take on many names, but they all aim to understand users, their environment, and the tasks they must perform. Here we describe them under the general term of *task analysis*.

Time motion studies identify ways to improve worker efficiency.

Contextual inquiry reveals user needs.

2.3 How to Perform a Task Analysis

Most generally, task analysis is a way of systematically describing human interaction with a system to understand how to match the demands of the system to human capabilities. Task analysis is a broad term that encompasses many other techniques such as use cases, user stories, and user journeys. All of these techniques focus on understanding the users' goals and motivations, the tasks and subtasks to achieve these goals, the ordering and timing of these tasks, and the location and situation where tasks occur. Task analysis consists of the following steps:

1. Define the purpose and identify the required data

2. Collect task data

3. Interpret task data

4. Innovate from task data

We describe this process as sequential, but in practice it is often iterative. As an example a hierarchical task diagram might be drawn during an interview and might be revised and adjusted as part of the interview process. This diagram might be further refined based on observations of work being performed. The observations and analysis that make up a task analysis help focus on the activity details that matter for the user. A deep, empathetic, and obsessive understanding of these details is what makes designs succeed.

2.3.1   Step 1: Define Purpose and Required Data

The first step of task analysis is to define the design considerations that the task analysis will address. Because a task analysis can be quite time consuming, it is critical to clearly define the purpose of the analysis. Typical reasons for performing a task analysis include:

- Redesigning processes

- Identifying software and hardware design requirements

- Identifying content of the human-machine interface

- Defining procedures, manuals, and training

- Allocating functions across teammates and automation

- Estimating system reliability

- Evaluating staffing requirements and estimating workload

As an example, a task analysis for entering a car, adjusting settings, and starting the engine could provide important information to re-imagine the car key and create a new system for entering and selecting vehicle settings. The task analysis could also be used to define features of the interface to adjust settings, and could even be used to define the content of the owner's manual.

Both the purpose of the analysis and the type of the task will influence the information gathered. Tasks can be physical tasks, such as opening the car door, or they can be cognitive tasks, such as selecting music to listen to while driving after entering the car. Because an increasing number of jobs have a large proportion of cognitive tasks, the traditional task analysis is being increasingly augmented to describe the cognitive processes, skills, strategies, and use of information required for task performance. There are methods specifically developed for *cognitive task analysis* [33, 34], but we will treat these as extensions of standard task analyses, referring to all as task analysis. However, designers should pay particular attention to the cognitive components in conducting the analysis if any of the following characteristics are present:

- Complex decision making, problem solving, diagnosis, or reasoning

- Much conceptual knowledge needed to perform tasks

- Large and complex rule structures that are highly dependent on the situation

- Performance depends on memory of information that needs to be retained for seconds or minutes

Whether the task analysis is focused on the physical or cognitive aspects of the activity, four categories of information are typically collected:

- **Hierarchical relationships:** What, why, and how tasks are performed

- **Information flow:** Who performs the task, with what indications, and with what feedback

- **Sequence and timing:** When, in what order, and how long tasks take to perform

- **Location and environmental context:** Where and under what physical and social conditions tasks are performed

Hierarchical relationships describe how subtasks combine into tasks, and how tasks combine to achieve people's goals. With the car example, a goal is to keep the car secure, a task is to lock the door, and subtasks include inserting the key, turning the key, or press the lock icon on a keyfob. Moving up the hierarchy describes *why* a task is performed—secure the car—and moving down the hierarchy describes *how* a task is performed—turn the key/click lock on keyfob. Describing the hierarchical relationships between goals, tasks, and subtasks makes the reason for the many subtasks understandable. These hierarchical relationships help us focus on the underlying goals of people and can prompt innovations by identifying new and more efficient ways of achieving people's goals, such as securing the car using a keyfob rather than a key.

Hierarchical relationships identify new ways of doing things

Information flow specifies the communication between people and the interactions between people and technology. This information flow can also include stored information needed to complete the task, such as knowledge and skills or information on a display. With the car example, the flow of information might include a signal to identify that the doors are unlocked. For some systems, a complex network of people and automation must be coordinated. In other systems, it may be only a single person and the technology. However, most systems involve coordination with multiple people. Defining their roles and their information needs often identifies important design considerations that might otherwise go unnoticed, such as how to get passenger preferences regarding music into the car.

> Information flow helps specify interface content.

Sequence and timing describe the order and duration of tasks. In the car example, the driver must first turn the key, then lift the door handle, and finally pull the door open. Performed in a different order, these tasks would not achieve the goal of opening the door. In other situations, tasks could be performed in parallel. Task sequence often determines how much time a set of tasks will take to complete. Specific task sequence information includes the goal or intent of task, sequential relationship (what tasks must precede or follow), trigger or event that starts a task sequence, results or outcome of performing the tasks, duration of task, number and type of people required, and the tasks that will be performed concurrently. Eliminating tasks, reducing their duration, or assigning them to different people can make systems easier to use and more efficient.

> Sequence and timing specifies efficient interactions.

Location and environmental context describe the physical and social world in which tasks occur. In the car example, important location information might be the physical layout of the vehicle interior that can make it difficult to insert the key to start the car. Specific location information might include places people work and the paths people take from one place to another, as well as the location of equipment and the physical barriers such as walls and desks. Location of equipment can greatly influence the effectiveness of people in production-line settings. The physical space can also have a surprisingly large effect on computer-based work, as anyone who has had to walk down the hall to a printer knows.

> Physical layout can strongly affect task difficulty.

These categories of information describe tasks from different perspectives and are all required for a comprehensive task analysis. Other useful information can be included such as the probability of performing the task incorrectly, the frequency with which an activity occurs, and the importance of the task. For example, the frequency of occurrence can describe an information flow between people or the number of times a particular path is taken. Most importantly, a task analysis should record instances where the current system makes it difficult for users to achieve their objectives; such data identify opportunities for redesigning and improving the system.

> Task analysis can identify frequent, error-prone, important tasks for careful consideration in design.

After the purpose of the task analysis is defined and relevant data identified, task data must be collected, summarized, and analyzed. There are a wide range of methods to complete these steps and here we review the most commonly used methods.

2.3.2 Step 2: Collect Task Data

Task data are collected by observing and talking with multiple users. The overall objective is to see the world through the eyes of the person the system is being designed for, and to develop empathy for the challenges, demands, and responsibilities they face. This empathy helps focus attention to the details of the system that matter to the user. These details might be very different than those that might be noticed by the engineer implementing the design. The particular way to understand users' tasks depends on the information required for the analysis. Ideally, human factors specialists observe and question users as they perform tasks in the place where they typically perform those tasks. This is not always possible, and it may be more cost effective to collect some information with other techniques, such as surveys or questionnaires. Task data collection techniques include:

1. Observations and questions of people as they use an existing system

2. Retrospective and prospective verbal protocol analysis

3. Unstructured and structured interviews including focus groups

4. Surveys and questionnaires

5. Automatic data recording

Probe questions for the Master/Apprentice observation approach:

Who and what is needed to perform the task?

What happens before, what after?

What does the task change, how is this detected?

What has to be remembered?

What is the consequence of failure to complete the task?

When in the day, and when relative to other events, is the task performed?

How do people communicate and coordinate their activity?

Table 2.2 Observing people to guide design. (Photo: Seaman Apprentice Karolina Oseguera (https://www.dvidshub.net/image/1415836) [Public domain], via Wikimedia Commons)

Observations of users as they interact with existing versions of the product or system is one of the most useful data collection methods. If we were interested in car design, we would find drivers who represent the different types of people the car is to be designed for and then observe how they use their cars. People are asked to perform the activities under a variety of typical scenarios, and the analyst observes the work, asking questions as needed. Observation should be performed in the environment that the person normally accomplishes the task.

The meaning behind users' tasks is often revealed in their thoughts, goals, and intentions, and so observations of physical activity may not be sufficient to understand the tasks. This is particularly true with primarily cognitive tasks that may generate little observable activity and so it can be useful for people to think out loud as they perform various tasks. One approach is to adopt a *master-apprentice relationship*, where the observer acts as an apprentice trying to learn how tasks are performed [32]. Adopting this relationship makes it easy for observers to ask questions that help users to describe their underlying goals, strategies, decisions (See Table 2.2).

Retrospective and prospective verbal protocol address important limits of direct observations. Direct observations disrupt ongoing activity or they can fail to capture rarely occurring situations. For example, trying to understand how people deal with being lost as they drive would be difficult to observe because talking to the driver could be distracting and the analyst would have to ride with the driver for many trips to observe the rare case that they get lost. Talking about tasks is termed verbal protocol, and retrospective verbal protocols require that people describe past events and prospective verbal protocols require that people imagine how they would act in future situations.

Video recordings of users' activity are an effective way to prompt retrospective protocols. People can describe what they were thinking as they performed the tasks, the human factors specialist can pause the playback and ask probe questions. Because users do not have to perform the task and talk about it at the same time retrospective protocols can be easier on the user. Retrospective protocols can even yield more information than concurrent protocols.

Structured and unstructured interviews involve the human factors specialist asking the user to describe their tasks. Structured interviews use a standard set of questions that ensure the interview captures specific information for all interviewees. Unstructured interviews use questions that are adjusted during the interview according to the situations. The analyst might ask about how the users go about the activities and also about their preferences and strategies. Analysts should also note points where users fail to achieve their goals, make errors, show lack of understanding, and seem frustrated or uncomfortable.

Unstructured interviews use probe questions similar to those used for direct observation (See Table 2.2). These question probes address when, how, and why a particular task is performed, as well as the consequences of not performing the task. *Critical incident technique* is a particularly useful approach for understanding how people respond to accident and near accident situations in high-risk systems. Because accidents are so rare direction observation is not feasible. With the critical incident technique, the analyst asks users to recall the details of specific situations and relive the event. By reliving the event with the user, the analyst can get insights similar to those from direct observation [35].

Critical incident technique makes it possible to "observe" past events.

Focus groups are interviews with small groups of users, rather than individuals [36, 37]. Focus groups typically consist of between six and ten users led by a facilitator familiar with the task and system. The facilitator should be neutral with respect to the outcome of the discussion and not lead the discussion to a particular outcome. One benefit of focus groups is that they cost less than individual interviews (less time for the analyst), also discussion among users often draws out more information because the conversation reminds them of things they would not otherwise remember.

Observations are typically more valuable than interviews or focus groups because what people say does not always match what they do. In addition, people may omit critical details of their work, they may find it difficult to imagine new technology, and they may distort their description to avoid appearing incompetent. It is often difficult for people to describe how they would perform a given task without actually doing it—describe how you tie your shoes without being able to touch and see your shoes.

Surveys and questionnaires are typically used after designers have obtained preliminary descriptions of activities or basic tasks. Questionnaires are often used to affirm the accuracy of the information, determine the frequency with which various groups of users perform the tasks, and identify any user preferences or biases. These data help designers prioritize different design functions or features. See Chapter 3 for a more complete discussion of surveys and their limits.

Automatic data recording with smartphones and activity monitors make it possible to record people's activities unobtrusively. An example of such a system is a data logging device that records the position of the car, its speed, and any hard braking or steering events. Such a device can show where drivers go, when they choose to travel, and whether they travel safely. Such data has the benefit of providing a detailed and objective record, but it lacks information regarding the purpose behind the activity. This limitation might be avoided by using automatically recorded data to prompt retrospective verbal protocol.

Limitations plague all of these methods, but combinations of the methods, such as automatic data recording and retrospective protocols, can compensate to some extent. A more general limit is that all of these methods document existing behavior. Designing to support existing behavior means that new controls, displays, or other performance aids might simply enable people to do the same tasks better, but might not produce dramatic innovation. Innovation requires the analysis to focus on underlying goals and needs, and identify different ways of accomplishing these goals.

One way to go beyond describing existing tasks is to evaluate the underlying characteristics of the environment and the control requirements of the system. In driving, this would mean examining the reason why people get into their cars. Often, such an analysis reveals new ways to doing things that might not be discovered by talking with people in a focus group.

Innovation from observations requires analysts to go beyond current activities and identify better ways to achieve users' goals.

2.3.3   Step 3: Interpret Task Data

Once task-related information has been collected, it must be organized, summarized, and analyzed. At the simplest level, the task analysis might be summarized as a list of challenges faced by people. As an example, observing drivers trying to unlock their cars during a Wisconsin winter could show how fumbling for keys might threaten drivers with frostbite. Sometimes these challenges can inspire important innovations, but often a more detailed analysis of tasks is needed to identify solutions and to avoid unintended consequences. Some of the most common ways to organize task data include:

1. **Task hierarchy:** Goal, task, subtask decomposition

2. **Task flow:** Control, decisions regarding the flow from one task to another

3. **Task sequence:** Task duration and sequence, as well as communication between system components

Task hierarchy can be shown as an arrangement of tasks where tasks are broken into more specific subtasks. Goals are at the top of the hierarchy and the tasks at the bottom represent detailed actions needed to accomplish those goals. The tasks higher in the hierarchy are why the ones below are performed, and the tasks lower in the hierarchy describe how the tasks above are achieved. A task hierarchy makes it possible to organize a complex array of many actions into a few general tasks, linking a detailed description to a more general description.

Figure 2.3 shows a task hierarchy for driving a car. General tasks, such as "Enter and start the car" are broken into more specific tasks, such as "Unlock the door" and "Starting the engine". These are further broken into very specific actions. For unlocking the door this might include "Find key", "Removing key from pocket", "Insert key in door" and so on. The level of task hierarchy should be aligned with the purpose of the analysis and should avoid unnecessary detail. Figure 2.3 shows how the purpose of the analysis focuses attention on describing entering and exiting the car and so the "Drive" activity is not broken into subtasks.

A task hierarchy prompts innovation by identifying different ways of achieving the same overall task with different subtasks. For example, considering different ways to "Enter and start car" could lead to using a smartphone to rather than keys. A task hierarchy also makes it easy for the analyst to develop spreadsheets to record information for each task or subtask. The spreadsheet contains a row for each task, and columns for information describing the tasks. This information might include task duration, conditions that must be met to perform the tasks, why the task is difficult, common errors, strategies, skills, or knowledge. One simple analysis that might be part of a time-motion study is to use a spreadsheet to calculate total task time and compare it with a new design with different tasks.

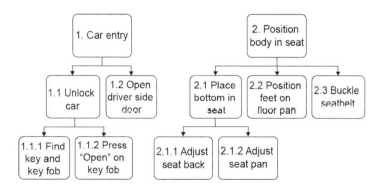

Figure 2.3 Task hierarchy for driving a car with a focus on unlocking the door.

Task flow from one task to another is captured by a flow chart. Activity diagrams build on flow charts and also show tasks that are performed concurrently (Figure 2.4). This diagram shows the flow from one task to another and the decision points that determine which task should follow. The flow from one task to another can be sequential, shown as an arrow connecting tasks, which are shown as rounded rectangles. The flow can also be branched, where the decision to flow to one task or another is indicated by a diamond. The flow can also be concurrent where tasks can occur in parallel, which is indicated by a set of tasks bounded on the top and bottom by horizontal bars.

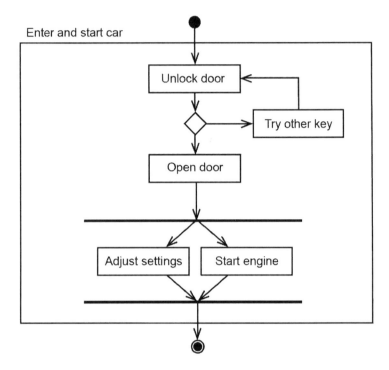

Figure 2.4 An activity diagram shows task flow for entering and starting a car.

Figure 2.4 shows an activity diagram for the entering and starting the car. It begins with unlocking the door and finishes when the settings are adjusted and the engine is started. The diamond after the unlock door indicates that the door can be opened only if the correct key is turned and loop from the diamond back to unlock door indicates that keys are tried until the correct one is inserted. After opening the door, the horizontal bar indicates that the settings can be adjusted and the engine can be started in any order.

Activity diagrams highlight decisions and the information required to make them. Sometimes this information is trivial and built into the interaction, such as the resistance experienced with the wrong key is used to open a car door. In other situations, identifying the cues that guide decisions can specify critical information for an interface. Activity diagrams also indicate mandatory ordering of tasks that need to be conveyed through the physical configuration of the device, the interface, or through instructions. In the case of a car, the physical configuration of the door and key makes it impossible to start the engine before opening the door. Positive locking of the door by a keyfob only outside the car makes it impossible to lock ones keys in the car. Beyond inspecting these diagrams, it is possible to "run" these diagrams as a computer simulation and estimate the time it takes to perform these tasks.

Task sequence is described in sequence diagrams that show the order and duration tasks for each object and person in the system. The activity of each person and object is represented as a timeline that runs from the top of the diagram to the bottom and rectangles on this timeline indicate when the person or object is active in responding to other elements of the system. Horizontal arrows indicate communication between people and objects. Solid lines and arrows indicate synchronous messages, where a response is required before other activities can proceed. Dashed lines with open arrows indicate asynchronous messages, where activities can proceed without a response. Responses are indicated by dashed lines with open arrows at the of a rectangle.

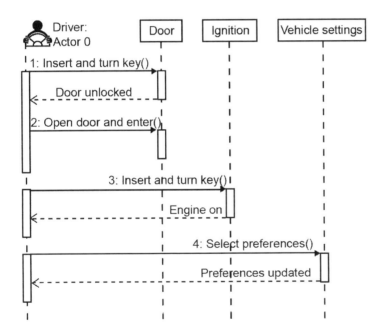

Figure 2.5 Sequence diagram for unlocking and entering a car.

Figure 2.5 shows a sequence diagram for entering and turning on a car. The left-most timeline begins with the driver inserting a key into the door. The next timeline indicates the feedback provided by the door that the door is unlocked. The driver then inserts the key into the ignition and tries to start the car, which is indicated by the engine being on. This figure shows that sequence diagram focuses a particular sequence of tasks of the many possible sequences—it shows the sequence associated with inserting the correct key and not what would have happened with the incorrect key. Fault tree analysis (see Chapter 16 on system safety) directly addresses how the probability of failure associated with many tasks combine to influence safety. Sets of tasks with many decision points are better represented by activity diagrams than by sequence diagrams.

Sequence diagrams highlight communication, particularly the responses that provide people with feedback regarding the success or failure of their actions. Interaction design should ensure clear and timely feedback. Diagrams that have many messages that cross several timelines indicate a need to reorganize the communication and simplify so that each person or object communicates with neighbors and that messages only occasionally cross timelines. One way to avoid messages crossing multiple timelines is to ensure that each object and person has a clear role that describes how it is responds to messages. High activity for one person and low activity for another might indicate workload that could be balanced by adjusting the roles of each.

Representing task analysis data as hierarchies, flows, and sequences have advantages and disadvantages, and choosing the most appropriate method depends on the type of activity being analyzed. If the tasks are basically linear and usually done in a particular order, as is entering and starting a car, it is appropriate to use a sequence diagram. If there are many decision points and conditions for choosing among actions, then an activity diagram might be more useful. There is a major disadvantage to activity and sequence diagrams that is often not readily apparent. People think about goals and tasks in hierarchies. To be consistent with such thinking, the design of controls and displays should reflect these hierarchies. However, when describing or performing a task, the actions will appear as a linear sequence. If the task analysis is represented in an activity or sequence diagram, the importance of cognitive groupings described in the task hierarchy might be lost. This makes it harder for the designer to match the interface to how the person thinks about the system. To develop efficient interfaces, designers must consider the hierarchical structure, decision points, and the linear sequence of tasks.

> Match the representation of tasks—hierarchy, flow, and sequence—to the design issue.

2.3.4   Step 4: Innovate from Task Data

Task analysis reveals the potential to help people by creating new systems or revising existing systems. Sometimes these insights come immediately from observations of people interacting with an existing system, such as a Wisconsin driver getting cold hands trying to unlock her car. But often these insights come from careful analysis of task hierarchy, flow or sequence, such as feedback indicating when the door has been unlocked. This analysis can be qualitative, with a focus on empathy and general insights about the users' experiences. It can also be quantitative, where task performance is described in terms of frequency of occurrence, probability of a failure, and task duration. This focus on task details needs to be placed in the broader user experience and then linked to design solutions. Here we discuss developing personas and use scenarios as first steps in linking task analysis results to system specifications.

User identification and persona development describes the most important user populations of the product or system. For example, designers of a more accessible ATM might characterize the primary user population as people ranging from teenagers to senior citizens with an education ranging from junior high to PhD and having at least a third-grade English reading level, or possible physical disabilities. After identifying characteristics of the user population, designers should also specify the people who will be installing or maintaining the systems.

It is important to create a complete description of the potential user population. This usually includes characteristics such as age, gender, education level or reading ability, physical size, physical abilities (or disabilities), familiarity with the type of product, and task-relevant skills. For situations where products or systems already exist, one way that designers can determine the characteristics of primary users is to sample the existing population of users. For example, the ATM designer might measure the types of people who currently use ATMs. Notice, however, that this will result in a description of users who are capable of using, and do use, the existing ATMs. This is not an appropriate analysis if the goal is to attract, or design for, a wider range of users.

A simple list of user characteristics often fails to influence design. Disembodied user characteristics may result in an "elastic user" whose characteristics shift as various features are developed. Designing for an elastic user may create a product that fails to satisfy any real user. Cooper [21] developed the concept of personas to represent the user characteristics in a concrete and understandable manner. A persona is a hypothetical person developed through interviews and observations of real people. Personas are not real people, but they represent key characteristics of the user population in the design process. The description of the persona includes not only physical characteristics and abilities, but also the persona's goals, work environment, typical activities, past experience, and precisely what he or she wishes to accomplish. The persona should be specific to the point of having a name.

Sequence diagrams help define personas by identifying roles, tasks, and communications. The task hierarchy specifies goals and motivations. For most applications, three or four personas can represent the characteristics of the user population. Separate personas may be needed to describe people with other roles in the system, such as maintenance personnel. The personas exist to define the goals that the system must support and describe the capabilities and limits of users in concrete terms. Personas describe who the design is for and act as the voice of the user, preventing the natural tendency of the design team to assume users are like themselves.

Because persona define several "typical" users, this method runs the risk of neglecting the extremes, such as the 5th and 95th percentiles of a population. Techniques to systematically accommodate these extremes are discussed in the context of fitting designs to the physical dimensions of people in Chapter 12.

Scenarios, user journeys, and use cases complement personas. Personas are detailed descriptions of typical users and scenarios

Design Exercise

Stanford wallet design The goal is to practice designing and recognizing a user's need. This exercise is an abbreviated version of the Stanford Wallet Project [38]. Depending on time, each step can be completed in less than five minutes.

Step 1 Everyone: Design/draw the ideal wallet.

Step 2: Pair up in teams of two. Person 1 acts as the designer. Person 2 acts as the user.

Step 3. Show NOT tell. Designer asks user about the ideal wallet. Some questions to consider: What should it look like? How should it feel? What should it be able to hold? How do you want to carry it? What functions do you want?

Step 4. Switch roles. Person 1 is now the user. Person 2 is now the designer. **Step 5. Continue Show NOT tell.**

Repeat Step 3.

Step 6. Self reflection Designer explains the users needs in one sentence: "[user] needs a way to [user's needs] because... (or "but"... or "surprisingly"...)

Table 2.3 Stanford design exercise.

are stories about these personas in a particular context. Scenarios, also termed user journeys, describe situations and tasks relevant to the use of the system or product being developed. Scenarios are a first step in creating the sequence of screens in software development, and they also define the tasks users might be asked to complete in usability tests. In creating a scenario, tasks are examined, and only those that directly serve users' goals are retained. Two types of scenarios are useful for focusing scenario specification on the design. The first is daily use scenarios, which describe the common sets of tasks that occur daily. In the car example, this might be the sequence of activities associated with entering the car when parked in the owners' garage. The second is necessary use scenarios, which describe infrequent but critical sets of tasks that must be performed. In the car example, this might be the sequence of activities associated with entering the car during a snowstorm. Scenarios can be thought of as the script that the personas follow in using the system [21].

Scenarios typically support conceptual design, where the general activities of people are described independent of technology. Use cases help move from conceptual design to prototypes. Use cases are a user-centered description of what the technology is meant to do. At the simplest level, a use case is a sequence of tasks that produce a meaningful outcome, such as entering and starting a car. These tasks can be described in a more formal way in a flow diagram and implemented in a software or hardware prototype.

Observations organized in task hierarchies, flows, and sequences help define personas and scenarios. Personas and scenarios, in turn, help define new task hierarchies, flows and sequences that the new design will make possible. As an example, a flow diagram associated with the new system, such as a keyless entry system for a car, would document the intended interactions between the person and new system. Often it is possible to use scenarios, use cases and personas to create prototypes. Personas and scenarios also provide a starting point for more specific task analysis, such as those that focus on the environment, workload, safety, and automation. The type of analyses needed depends on the scope of the design and the particulars of the system.

Environment and context analysis describes where the tasks, scenarios, and personas live. For example, if ATMs are to be placed indoors, environmental analysis would include a somewhat limited set of factors, such as type of access (e.g., will the locations be wheelchair accessible?), weather conditions (e.g., will it exist in a lobby type of area with outdoor temperatures?), and type of clothing people will be wearing (e.g., will they be wearing gloves?), issues considered in more detail in Chapter 14 where we discuss the physiology of work. Beyond the physical environment, the culture and norms of workplace should be considered as discussed in Chapter 18.

Workload analysis considers whether the system is going to place excessive mental or physical demands on the user, either alone or in conjunction with other tasks, an technique we discuss in Chapter 15.

Safety and hazard analyses should be conducted any time a product or system has implications for human safety. Such analyses identify potential hazards (e.g., electrocution, chemical exposure, or falls) or the likelihood of human error. Several standard methods for performing such analyses are covered in more detail in Chapter 14, which focuses on safety.

Function allocation considers how to distribute tasks between the human operator and technology. To do this, the we first evaluate the basic functions that must be performed by the human-machine system to accomplish the activities. Then we determine whether each function is to be performed by the technology (automation), the person (manual), or some combination. This process is termed function allocation and is covered in Chapter 11, which focuses on automation.

2.4 Iterative Design and Refinement

Once the front-end analysis has been performed, the designers have an understanding of the user's needs. This understanding can be used to identify initial system specifications and create prototypes. Creating prototypes depends on two types of understanding: understanding tasks and understanding general human capabilities. Prototypes must support user tasks in a way that is consistent with how people see, hear, feel, comprehend and act on the world. This understanding is often distilled into principles or design heuristics and Chapters 4 through 18 describe these in detail. As initial prototypes are developed, the design team begins to characterize the product in more detail, and evaluates how people respond to the evolving product. The human factors specialist usually works to ensure that the tasks people will perform fall within the limits of human capability. In other words, can people perform the tasks safely and easily with the proposed system?

2.4.1 Providing Input for System Specifications

Design heuristics help human factors professionals provide design teams with quick input on whether the design alternatives are consistent with human capabilities. Design heuristics or principles provide a response that is grounded in years of design practice and much research on human behavior. Table 2.4 shows 11 design heuristics derived from those of Rams, Nielsen, and Tognazzini [39, 40, 41]. The table also shows the chapters that contain the specific information on cognitive, physical and organizational characteristics that underlies these heuristics.

These heuristics suggest promising design alternatives, but are not simple rules that can be applied without thought. In fact, in many instances, the heuristics conflict. As an example, providing flexibility is needed to accommodate differences between people and environments, such as hearing impairment making it impossible to define an optimal volume setting. At the same time, providing too much flexibility burdens the user with finishing the

---

Evaluation Heuristics

Create useful innovation: Address a need, solve a problem (Chapter 2).

Attend to details: Small changes to the design can have a big effect on people.

Simplify: Remove irrelevant information, but do not mask essential indicators and feedback (Chapter 8).

Honest and understandable: Functions should be reflected in forms that make their states visible, changes predictable, and interactions intuitive (Chapter 4, 9, and 11).

Provide flexibility: People should be able adjust, navigate, undo and redo, adopt shortcuts (Chapter 10, 12).

Consistency: The same label or action should mean the same thing in different situations—don't deviate from well-defined conventions (Chapter 6, 10 and 11).

Anticipate needs: Provide options rather than require people to recall them. Choose thoughtful defaults because people often adopt initial settings (Chapter 6 and 10).

Minimize memory demands: Interactions with technology should not disrupt the flow of activities unless necessary (Chapter 6).

Consider adaptation: Adopt a systems perspective to identify otherwise unanticipated outcomes, particularly as people adapt to the changes in the system (Chapter 18).

Fit the task to the person rather than the person to the task

Table 2.4 General design heuristics.

design, and might lead to the user designing a poor system. As a consequence, the 11 heuristics are not a set of strict rules, but instead should be thought of as a checklist to provoke conversation. Chapter 3 describes how these and other principles can be used as part of *heuristic evaluations* and cognitive walkthroughs. In a heuristic evaluation, you assess whether the system design is consistent with the heuristics. To apply any of these principles effectively requires an understanding of the underlying human characteristics described in forthcoming chapters as indicated for each heuristic.

Design patterns are solutions to commonly occurring design problems and are most typically associated with software, but also apply to physical systems. A number pad for a phone is an example of a design pattern. Using design patterns, such as a conventional number pad for entering phone numbers has benefit of eliminating many design decisions and it is likely to present people with a familiar interaction that is consistent with other systems they might use. Design patterns for user interfaces include common ways of navigating multi-page applications, getting user input, and browsing data (http://ui-patterns.com). Design patterns provide a ready-made shortcut to the final design, but should be carefully assessed to determine whether previously developed patterns fit the current situation.

Human Factors requirements and system specifications need to be defined where design patterns don't fit the current situation. These requirements include the system characteristics that are needed to achieve the desired levels of safety, performance, and satisfaction. For software design, human factors requirements might include error recovery, or the ability to support people performing more than one task at a time. As an example, for an ergonomic keyboard design, McAlindon [42] specified that the new keyboard must eliminate excessive wrist deviation, eliminate excessive key forces, and reduce finger movement. The design that resulted from these requirements was a "keybowl" that is drastically different from the traditional QWERTY keyboard currently in use, but a design that satisfied the ergonomic criteria.

Detailed human factors requirements are most critical to the Vee design approach.

Identifying system requirements is a logical extension of the task analysis that draws on the task data to specify (1) the overall objectives of the system, (2) performance requirements and features, and (3) design constraints. The challenge is to generate system specifications that identify possible features and engineering performance requirements that best satisfy user objectives and goals. The objectives should be written to avoid premature design decisions. They should describe what must be done to achieve the user's goals, but not how to do it. The system objectives should reflect the user's goals and not the technology used to build the system. The objectives do not specify any particular product configuration and should not state specifically how the user will accomplish goals or perform tasks.

After the objectives, designers determine the means by which the system will help the user achieve the goals. These are termed performance requirements and features. The features define what the system will be able to do and under what conditions. The performance requirements and system features provide a design space in which the team develops various solutions.

Finally, in addition to the objectives and system features, the specifications document identifies design constraints, such as weight, speed, cost, abilities of users, and so forth. More generally, design constraints include cost, manufacturing, development time, and environmental considerations. The constraints limit possible design alternatives.

Translating the user needs and goals into system specifications requires the human factors specialist to take a *systems thinking* approach, analyzing the entire system to determine the best configuration of features. The focus should not be on the technology or the person, but on the person-technology system as a unit. The systems design approach draws upon several tools and analyses, that we highlight in the following discussion.

Quality Function Deployment addresses critical questions. What is the role of the human factors specialist as the system specifications are written? He or she compares the system features with user characteristics, activities, environmental conditions, and especially the users' preferences or requirements. This ensures that the design specifications meet the needs of users and avoids adding features that people do not want. Human factors designers often use a simple yet effective method for this process known as the QFD (quality function deployment), which uses the "house of quality" analysis tool [43]. This tool uses a decision matrix to relate user needs to system features, allowing designers to see which features will satisfy user needs.

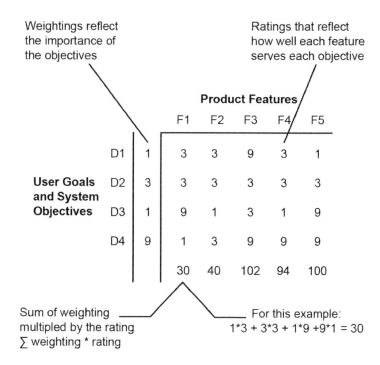

Weightings reflect
the importance of
the objectives

Ratings that reflect
how well each feature
serves each objective

Product Features

			F1	F2	F3	F4	F5
	D1	1	3	3	9	3	1
User Goals	D2	3	3	3	3	3	3
and System							
Objectives	D3	1	9	1	3	1	9
	D4	9	1	3	9	9	9
			30	40	102	94	100

Sum of weighting
multipled by the rating
∑ weighting * rating

For this example:
1*3 + 3*3 + 1*9 +9*1 = 30

Figure 2.6 Simplified house of quality decision matrix for evaluating the importance of features (F) relative to objectives (O).

Figure 2.6 shows a simplified house of quality for the car door design. The rows represent the user needs. The columns represent system features. The task analysis identifies the importance or weighting of each need, which is shown in the left-most column. These weightings are often determined by asking people to assign numbers to the importance of each user need, 9 for very important, 3 for somewhat important, and 1 for marginally important objectives. The rating in each cell in the matrix represents how well each system feature satisfies each user need. These ratings are typically defined using the 9/3/1 rating scale, where 9 is most important, 3 is moderately important, and 1 is least important. The importance of any feature can then be calculated by multiplying the ratings of each feature by the weighting of each user need and adding the result. This result identifies the features that matter most for the users, separating technology-centered features from user-centered features.

Cost/benefit analysis builds on the QFD analysis, which calculates the importance of features that best serve the user needs. This importance serves as the input to cost/benefit analysis, which compares different designs according to their costs relative to their benefits. Costs and benefits can be defined monetarily or by a 9/3/1 rating scale. A decision matrix similar to Figure 2.6 can support the cost/benefit analysis. The features are listed as rows on the left side of a matrix, and the different design alternatives are listed as columns. Each feature is given a weight representing importance of the feature—the result of the QFD analysis. For the features in

Figure 2.6 this would be the total importance shown in the bottom row of the decision matrix. Then, each design alternative is assigned a rating representing how well it addresses each feature. This rating is multiplied by the weighting of each feature and added to determine the total benefit of a design. The cost for each each design is divided by this number to determine the cost/benefit ratio. The design with the lowest cost/benefit ratio represents the greatest value.

Tradeoff analysis identifies the most promising way to implement a design. If multiple factors are considered (e.g., effort, speed, and accuracy), design tradeoffs might be based on the design that has the largest number of advantages and the smallest number of disadvantages. Alternatively, a decision matrix can be constructed. The matrix would assess how well systems, represented as columns, compare according to the performance criteria, represented as rows. For example, for the design of a new car, the performance criteria of a key-less entry system could be represented in one row and an existing key entry system could be another row. The columns would be time to enter the car, likelihood of errors, and ease of use.

Although the decision matrix analyses can be very useful, they tend to consider each product's features independently. Focusing on individual features may fail to consider global issues concerning the interactions of each feature on the overall use of the product. People use a product, not a set of features—a product is more than the sum of its features. Because of this, matrix analyses should be complemented with other approaches, such as scenario specification and user journeys, so that the product is a coherent whole that supports the user rather than simply a set of highly important but disconnected features. The overall objective of these analyses is to identify a small set of the most promising alternatives for implementing in prototypes for further evaluation. Chapter 7 on decision making provides a more detailed discussion for the strengths and weaknesses of decision matrix analysis.

2.4.2 Prototypes, Wireframes, and Mockups

Early prototypes for software development are created by drawing dialog boxes and other interface elements to create a paper prototype as shown in the sidebar table 2.5. Paper prototypes of software systems are useful because screen designs can be sketched, then modified with little effort, making it possible to try out many design alternatives. For this reason, they are useful early in the design process. Because paper prototypes are sketchy versions of the system, users feel more open to identifying flaws. Paper prototypes can even be created during the interviews and used as props to clarify conversations with users. The main purpose of paper prototypes is to guide interaction design and ensure that the structure of the system meets the users' needs (See sidebar Table 2.5).

After paper prototypes, wireframes are created, which are simple layouts that show grouping and location of content, but which omit graphics and detailed functionality. Wireframes are primarily used to communicate with the design team (see Chapter 10 for

Rough, easily created and easily changed paper prototypes invite changes.

A refined, high-fidelity prototype provides an experience that more precisely matches that of the final product.

Table 2.5 Paper prototype and high-fidelity prototype (Permission from designer, Xiaoxia Lu) .

Dieter Rams (1932-) Highly influential designer who asked and answered the question : What is good design [9]? Source: Vitsoe at English Wikipedia/CC BY-SA 3.0 https://commons.wikimedia.org/wiki/File:606-Universal-Shelving-System.jpg

Good design is innovative: The possibilities for innovation are not, by any means, exhausted. Technological development is always offering new opportunities for innovative design. But innovative design always develops in tandem with innovative technology, and can never be an end in itself.

Good design makes a product useful: A product is bought to be used. It has to satisfy certain criteria, not only functional, but also psychological and aesthetic. Good design emphasizes the usefulness of a product whilst disregarding anything that could possibly detract from it.

Good design is aesthetic: The aesthetic quality of a product is integral to its usefulness because products we use every day affect our person and our well-being. But only well-executed objects can be beautiful.

Good design makes a product understandable: It clarifies the product's structure. Better still, it can make the product talk. At best, it is self-explanatory.

Good design is unobtrusive: Products fulfilling a purpose are like tools. They are neither decorative objects nor works of art. Their design should therefore be both neutral and restrained, to leave room for the userâfts self-expression.

Good design is honest: It does not make a product more innovative, powerful or valuable than it really is. It does not attempt to manipulate the consumer with promises that cannot be kept.

Good design is long-lasting: It avoids being fashionable and therefore never appears antiquated. Unlike fashionable design, it lasts many years—even in today's throwaway society.

Good design is down to the last detail: Nothing must be arbitrary or left to chance. Care and accuracy in the design process show respect towards the user.

Good design is environmentally-friendly: Design makes an important contribution to the preservation of the environment. It conserves resources and minimizes physical and visual pollution throughout the lifecycle of the product.

Good design is as little as possible: Less, but better—because it concentrates on the essential aspects, and the products are not burdened with non-essentials.

Back to purity, back to simplicity.

more details), and are helpful in documenting decisions and communicating the essential interactions people might have with the product. Wireframes lack details for the look and feel of the interface or product; these are elements that are the focus of mockups. Mockups focus on the look and feel, and include color, font, layout, and choices of the final product. Wireframes are limited to software systems, but mockups are often created for hardware systems. Wireframes communicate the system's functional characteristics to the design team, and mockups are used to communicate the system's physical features to the design team and other stakeholders, such as users and managers.

Building on wireframes and mockups, we create high-fidelity prototypes so that users can experience elements of the final design. Collecting information from these experiences leads to redesigning the prototype. One analysis showed that user performance improved 12% with each redesign iteration and that the average time to perform software-based tasks decreased 35% from the first to the final design iteration [44], another analysis showed 20-40% improvement per iteration [45]. This redesign and evaluation continues for many iterations, sometimes as many as 10 or 20, or more for complex products.

To summarize, using paper prototypes, wireframes, mockups, and prototypes in the design process has a number of advantages:

- Paper prototypes help understand user needs and if the early design concepts meet those needs

- Wireframes communicate and document ideas for the design team

- Mockups make ideas concrete to stakeholders and sponsors

- Prototypes support heuristic evaluation

- Prototypes support evaluation by giving users something to react to and use

Beyond these specific uses, prototypes help build empathy for the user by allowing designers to directly experience the use of their system. However, simply using the prototype is often insufficient for the designer to have the same experience as the actual user because designers are often very different from the users. One method for designers to have an experience that more closely matches that of actual users is to use empathy suits. Empathy suits can help a 30-year old feel what it might be like to be an 85-year old to get into car, or what it is like to get into a car when nine months pregnant.

Although this discussion has focused on software prototypes, prototypes of hardware are equally important, as are prototypes of new work processes. Important elements of the overall system design that the prototyping process might neglect include the support systems, such as instruction manuals, and the broader organizational design. Human factors professionals can help design these with a prototyping mindset, where they develop initial designs based on a task analysis and understanding of human capabilities and then evaluate and improve the designs in an iterative manner before a final version is delivered. Prototypes of support material, such as manuals and help systems, and of the team and organizational design are sometimes neglected if the team is too focused on the physical and software elements of the system.

2.4.3 Supporting Materials and Organizational Design

Support materials development accelerates when the product specifications become more complete. Frequently, these materials are developed only after the system design is complete. This is unfortunate. The design of the support materials should begin as part of the system specifications that begin with the front-end analyses. Products are often accompanied by manuals, assembly instructions, owner's manuals, training programs, and so forth. A large responsibility for the human factors member of the design team is to make sure that these materials are compatible with the characteristics and limitations of the human user. For example, the owner's manual accompanying a table saw contains very important information on safety and correct procedures. This information is critical and must be presented in a way that maximizes the likelihood that the user will read it, understand it, and comply with it.

Organization design reflects the need to consider user experience most broadly. This means going beyond just the characteristics or interface of a single product or piece of equipment. Organizational design elements include team structure, training, and selection process. Chapter 17 addresses the design of training and selection processes. Often an entire reengineering of the organization, including the beliefs and attitudes of employees, must be addressed to achieve the promised benefits of equipment redesign. This global approach to system redesign is often termed macroergonomics. We discuss macroergonomics in Chapter 18, which deals with social factors. New technology often changes roles of the users considerably, and ignoring the social and organization implications of these changes can undermine system success.

2.5 Evaluation

As described at the start of this chapter, design is an iterative cycle of understanding, creating, and evaluating. Understanding begins with observations of people, task analysis, and knowledge of human characteristics. This understanding informs the creation of mockups and prototypes, which are immediately evaluated as designers and users experience their creations. We have seen that the human factors specialist performs a great deal of informal evaluation during the system design phases. These evaluations produce a deeper understanding of the design problem which leads to revisions. More formal evaluations are also required. These evaluations

Informal evaluation guides the iterative design process, but more formal evaluation is essential to ensure design objectives have been met.

must carefully assess the match of the system to human capabilities, as well as the ability of the system to support the tasks of the person. Chapter 3 describes evaluation methods in detail.

2.6 Summary

In this chapter we described some of the techniques used to understand user needs and to create systems to meet those needs. Designers who skip the front-end analysis techniques that identify the users, their needs, and their tasks risk creating technology-centered designs that tend to fail. The techniques described in this chapter provide the basic outline for creating human-centered systems: develop an understanding of people's needs through observation and then test that understanding with prototypes that can be quickly adjusted to better meet people's needs. A critical step in designing human-centered systems is to define the human factors requirements. Many of these requirements depend on cognitive, physical, and social considerations. The following chapters describe these characteristics in detail.

"Indifference towards people and the reality in which they live is actually the one and only cardinal sin in design." (Dieter Rams) [39]

Additional Resources

One of the best resources is for task analysis is *Guidebook to Task Analysis* [46], which describes 41 different methods for task analysis with detailed examples. For a more general set of design methods an excellent source is *Universal Methods of Design: 100 Ways to Research Complex Problems, Develop Innovative Ideas, and Design Effective Solutions* [47], as is *Guide to Methodology in Ergonomics: Designing for human use* [34].

TaskArchitect is a computer-based tool for implementing some of these tasks analysis methods (http://www.taskarchitect.com). Human factors specialists usually rely on many sources of information to guide their involvement in the design process, including previous published research, data compendiums, human factors standards, and more general principles and guidelines.

Data compendiums provide detailed information concerning human factors aspects of system design. One example is the four-volume publication by Boff and Lincoln [48], *Engineering Data Compendium: Human Perception and Performance.*

Human Factors design standards are another form of information to support design. Standards are precise recommendations that relate to very specific areas or topics. One of the commonly used standards in human factors is the *Human Engineering Department of Defense Design Criteria Standard* MIL-STD-1472G [49]. This standard provides requirements for areas such as controls, visual and audio displays, labeling, anthropometry, workspace design, environmental factors, and designing for maintenance, hazards, and safety. Other standards include the ANSI/HFES-100 VDT standard and the *ANSI/HFES-200 ANSI/HFES 200 Human Factors Engineering of Software User Interfaces* [50].

Human Factors principles and guidelines provide more general information than standards. Standards do not provide solutions for all design problems. For example, there is no current standard to tell a designer where to place the controls on a camera. The designer must look to more abstract principles and guidelines for this information. Human factors principles and guidelines cover a wide range of topics, some more general than others. Rams, Nielsen, and Tognazzini provide general principles for design [39, 40, 41], and Van Cott and Kinkade provide human factors guidelines for equipment design [51]. The following chapters reference specific guidelines related to physical facilities medical devices, and vehicle design.

Questions

Questions for 2.1 Human Factors in Design and Evaluation

P2.1   What is the difference between user interface design and user experience design?

P2.2   Why is designing a beautiful interface often insufficient in creating a useful system?

P2.3   What type of product or system might be best suited to Vee, plan-do-check-act, and scrum development cycles?

P2.4   What is the difference between the plan-do-check-act, and scrum development cycles?

P2.5   Why are observations and interviews of actual users important for designers and engineers?

P2.6   Describe how the speed-accuracy tradeoff would lead you to apply different human factors methods to a smart phone app versus a commercial airliner.

P2.7   Describe the difference between user-centered system design and the user designing the system.

P2.8   Why are observations generally preferred over focus groups for front-end analysis?

P2.9   Give an example of how the lack of holistic, systems thinking could lead technology development to have unintended consequences.

P2.10   What alternatives to evaluation would be preferable to a comprehensive test and evaluation?

P2.11   Explain why the sequence of understand, create, and evaluate is cyclical.

P2.12   Discuss how post-release surveillance would be used differently for consumer products and high-risk systems.

Questions for 2.2 Understanding the User, Context, and Tasks

P2.13   Describe the role of the Five Whys in understanding the role of human error in system performance.

P2.14   How does the master-apprentice mindset influence how one might observe and interview people as part of a contextual inquiry?

P2.15   How does a task analysis help designers develop a deeper empathy for those they design for?

P2.16   What are the three basic elements of a task analysis?

P2.17   How does the iterative nature of task analysis affect how you would organize your data collection and interpretation?

Questions for 2.3 How to Perform a Task Analysis

P2.18   Why is it important to identify the focus and purpose of a task analysis before you begin?

P2.19   What four types of data are typically collected and how might they be more or less relevant to designing the layout of machines in a factory? Compare that to designing news feed on a smartphone.

P2.20   What data recording technique or techniques would you use to design a route planning and navigation app for pizza delivery?

P2.21   Why is the critical incident technique particularly useful for understanding the tasks performed in high-risk environments?

P2.22   What general limitation affects all task analysis data collection techniques?

P2.23   If the focus of your task analysis is on coordinating the communication of baristas for a local coffee shop, what task summary approach would you use and why: Task hierarchy, task flow, or task sequence?

P2.24   If the focus of your task analysis is on supporting decisions with a checklist, which task summary approach would you use and why: Task hierarchy, task flow, or task sequence?

P2.25   If the focus of your task analysis is on training operators on the concepts needed to control a nuclear power plant, which task summary approach would you use and why: Task hierarchy, task flow, or task sequence?

P2.26   How does defining a persona differ from identifying a users' role?

P2.27   What is the benefit of defining a persona compared to simply listing user characteristics?

P2.28   How do scenarios differ from use cases in the context of moving from task analysis to system design?

Questions for 2.4 Iterative Design and Refinement

P2.29   Is it possible to create prototypes based on use cases and scenarios? Explain.

P2.30   Describe a specific analysis that goes beyond the development of personas and use cases to address a particular issue.

P2.31   Describe the difference in the two types of understanding that guides prototype design: user tasks and general human capabilities.

P2.32   Apply two of the design heuristics to the re-design of the instrument cluster of a car.

P2.33   How might a design pattern speed the design of the payment system for the website of car rental company?

P2.34   How does a decision matrix help justify the selection of particular product features for inclusion in a product?

P2.35   You are designing a website for a local real estate agent, describe how you would use wireframes, mockups, and prototypes. Describe the primary audience and purpose of each.

P2.36   Identify two elements of a system beyond the obvious focus of a prototype that often merit

Chapter 3

Evaluation Methods

At the end of this chapter you will be able to...

1. recognize differences between evaluation methods and how they support the human factors design cycle

2. understand the role of formative and summative human factors evaluations

3. apply experimental design principles to create a controlled study

4. understand representative sampling and the implications for study design and generalization

5. interpret results and recognize the limitations of a study

6. identify the ethical issues associated with collecting data from people

A government official was involved in a car crash when another driver ran a stop sign while texting on a mobile phone. The crash led the official to introduce legislation that banned all mobile phone use while driving. However, the public challenged whether one person's experience could justify a ban on all mobile phone use while driving. A consulting firm was hired to provide evidence regarding whether or not the use of mobile devices compromises driver safety. At the firm, Erika and her team must develop a plan to gather evidence to guide the design of effective legislation regarding whether or not mobile devices should be banned.

Where and how should evidence be obtained? Erika might review crash statistics and police reports, which could reveal that mobile phone use is not as prevalent in crashes even though the prevalence of use of these devices for talking, texting, and calling while driving seems high when collected from a self-reported survey. But how reliable and accurate is this evidence? Not every crash report may have a place for the officer to note whether a mobile phone was or was not in use; and those drivers completing the survey may not have been entirely truthful about how often they use their phone while driving. Erika's firm might also perform their own research in a costly driving simulator study, comparing the driving performance of people while the smartphone was and was not in use. But do the conditions in the simulator match those on the highway? On the highway, people choose when they want to talk on the phone. In the simulator, people are asked to talk at specific times. Erika might also review previously conducted research, such as controlled laboratory studies. For example, a laboratory study might show how talking interferes with computer-based "tracking task", as a way to represent steering a car, and performing a "choice reaction task", as a way to represent responding to red lights [52]. But are these tracking and choice reaction tasks really like driving?

No one evaluation method provides a complete answer.

These approaches to evaluation represent a sample of methods that human factors engineers can employ to discover "the truth" (or something close to it) about the behavior of people interacting with systems. Human factors engineers use standard methods that have been developed over the years in traditional physical and social sciences. These methods range from the *true experiment* conducted in highly controlled laboratory environments to less controlled, but more representative, quasi-experiment or *descriptive studies* in the world. These methods are relevant to both the consulting firm trying to assemble evidence regarding a ban on mobile devices and to designers evaluating whether a system will meet the needs of its intended users. In Chapter 2 we saw that the human factors specialist performs a great deal of informal evaluation during the system design phases. This chapter describes more formal evaluations to assess the match of the system to human capabilities.

Given this diversity of methods, a human factors specialist must be familiar with the range of methods that are available and know which methods are best for specific types of design questions. It is equally important for researchers to understand how practitioners ultimately use their findings. Ideally, this enables a human factors specialist to work in a way that will be useful to design, thus making the results applicable. Selecting an evaluation method that will provide useful information requires that the method be matched to its intended purpose.

> Evaluation methods are not equally suited to all design questions.

3.1 Purpose of Evaluation

In Chapter 2 we saw how human factors design occurs in the cycle of understanding, creating, and evaluating. Chapter 2 focused on understanding peoples' needs and characteristics and using that understanding to create prototypes that are refined into the final system through iteration. Central to this iterative process is evaluation. Evaluation identifies opportunities to improve a design so that it serves the needs of people more effectively. In the understand-create-evaluate cycle, the evaluate step is both the final step in assessing a design and the first step of the next iteration of the design, which provides a deeper understanding of what people need and want.

Evaluation methods that serve as the first step of the next iteration of the design are termed formative evaluations. *Formative evaluations* help understand how people use a system and how the system might be improved. Consequently, formative evaluations tend to rely on *qualitative measures*—general aspects of the interaction that need improvement. Evaluation methods that serve as the final step in assessing a design are termed summative evaluations. *Summative evaluations* are used to assess whether the system performance meets design requirements and benchmarks. Consequently, summative evaluations tend to rely on *quantitative measures*—numeric indicators of performance. The distinctions between summative and formative evaluations can be described in terms of three main purposes of evaluation:

- **Understand how to improve (Formative evaluation):** does the existing product address the real needs of people? is it used as expected?

- **Diagnose problems with prototypes (Formative evaluation):** how can it be improved? why did it fail? why isn't it good enough?

- **Verify (Summative evaluation):** does the expected performance meet design requirements? which system is better? how good is it?

Each of these questions might be asked in terms of safety, performance, and satisfaction. For Erika's analysis, predicting the effect of mobile phones on driving safety is most important: how dangerous is talking on a phone and driving?

Table 3.1 shows the example evaluation techniques for three evaluation purposes. The first row of this table shows methods associated with diagnosing problems with qualitative data. Qualitative data are not numerical and include responses to open-ended

Purpose	Data used	Evaluation Methods
Understand	Qualitative	Open-ended survey
	Quantitative	Task analysis
Diagnose	Qualitative	Heuristic evaluation
		Cognitive walkthrough
		Usability test
Verify	Quantitative	Field test

Table 3.1 Purpose and data and for different Evaluation methods.

questions, such as "what features on the device would you like to see?" or "what were the main problems in operating the device?" Qualitative data also includes observations of behavior and interpretation of interviews. They are particularly useful for diagnosing problems and identifying opportunities for improvement. These opportunities for improvement make qualitative data particularly important in the iterative design process, where the results of a usability test might guide the next iteration of the design.

The second row of the table shows methods associated with verifying the performance of the system with quantitative data. Quantitative data include measures of response time, frequency of use, as well as subjective assessments of workload. Quantitative data include any data that can be represented numerically. The table shows that quantitative data are essential for assessing whether a system has met its objectives and if it is ready to be deployed. Quantitative data offer a numeric prediction of whether a system will succeed. In the evaluating of whether there should be a ban of mobile phones, quantitative data might include a prediction of the number of lives saved if a ban were to be adopted.

The last two rows show how both qualitative and quantitative data can support understanding people's needs and characteristics relative to the design. Chapter 2 was primarily concerned with understanding people's needs and using that understanding to guide design, which is highlighted in the last two rows of Table 3.1. Although methods for understanding and methods for evaluation are presented in separate chapters there is substantial overlap between them. In this chapter, we focus on diagnosing design problems and verifying its performance, but these evaluations often require data that can inform understanding and can guide the next generations of the design.

Beyond evaluating specific systems or products, human factors specialists also evaluate more general design concepts and develop design principles. Such concept evaluations include assessing the relative strengths of keyboard versus mouse or touchscreen or rotating versus fixed maps. *Concept evaluation* reflects the basic science that supports the design principles and heuristics that make it possible to guide design without conducting a study for every design decision.

3.2 Timing and Types of Evaluation

In Erika's evaluation of the effect of mobile phones on driving safety, a critical consideration is time. If there were two years to find an answer, she might conduct a comprehensive field test, but if she has to provide an answer in weeks, then collecting field data might be difficult. More generally, the time available to provide an answer and the point in the design process—understand, create, or evaluate—are critical considerations in selecting an evaluation method.

Methods used early in the design process must diagnose problems and guide iterative design and they must do so in a very rapid manner. Methods used later in the design process, just before the product is released often take more time and must be more thorough. As discussed in Chapter 2, there are many system design processes and the emphasis (safety, performance, satisfaction), can greatly affect what type of evaluation method and data collection tools the human factors engineer can use.

The inner cycles of Figure 3.1 require very rapid evaluation methods that diagnose problems in a matter of day. Similarly, some design processes such as the Scrum approach described in chapter 2 requires response in days or weeks, but the Vee process might require a precise answer that might be possible with evaluation studies taking months. A general challenge alluded to in Chapter 2 is matching the rapid response required in a scrum design context with the time to conduct a user study, particularly when such user studies are critical as in high-risk systems. More generally, it is critical that human factors practitioners work to identify the evaluation approach that fits the timeline and needs of design process.

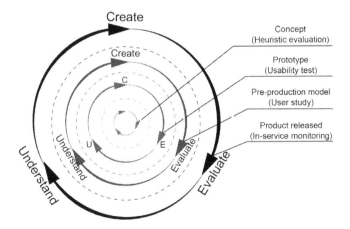

Figure 3.1 Role of evaluation at various points in the iterative design process.

3.2.1 Literature Review, Heuristic Evaluation, and Cognitive Walkthrough

Literature reviews can serve as a useful starting point for evaluation. A literature review involves reading journal articles, books, conference papers, and technical reports on previously completed studies that describe how people behave in similar situations. As an example, a good place to start such a review would be the literature cited in this textbook that relates to the focus evaluation. A good literature review can often substitute for a study itself if other researchers have already answered the question. In Erika's case, hundreds of studies have addressed various aspects of driver distraction. One particular form of literature review, known as a meta-analysis, integrates the statistical findings of many experiments that have examined a common independent variable in order to draw a very reliable conclusion regarding the effect of that variable [53, 54].

Like literature reviews, *heuristic evaluations* build on previous research and do not require additional data collection. In a heuristic evaluation a human factors specialist applies heuristics—rules of thumb, principles, and guidelines—to identify ways to improve a design. It is important to point out that many guidelines are just that: guides rather than hard-and-fast rules. Guidelines require careful consideration rather than blind application. For a computer application, heuristic evaluation might mean examining every aspect of the interface to make sure it meets usability standards [55, 26]. However, there are important aspects of a system that are not directly related to usability, such as safety and satisfaction. Thus, the first step of a heuristic evaluation would be to select human factors principles that are particularly applicable to the design, such as those listed at the end of Chapters 4-18. We will discuss below the research methods necessary to establish the validity of the principles.

The second step of a heuristic evaluation is to carefully inspect the design and identify where it violates the design principles. A simple example might be a violation of control-display compatibility, where a control is located far from the display that it changes; or violation of font size, where digits are too small to be read at a glance. While an individual evaluator can perform the heuristic evaluation, the odds are great that this person will miss most of the usability or other human factors problems. Nielson [55] reports that, averaged over six projects, only 35 percent of the interface usability problems were found by individual evaluators. Because different evaluators find different problems, the difficulty can be overcome by having multiple evaluators perform the heuristic evaluation. Nielson recommends using at least three evaluators, preferably five. Each evaluator should inspect the design in isolation from the others. After each has finished the evaluation, they should communicate and aggregate their findings.

Once the heuristic evaluations have been completed, the results should be conveyed to the design team. Often, this can be done in a group meeting, where the evaluators and design team

Heuristic evaluation should include at least three evaluators.

members discuss the problems identified and brainstorm to generate possible design solutions. Heuristic evaluation has been shown to be very cost effective. For example, Nielson [56] reports a case study where the cost was $10,500 for the heuristic evaluation, and the expected benefits were estimated at $500,000 (a 48:1 benefit-cost ratio).

A heuristic evaluation provides a relatively broad and informal assessment of the design, compared to the more structured approach of a *cognitive walkthrough*. A cognitive walkthrough considers each task associated with a system interaction, and poses a series of questions for each task to highlight potential problems that might confront someone trying to actually perform the sequence of tasks. Cognitive walkthroughs are best suited for interaction design and heuristic evaluations are best suited for interface design. Questions that guide the cognitive walkthrough include [57]:

- Is it likely that the person will perform the right action?

- Does the person understand what task needs to be performed?

- Will the person notice that the next task can be performed?

- Will the person understand how to perform the task?

- Does the person get feedback after performing the task indicating successful completion?

Walking through each task with these questions in mind will identify places where people are likely to make mistakes or get confused, which can be noted for discussion with the design team in a manner similar to the results of a heuristic evaluation.

Literature reviews, heuristic evaluations, and cognitive walkthroughs do not involve collecting data from people interacting with the system, which makes them fast to apply to a system, and so are particularly useful early in the design. One important limitation of these approaches is that the analysts might suffer from learned intuition and the curse of knowledge about how the system works. In these situations, even with the help of the heuristics and the walkthrough questions, they might not notice problems that might frustrate a less familiar person. For this reason, usability testing with people who are similar to those who will eventually use the system is essential. For example, a team of engineers in their 30's might not understand how an 85 year old woman does could be confused by what to do with a computer mouse. The following sections describe usability testing and other evaluation techniques that collect data from likely users.

Learned intuition can undermine heuristic evaluations and cognitive walkthroughs.

3.2.2 Usability Testing

Usability testing is a formative evaluation technique—it helps diagnose problems and identify opportunities for improvement as part of an iterative development process. Usability testing involves users interacting with the system and measuring their performance as ways to improve the design. Usability is primarily the degree to which the system is easy to use, or "user friendly." This translates into a cluster of factors, including the following five variables (from Nielson[55]):

- **Learnability:** The system should be easy to learn so that the user can rapidly start getting some work done.

- **Efficiency:** The system should be efficient to use so that once the user has learned the system, a high level of productivity is possible.

- **Memorability**: The steps in system use should be easy to remember so that the casual user is able to return to the system after some period of not having used it, without having to learn everything all over again.

- **Errors:** The system should have a low error rate so that users make few errors during the use of the system and so that if they do make errors, they can easily recover from them. Further, catastrophic errors must not occur.

- **Satisfaction:** The system should be pleasant to use so that users are subjectively satisfied when using it; they like it.

Usability testing identifies how to improve a design on each of these usability dimensions, which differs substantially from typical experiments that can have anywhere from 20 to 100 participants. Usability testing typically includes just five participants, and each test is part of a sequence. After each usability test, the results are shared with the design team, the design is refined, and another usability test is conducted with a new set of users [45]. A single usability test is not enough—a minimum of two and ideally five or more tests and refinement iterations are needed.

Figure 3.2 shows a powerful way to enhance usability testing. Rather than focusing on a single design, three are developed in parallel and tested. The best elements of the three designs are merged and the resulting design is assessed and refined in a series of two to five tests and refinement iterations.

3.2.3 Comprehensive Evaluations and Controlled Experiments

Comprehensive system evaluation provides a more inclusive, summative assessment of the system than a usability evaluation. The data source for a comprehensive system evaluation often involves controlled experiments. Similarly, user studies aimed at understanding more general factors affecting human behavior, such as how voice control compares to manual operation of a mobile device while driving, also require controlled experiments. So does research required to establish the validity of general human factors principles, such as control-display compatibility. The experimental method consists of deliberately producing a change in one or more causal or *independent variables* and measuring the effect of

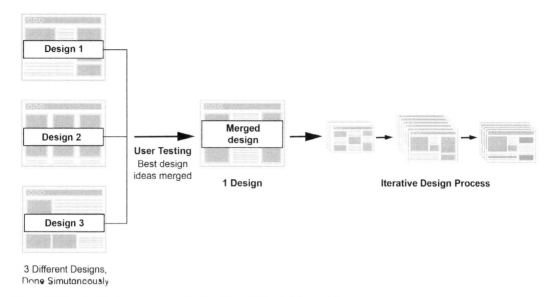

Figure 3.2 Parallel design process with iteration. (Adapted from: https://www.nngroup.com/articles/parallel-and-iterative-design/).

that change on one or more *dependent variables*. An experiment should change only the independent variables of interest while all other variables are held constant or controlled. However, for some human factors studies, participants need to perform the tasks in various real-world contexts for a comprehensive system evaluation. In such cases, control can be difficult. As control is loosened, the researcher will need to depend more on quasi-experiments and *descriptive methods*: describing relationships even though they could not actually be manipulated or controlled. For example, the researcher might describe the greater frequency of mobile phone crashes in city driving compared to freeway driving to help draw a conclusion that mobile phones are more likely to distract drivers when they are dealing with complex traffic situations.

3.2.4 In-service Evaluation

In-service evaluation refers to evaluations conducted after a design has been released, such as after a car has been on the market, after a modified manufacturing line has been placed in service, or after a new mobile phone operating system has been released. Descriptive studies are critical for in-service evaluation because experimental control is often impossible. In the vignette presented at the beginning of this chapter, an in-service evaluation of existing mobile phone use might start by examining crash records, or moving violations. This will give us some information regarding road safety issues, but there is a great deal of variation, missing data, and underreporting in such databases. Like most descriptive studies, such a comparison of crashes is a challenge because each crash involves many different conditions and important driver-related activities (e.g., eating, cell-phone use, looked but did not see) might go unreported.

A-B testing is a type of in-service evaluation where one version of a system (A) is compared to another version of the system (B), where one is typically an improvement over the existing system [47]. A-B testing is very common for internet applications, where thousands of A-B tests provide data to guide screen layout and even shades of color[58]. A-B testing typically collects data from many thousands of people, compared to the 3-5 for usability testing or the 20-100 participants in a typical experiment.

Collecting data, whether in an experimental or descriptive study, is only half of the process. The other part is inferring the meaning or message conveyed by the data, and this usually involves generalizing or predicting from the particular data sampled to the broader population. Do mobile phones compromise (or not) driving safety in the broad section of automobile drivers, and not just in the sample of drivers from a driving simulator experiment, from crash data in one geographical area, or from self-reported survey data? The ability to generalize involves care in both the design of experiments and in the statistical analysis.

Although the details depend on the type of study, the general steps are similar. A descriptive study, a usability evaluation, and a controlled experiment might differ substantially in the amount and type of data collected and how it would be analyzed, but the general steps would be similar. In Table 3.2 we outline the five general steps with more detail to follow.

The following sections expand on steps associated with conducting a study that are particularly complicated: study design, measurement, data analysis, and drawing conclusions.

3.3 Study Design

An experiment involves examining the relationship between independent variables and the resulting changes in one or more dependent variables, which are typically measures of performance, workload, situation awareness, or preference. The goal is to show that manipulations of the independent variable, and no other variable, causes changes in behavior and attitudes measured by the dependent variables.

An *experimental design* identifies the independent variables that will be controlled in the study, and the dependent variables or outcomes of interest to be measured and observed. The key to good experiments is control. That is, only the independent variable should be manipulated, and all other variables should be held constant. However, control becomes increasingly difficult as the tasks being examined need to be considered in the context of the environment to which the research results are to generalize. Furthermore, controlling everything in an experiment may lead to results that do not *generalize*—hold up in the messy real world when other variables that we cannot control affect performance.

In designing a study, consider all variables that might affect the

---

⸎Google used A-B testing to pick one of 41 shades of blue [58].

Steps in Conducting a Controlled Experiment

Step 1. Define research questions and hypotheses. The researcher must clearly state the questions addressed by the study, often in terms of hypotheses about how the experimental conditions will affect the outcomes. For example, does using a smartphone while driving create more driving errors?

Step 2. Specify the experimental design by defining how the independent variables that will be manipulated, as well which dependent variables will be recorded. For the example of using a smartphone while driving, one independent variable is using the smartphone (yes, no) and the dependent variable might be driving errors.

Step 3. Conduct the study by collecting pilot data and then collecting the main experiment data. Collecting pilot data prior to the main experiment is essential. Such data ensures that the experimental conditions are implemented properly, that participants understand the instructions, and that the data are recorded properly.

Step 4. Analyze the data Data for each participant are summarized. For our example, the number of lane deviations. This summary indicates whether there are meaningful differences among the groups.

Step 5. Draw conclusions and communicate results by properly interpreting strength of effects and uncertainty of the results. Variability of drivers' response means that any effect of using the smartphone on driving might be due to difference between drivers.

Table 3.2 Steps for conducing a controlled experiment.

dependent variables. Extraneous variables have the potential to interfere in the causal relationship of interest and must be controlled. If these extraneous variables do influence the dependent variable, we say that they are *confounding variables*. One group of extraneous variables is the wide range of differences between people, so it is important that the people in each experimental condition differ only with respect to the treatment condition and not on any other variable. For example, in the smartphone study, you would not want elderly drivers using the car phone and young drivers using no phone, otherwise age would be a confounding variable. One way to make sure all groups are equivalent is to randomly assign people to each experimental condition. If the sample is large enough, this *random assignment* of participants to experimental conditions ensures that differences between people, such as age, even out across the groups.

Another way to avoid the confounding effect of differences between groups is to use a *within-subjects or repeated measures design*: the same person drives with and without a cell phone. However, this design creates a different set of challenges for experimental control. For within-subjects designs, there is another variable that must be controlled: the order in which the participants experience the experimental conditions, which creates what are called order effects. When people participate in several conditions, the dependent measure may show differences from one condition to the next simply because the conditions are experienced in a particular order. For example, if participants use five different cursor-control devices in an experiment, they might be fatigued by the time they are tested on the fifth device and therefore exhibit more errors or slower times. This would be due to the order of devices used rather than the device. Alternatively, if the cursor-control task is new to the participant, he or she might show learning and actually do best on the fifth device tested, not because it was better, but because the cursor-control skill was more practiced. These order effects of fatigue and practice effects in between-subjects designs are both confounding variables; while they work in opposite directions, to penalize or reward the late-tested conditions, they do not necessarily balance each other out.

We can use a variety of methods to keep order from confounding the independent variables. For example, extensive practice can reduce learning effects. Time between conditions can reduce fatigue. More commonly, the order of conditions is counterbalanced. Counterbalancing simply means that different participants receive the treatment conditions in different orders. For example, half of the participants in a study would use a trackball and then a mouse. The other half would use a mouse and then a trackball. There are various techniques for counterbalancing order effects; the most common is a Latin-square design. Research methods books describe how to use these designs [59, 47, 60].

Other variables in addition to participant variables must be controlled. For example, it would be a poor experimental design to have one condition where smartphones are used in a Jaguar and another condition where no phone is used in an Oldsmobile.

Good experimental design avoids confounding the independent variables with extraneous variables.

Differences in vehicle dynamics affect driving performance. The phone versus no-phone comparison should be carried out in the same vehicle (or same type of vehicle). We need to remember, however, that in more applied research, it is sometimes impossible to exert perfect control.

In summary, the researcher must control extraneous variables by making sure they do not covary with the independent variable. If they do covary, they become confounds and make interpretation of the data impossible. This is because the researcher does not know which variable caused the differences in the dependent variable. Sometimes confounding is hard to avoid: in driving, age and experience are confounded. Younger drivers usually have less experience and it is not always clear if they driving performance is worse because they are young or because they have less experience.

3.3.1 One-factor Designs

One-factor (2 levels)

The simplest experimental designs involve one independent variable, and at their very simplest level involve only 2 levels (or 2 conditions) to be compared (two groups in a between-participant study). These two may involve a control and a treatment, like driving only (control) and driving with a cell phone (treatment). Alternatively they may involve two different levels of a treatment, such as comparing a keyboard versus a voice control on the cell phone.

Two-factor (2×2)

Sometimes the one factor with two-level design does not adequately test our hypothesis of interest. For example, if we want to assess the effects of display brightness on reading speed, we might want to evaluate several different levels of brightness. We would be studying one independent variable (brightness) but would want to evaluate many levels of the variable. If we used five different brightness levels and therefore five levels or conditions, we would still be studying one independent variable but would gain more information than if we used only two levels or conditions. With this design, we could develop a quantitative model or equation that predicts performance as a function of brightness. In a different multilevel design, we might want to test four different input devices for cursor control, such as trackball, thumbwheel, traditional mouse, and key-mouse. We would have four different experimental conditions but still only one independent variable (type of input device).

Three-factor (2×2×2)

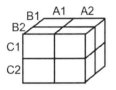

Figure 3.3 Examples of different experimental designs.

3.3.2 Multiple-factor Designs

In addition to increasing the number of levels of a single independent variable, we can increase the number of independent variables or factors. This makes it possible to evaluate more than one independent variable or factor in a single experiment. In human factors, we are often interested in complex systems and therefore in simultaneous influence of many variables rather than just two. In the case of the cell phone and driving, we may wish to determine if using the cell phone (Factor A) has the same or different effects on older versus younger drivers (Factor B).

A multifactor design that evaluates two or more independent variables by combining the different levels of each independent variable is called a *factorial design*. The term factorial indicates that all possible combinations of the independent variable levels are combined and evaluated. Factorial designs allow the researcher to assess the effect of each independent variable by itself and also to assess how the independent variables interact with one another. That is, for example whether night shifts may have a more adverse effect on older than younger workers. Because much of human performance is complex and human-machine interaction is often complex, factorial designs are the most common research designs used in both basic and applied human factors research. Factorial designs can be more complex than a $2X2$ design in a number of ways. First, there can be more than two levels of each independent variable. For example, we could compare driving performance with two different smartphone designs (e.g., hand-dialed and voice-dialed), and also with a "no phone" control condition. Then we might combine that first three-level variable with a second variable consisting of two different driving conditions: city and freeway driving. This would result in a $3X2$ factorial design. Another way that factorial designs can become more complex is by increasing the number of factors or independent variables. Suppose we repeated the above $2X3$ design with both older and younger drivers. This would create a $2X3X2$ design with 12 different conditions. A design with three independent variables is called a three-way factorial design.

Combining independent variables has three advantages: (1) It is efficient because you can vary more system features in a single experiment: (2) It captures more of the complexity found in the real world, making experimental results more likely to generalize. (3) It allows the experimenter to detect interactions between independent variables.

3.3.3 Between-subjects Designs

As we have seen, a between-subjects design is a design in which all of the independent variables are between-subjects, and therefore each combination of independent variables is administered to a different group of subjects. Between-subjects designs are most commonly used when having participants perform in more than one of the conditions would be problematic. For example, if you have participants receive one type of training (e.g., on a simulator), they can not start over for another type of training because they already know the material. Between-subjects designs also eliminate certain confounds related to order effects, which we discussed earlier.

3.3.4   Within-subjects Designs

This is a design where the same participant is used in multiple treatment conditions, many experiments, it is feasible to have the same people participate in all of the experimental conditions. For example, in the driving study, we could have the same subjects drive for periods of time in each of the four conditions. In this way, we could compare the performance of each person with him- or herself across the different conditions. This within-subject performance comparison illustrates where the methods gets its name. An experiment where all independent variables are within-subject variables is termed a within-subjects design. Using a within-subjects design is advantageous in two major respects: it is more sensitive and easier to find statistically significant differences between experimental conditions, and it is also advantageous when the number of people available to participate in the experiment is limited.

3.3.5   Mixed Designs

In factorial designs, each independent variable can be either between-subjects or within-subjects. If both types are used, the design is termed a mixed design. If one group of subjects drove in heavy traffic with and without a smartphone, and a second group did so in light traffic, this is a mixed design.

3.3.6   Sampling People, Tasks, and Situations

Once the experimental design has been specified with respect to independent variables, the researcher must decide what people will be recruited, what tasks the people will be asked to perform, and in what situations. The concept of *representative sampling* guides researchers to select people, tasks, and situations that they are designing for.

Participants should represent the population or group in which the researcher is interested in studying. For example, children under 18 may not be the appropriate sample for studying typical driver behavior—unless you are interested in studying new drivers in Iowa where they start driving at 14.5 years old. If we are studying systems that will be used by the elderly, the target population should be those aged 65 and older. Depending on the aims of the design, this target population might be defined more narrowly: English speakers living in the United States who are healthy, and read at a certain grade level. Importantly, a sample representative of typical users is probably not the population of university students or engineers designing the product, which are often used.

Just as you would not conduct a study with a single study participant, it is also important to include more than one task that people might encounter with the design. In the example of the mobile phone evaluation, these tasks might include placing a call, answering a call, reading a text, and sending a text. Each of these would define a different dependent variable, and hence, perhaps, a different unit of measurement. Just as the sample of people needs

Between-subject Design

	A1	A2
B1	S1	S2
B2	S3	S4

Within-subject Design

	A1	A2
B1	S1, S2 S3, S4	S1, S2 S3, S4
B2	S1, S2 S3, S4	S1, S2 S3, S4

Mixed Design

	A1	A2
B1	S1, S2	S1, S2
B2	S3, S4	S3, S4

Figure 3.4 Examples of between-subject, within-subject, and mixed subject designs.

to be representative of the population using the design, the sample of tasks should be representative of tasks people are likely to perform.

For applied research, we try to identify tasks and environments that will give us results that generalize to those situations and allow us to predict outcomes of our human factors interventions precisely. This often means conducting the experiments in situations that are most representative of those actually encountered with the design. For basic research we try to identify tasks and environments that generalize across many situations beyond those included in the study. Consider these somewhat conflicting aims as you work through the exercise in Table 3.3. All elements of the study design will likely differ if you intend to test a specific intervention for a specific situation (*applied research*), or if you hope your results might help us understand human behavior in a broad range of situations (*basic research*).

3.4 Measurement

Because an experiment involves examining the relationship between independent variables and changes in one or more dependent variables, defining what is measured—the dependent variables—is crucial. The dependent variables are what can be measured and relate to the outcomes described in the research questions. The research questions are often stated in terms of *theoretical constructs*, where constructs describe abstract entities that cannot be measured directly. Common constructs in human factors studies include: workload, situation awareness, fatigue, safety, acceptance, trust, and comfort. These constructs cannot be measured directly and the human factors researcher must select variables that can be measured, such as subjective ratings and response times that are strongly related to the underlying constructs of interest. To assess how smartphones affect driving, the underlying construct might be safety and the measure that relates to safety might be error in lane keeping where the car's tire crosses a lane boundary. Safety might also be measured by ratings from the drivers indicating how safe they felt

Subjective ratings are often contrasted with objective performance data, such as error rates or response times. The difference between these two classes of measures is important, given that subjective measures are often easier and less expensive to obtain, with a large sample size. Both objective and subjective measures are useful. For example, in a study of factors that lead to stress disorders in soldiers, objective and subjective indicators of event stressfulness and social support were predictive of combat stress reaction and later posttraumatic stress disorder. The subjective measure were the stronger predictors than the objective measure [61]. In considering subjective measures, however, what people rate as "preferred" is not always the system feature that supports best performance [62]. For example, people almost always prefer a color display to a monochrome one, even when color undermines

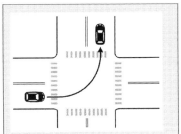

Study Design Exercise

Design a study to address the following problem statement:"Making left turns is challenging for older drivers. What design changes would you recommend to make such turns safer for older drivers?"

Step 1: Define a research question
Write out a testable research question to address the problem statement.

Step 2: Independent variables Identify two factors you can control for given your study protocol and identify the number of levels for each factor.

Step 3: Participants Who will you include in the study? What screening criteria would you use?

Step 4: Dependent variables What dependent variables will you measure? That is, what are the outcomes of interest?

Step 5: Study protocol What type of experimental set up will allow you to capture these outcomes?

Table 3.3 An exercise in study design to assess an intervention to make left turns safer.

performance.

Furthermore, people cannot always predict how they would respond to surprising events in different conditions, like during system failures. Human factors psychology is much more than intuitive judgment (of either the designer OR the participant). It is for this reason that objective experiments must go beyond the data offered by heuristic evaluations.

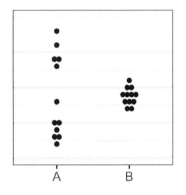

Subjective and objective dependent variables provide important and complementary information. We often want to measure how causal variables affect several dependent variables at once. For example, we might want to measure how use of a smartphone affects a number of driving performance variables, including deviations from the lane, reaction time to cars or other objects in front of the vehicle, time to recognize objects in the driver's peripheral vision, speed, acceleration, and so forth. Using several dependent variables helps triangulate on the truth—if all the variables indicate the same outcome then one can have much greater confidence in that outcome.

3.5 Data Analysis

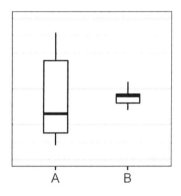

Collecting data, whether from an experiment or descriptive study, is only part of the process. Another part is inferring the meaning or message conveyed by the data, and this usually involves generalizing or predicting from the particular sample of data to the broader population of people and context of use. Do mobile phones compromise driving safety for most automobile drivers, or are our findings specific to the sample of drivers used in the simulator experiment or the sample represented in crash statistics? Do all mobile phones comprise safety in a similar way? The ability to generalize involves care in both the design of experiments and in the statistical analysis.

Data visualization is a first step in any data analysis. Figure 3.5 shows three graphs of the same data for two conditions, such as driving with and without a smartphone. The top graph is a dot plot, which shows the response of each person in the study and gives a sense of how each individual compares to the others. Here it shows that the condition on the left generates a wide distribution of responses compared to that on the right. The middle graph is a box plot, which shows the median value as a horizontal line and the 25th and 75th percentiles of the distribution as the upper and lower edges of the box. The box plot is useful when you have too many data points to plot individually. The bottom graph is a bar chart, which shows the mean as the height of the bar and error bars indicating variability. The error bars shows there is much greater variability for the condition on the left than on the right. Without the error bars, the bar chart might mislead you into thinking the conditions were very similar. More generally, visualizations that show individual data points, such as the dot plot, are critical in understanding whether summary statistics such as the mean value are a good representation of the underlying data, which is not the

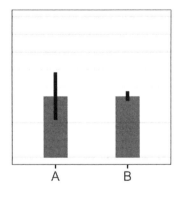

Figure 3.5 Three views of the same data. Upper graph shows individual data points, middle graph shows box plots, and the bottom graph shows mean and confidence interval. Showing individual data points identifies odd responses. (Created in R with ggplot2.)

case in this example.

3.5.1 Analysis of Controlled Experiments

Once the experimental data have been collected, the researcher must determine whether the dependent variable(s) actually did change as a function of experimental condition. For example, was driving performance really "worse" while using a smartphone? To evaluate the research questions and hypotheses, the experimenter calculates two types of statistics: descriptive and inferential statistics. Descriptive statistics are a way to summarize the dependent variable for the different treatment conditions, while inferential statistics tell us the likelihood that any differences between our experimental groups are "real" and not just random fluctuations due to chance .

Differences between experimental groups are usually described in terms of averages or means. Research reports typically describe the mean value of the dependent variable for each group of subjects. This is a simple way of conveying the effects of the independent variable(s) on the dependent variable. Standard deviation for each group convey the spread of scores. Standard errors or confidence intervals, as described later in this section are preferred to simply reporting the mean values [63].

Report uncertainty along with your estimate of the mean value.

While experimental groups may show different means for the various conditions, it is possible that such differences occurred solely due to chance. Humans almost always show random variation in performance, even without manipulating any variables. It is quite possible to get two groups of participants who have different means on a variable, without the difference being due to any experimental manipulation; in the same way that you are likely to get a different number of "heads" if you do two series of 10 coin tosses. In fact, it is unusual to obtain means that are exactly the same. So, the question becomes: are the difference big enough that we can rule out chance and assume the independent variable had an effect? Inferential statistics give us, effectively, the probability that the difference between the groups is due to chance. If we can rule out the "chance" explanation, then we infer that the difference was due to the experimental manipulation.

A comparison of two conditions is usually conducted using the t-test. Comparison of proportions is done using a χ^2 test. For more than two groups, we use an analysis of variance (ANOVA). All three tests yield a score; for a t-test, we get a value for t, and for ANOVA, we get a value for F. Most important, we also identify the probability, p, that the t or F value would be found by chance for that particular set of data if there was no effect or difference. The smaller that p is, the more significant our result becomes and the more confident we are that our independent variable really did cause the difference. This p value will tend to be smaller when the difference between means is greater, and when the variability between our observations within a condition (standard deviation) is smaller, and, importantly, when the sample size (N) increases (more participants, or more measurements per participant). A

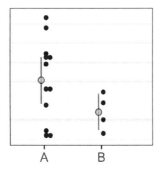

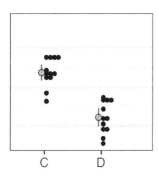

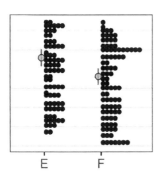

Figure 3.6 Differences in sample size and standard deviation for six conditions. Black dots show individual data points and the gray dots and lines show means and confidence intervals. (Created in R with ggplot2.)

larger sample gives an experiment greater statistical power to find true differences between conditions.

Although p values are a common inferential statistic, a more useful approach is to report confidence intervals [64]. Confidence intervals show the range of the mean value that might be expected if the study were to be repeated. The confidence interval is more informative than the p value because it can show the difference between a situation where the mean values of two conditions were very large but the variability was also large and a situation where the difference in the mean values was small but the variability was very small. These conditions might produce the same p values, but considering the means and confidence intervals would suggest a very different interpretation of the results.

Figure 3.6 shows how the variability of the data and the sample size contribute to uncertainty about differences between conditions. The upper graph shows that Condition A has a larger number of samples, but a large standard deviation. Condition B has a smaller standard deviation but a smaller sample. Hence both have a wide confidence interval, and they so are not statistically different from each other. In the lower graph, both conditions C and D have a smaller standard deviation and larger sample. Hence their smaller confidence interval. Thus, even though the difference between means is the same in the two graphs the difference will not be significant in the top, but it will be in the bottom.

The size of the confidence interval is conventionally set at 95%. This means that if the mean of one condition lies within the 95% confidence interval of another, it cannot be firmly concluded that the two conditions differ as a result of the experimental manipulation. This is the case on the upper graph of Figure 3.6.

To complement the p values and confidence intervals, the *effect size* should also be reported. The effect size describes the how much effect the independent variable has relative to the variability of sample. Typically this is expressed as the mean difference divided by the standard deviation. Effect size is important because even a small difference between mean values will show a high degree of statistical significance—a very low p value—if a very large sample is collected. Conditions E and F in Figure 3.6 show such a situation. A small effect size indicates the difference between conditions may be statistically significant, but not practically significant, a topic we return to later in the chapter.

In our examples so far, we have considered two conditions, levels or groups. But inferential statistics apply as well to multi-level experiments (e.g., three levels of one IV) or to multi-factor experiments (e.g., more than two independent variables). In these cases the ANOVA (using an F-test), rather than the t test will return an inference measure (a p value) which is, essentially, a collective measure of the comparison between all conditions in the experiment. Each condition will of course have its own standard error and confidence interval.

3.5.2 Analysis of Continuous Variables in Descriptive Studies

Descriptive studies do not place people into experimental conditions defined by discrete levels of independent variables. Instead, the independent and dependent variables are observed in naturally occurring situations and the independent variables or predictor variables are often continuous. Relationships between these variables are often assessed with correlational or regression analyses.

For example, a correlational analysis could assess the relationship between job experience and safety attitudes within an organization. The correlational analysis measures the extent to which two variables covary, such that the value of one can be somewhat predicted by knowing the value of the other. In a positive correlation, one variable increases as the value of another variable increases; for example, the amount of illumination needed to read text will be positively correlated with age. In a negative correlation, the value of one variable decreases as the other variable increases; for example, the frequency of car crashes is negatively correlated with experience. By calculating the *correlation coefficient*, *r*, we get a measure of the strength of the relationship–the stronger the relationship the more precise the predictions.

Similar to comparing mean values between experimental conditions, statistical tests can assess the probability that the relationship is due to chance fluctuation in the variables. The *p-value* indicates whether a relationship exists and a measure of the strength of the relationship (*r*). As with other statistical measures, the likelihood of finding a significant correlation increases as the sample size N—the number of measurements—increases.

Correlational analysis often goes beyond reporting a correlation coefficient, and typically describes the relationship with a regression equation. This equation uses the observed data to show how much change one variable will change another variable. This statistical model can even be used to predict future outcomes and might suggest optimal values for a design.

One caution should be noted. When we find a statistically significant correlation, it is tempting to assume that one of the variables caused the changes seen in the other variable. This causal inference is unfounded for two reasons. First, the direction of causation could actually be in the opposite direction. For example, we might find that years on the job is negatively correlated with risk-taking. While it is possible that staying on the job makes an employee more cautious, it is also possible that being more cautious results in a lower likelihood of injury or death. This may therefore cause people to stay on the job. Second, a third variable might cause changes in both variables. For example, people who try hard to do a good job may be encouraged to stay on and may also behave more cautiously as part of trying hard.

Don't confuse correlation with causation, such as the frequency of death by bed-sheet entanglement and per capita cheese consumption (http://tylervigen.com/spurious-correlations).

3.6 Drawing Conclusions and Communicating Results

Statistical analysis provides an essential method to differentiate between systematic effects of the independent variables and random variation between people and conditions. This analysis often seems to provide a clear decision: if the p value is less than 0.05 there is an important difference between conditions. This clarity is an illusion and drawing conclusions from statistical results requires careful judgment and communication.

3.6.1 Statistical Significance and Type I and Type II Errors

Researchers often assume that if the computed statistics show a *p-value* less than 0.05, that there is high chance that the independent variable had an effect on the dependent variable. This assumption is based on a 0.05 cutoff that suggests that you will be wrong approximately one in 20 times. Concluding that the independent variable had an effect when it was really just chance (that one in 20 times) is referred to as making a *Type I error*, often denoted as α.

In general, our goal of minimizing Type I errors and using a criterion of $\alpha < 0.05$. is reasonable when developing cause-and-effect models of the world. Type I errors can lead to the development of false theories and misplaced expectations about the benefits of design changes. Minimizing Type I errors is often reasonable in applied settings as well, where you want to minimize the chance of recommending new equipment that is actually no better than the old equipment.

We tend to accept the implicit assumption that always minimizing Type I errors is desirable and ignore the cost of a *Type II errors*. A Type II error (denoted as β) is concluding there was no effect when there was actually an effect. In evaluating a new piece of equipment, a Type II would be to conclude the new equipment is no better, based on $\beta=0.10$, when in fact it is better. Type II errors can lead to improvements being rejected.

The likelihood of making Type I and Type II errors are inversely related. If the Type I error is set at $\alpha=0.10$ instead of 0.05, we would conclude the new equipment to be better. However, increasing α would also increase the risk of recommending new equipment that might not actually be better than the old one.

Focusing on only the Type I error without considering the Type II error is particularly problematic in human factors evaluation. We frequently must conduct experiments and evaluations with relatively few participants because of expense or the limited availability of certain highly trained professionals. Using a small number of participants makes the statistical test less powerful and more likely to show no significance with a computed $p > .05$, even when there is a difference. The smaller sample size will also show greater In addition, the variability in performance between different participants or for the same participant over time and conditions is likely

Type I errors indicate that you falsely concluded a difference between conditions, and Type II errors indicate that you falsely concluded a difference does not exist.

to be great when we try to do our research in more applied environments. Again, these factors make it more likely that the results will show no significance when examined at $\alpha = 0.05$. The result is that human factors specialists might be particularly likely to commit Type II errors.

Type II errors can have practical consequences. For example, will a safety-enhancing device fail to be adopted? In the smartphone study, suppose that performance really was worse when using a smartphone than without, but the difference was not big enough to achieve statistical significance. Might the legislature conclude, in error, that cell phone use was "safe"? There is no easy answer to the question of how best to balance Type I and Type II errors [59]. The best advice is to realize that the larger the sample size, the greater is the power $(1-\beta)$ of our study and the tighter is our sample around the true mean. With a larger sample size, we will be less likely to commit a Type I or Type II error, as shown in Figure 3.7.

Low statistical power can sometimes be compensated for by building one's findings on top of prior research that have yielded similar effects in similar conditions. The formal way of compiling such research in quantitative form is through a meta-analysis and Bayesian statistics.

3.6.2 Statistical and Practical Significance

Once chance is ruled out, often meaning that the computed *p-value* is less than 0.05, researchers will examine the differences between groups. It is important to remember that two groups of numbers can be statistically different from one another without the difference being very large if statistical power is quite high. Suppose we compare two groups of Army trainees. One group is trained in tank gunnery with a low-fidelity personal computer. Another group is trained with an expensive, high-fidelity simulator. We might find that when we measure performance, the mean percent correct for the computer group is 80, while the mean percent correct for the simulator group is 83. Although the mean percent correct is only three, this could be considered statistically different if the variation was quite small. If they are significantly different when examined at $\alpha = 0.05$, we would conclude that the more expensive simulator is a better training system. However, especially for applied research, we must also look at the difference between the two groups in terms of practical significance. Is it worth spending millions to place simulators on every military base to get this three percent increase? This illustrates the tendency for some researchers to place too much emphasis on statistical significance and not enough emphasis on practical significance (sometimes called "engineering significance"). Focusing on mean values and confidence intervals (uncertainty about the mean value) can help avoid misinterpreting statistical significance as practical significance.

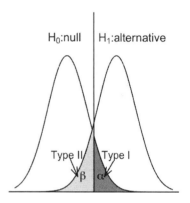

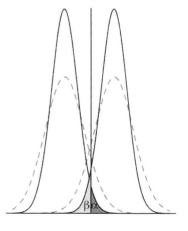

Figure 3.7 Type I and Type II errors and the effect of sample variability on their likelihood. (Created in R.)

3.6.3   Generalizing and Predicting

No single study proves anything. As we will see in the following chapters, despite substantial regularity in human behavior, individual differences are substantial as are the effect of expectations and context. A different sample of people, different instructions, and different tasks might produce a different outcome. Communicating the results of a study to the design team should reflect this uncertainty. When Erika interprets the results of a study that shows a statistically significant effect of smartphone use on lane keeping error she must consider the degree to which that effect depends on the specific people, tasks, and situations that she included in her study. Would her prediction about safer driving without a smartphone materialize if the government enacted a ban? This uncertainty makes it important to consider any study as part of the cycle where evaluation feeds back to understanding of human behavior that improves in an iterative manner after many studies.

3.7   Driver Distraction: Example of a Simple Factorial Design

To illustrate the logic behind controlled experiments we consider an example of a simple factorial design. This is where two levels of one independent variable are combined with two levels of a second independent variable. Such a design is called a $2X2$ factorial design. Imagine that a researcher wants to evaluate the effects of using a smartphone on driving performance (and hence on safety). The researcher manipulates the first independent variable by comparing driving with and without a smartphone. However, the researcher suspects that the smartphone might only affect people when they are driving in heavy traffic. Thus, the researcher adds a second independent variable consisting of light versus heavy traffic, resulting in the experimental design shown in Figure 3.8.

DRIVING CONDITIONS

	Light traffic	Heavy traffic
No mobile device	No mobile device while driving in light traffic	No mobile device while driving in heavy traffic
Mobile device	Use mobile device while driving in light traffic	Use mobile deive while driving in heavy traffic

Figure 3.8 The four experimental conditions for a 2X2 factorial design).

Imagine that we conducted the study, and for each of the drivers in the four groups shown in Figure 3.8, we counted the number of times the driver strayed outside of the lane as the dependent variable. We can look at the general pattern of data by evaluating the cell means; that is, we combine the performance of all drivers within each of the four groups.

If we look only at the effect of smartphone use (combining the light and heavy traffic conditions), we might be led to believe that use of cell phones impairs driving performance in all situations. But looking at the entire picture, as shown in Figure 3.9, we see that the use of a cell phone impairs driving only in heavy traffic conditions. When the lines connecting the cell means in a factorial study are not parallel, as in Figure 3.9, we know that there is an interaction between the independent variables: The effect of phone use depends on driving conditions. Factorial designs are popular for both basic research and applied questions because they allow researchers to evaluate interactions between variables.

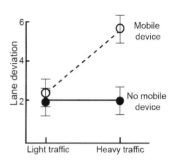

Figure 3.9 The interaction of mobile phone use and traffic.

3.8 Ethical Issues

The majority of human factors studies involve people as participants. Many professional affiliations and government agencies have written specific guidelines for the proper way to involve participants in research. Federal agencies rely on the guidelines found in the Code of Federal Regulations HHS, Title 45, Part 46; Protections of Human Subjects (http://www.hhs.gov/ohrp/regulations-and-policy/regulations/45-cfr-46/). The National Institute of Health has a web site where students can be certified in data collection with human subjects (https://humansubjects.nih.gov/resources). Anyone who conducts research using human participants should be familiar with the federal guidelines as well as American Psychological Association (APA) guidelines for ethical treatment of human subjects [65]. These guidelines fundamentally advocate the following *principles for protecting participants*:

- Protection of participants from mental or physical harm

- The right of participants to privacy with respect to their behavior

- The assurance that participation in research is completely voluntary

- The right of participants to be informed beforehand about the nature of the experimental procedures

When individuals are asked to participate in an experiment, or asked to provide data for a study, they are provided information on the general nature of the study. Often, they cannot be told the exact nature of the hypotheses because this may bias their behavior. Participants should be informed that all results would be kept anonymous and confidential. This is especially important in human factors research because participants should be at ease during

Hawthorne Effect (1924-1932) Studies conducted at the Western Electric Hawthorne Works seemed to show how various interventions, such as light levels, improved worker productivity. These effects did not last and were not a consequence of the interventions, but due to researchers failing to consider how their attention to the workers might affect their behavior—the Hawthorne effect— as well as many other uncontrolled variables [66]. How can you avoid these sorts of problems and create a good study?

Get up to speed: Collect background information. Meet with the project team in person and understand the need.
Clarify research questions: Make sure the question can, and should be answered. Find the real question, which sometimes differs from the one initially asked.
Avoid confounding: Good experimental design controls variables affecting behavior and ensures that you can draw conclusions.
Use representative sampling: Seek out a broad sample of people and tasks. Your conclusions are only as robust as your sample of tasks and people.
Triangulate with multiple measures: Answer the questions with objective data and complementary subjective data. Collect what you need and a little more. Be opportunistic.
Conduct pilot tests: Verify all elements of data collection before starting your full data collection. Use frequent internal and external checkpoints.
Look beyond the initial question: Think beyond what you set out to study. As you see issues, consider pivoting and adjusting.
Plot raw and summary data: Don't blindly summarize data. Report central tendency (e.g., mean) *and* variability (e.g., standard deviation). Show limits of the data.
Thoroughly analyze and interpret: Consider alternate explanations and possible confounding variables. Get feedback and discuss with colleagues.
Treat participants like gold: Show them they're appreciated, respect them. Without the participants you have no data.

a study and not fear that their performance might affect their job, their ability to drive, or their overall health. Finally, participants are generally asked to sign a document, an informed consent form, stating that they understand the nature and risks of the experiment, or data gathering project, that their participation is voluntary, and that they understand they may withdraw at any time. In human factors field research, the experiment is considered to be reasonable in risk if the risks are no greater than those faced in the actual job environment. Research boards in the university or organization where the research is to be conducted certify the adequacy of the consent form and that any risks to the participant are outweighed by the overall benefits of the research to society.

As one last note, experimenters should always treat participants with respect. Participants are usually self-conscious because they feel their performance is being evaluated (which it is, in some sense) and they fear that they are not doing well enough. It is the responsibility of the investigator to put participants at ease, assuring them that the system components are being evaluated and not the people themselves. This is one reason that the term user testing has been changed to usability testing to indicate the system, not the person, is the focus of the evaluation.

3.9   Summary

Evaluation completes the understand-create-evaluate cycle by providing an indication of how well the design meets the users' needs. Evaluation also provides the basis for understanding how design can be improved, and also serves as the beginning of the cycle. Evaluation in its various forms is a core element of iterative design.

Additional Resources

This chapter provides a very initial introduction to usability testing and user studies. Many excellent books provide much more detail. Some useful resources include:

Tullis, T., & Albert, B. (2013). *Measuring the User Experience, Second Edition: Collecting, Analyzing, and Presenting Usability Metrics (Interactive Technologies)* (Second edition). Waltham, MA: Morgan Kaufmann.

Wickham, H., & Grolemund, G. (2017). *R for Data Science: Visualize, Model, Transform, Tidy, and Import Data*. O'Reilly.

Keppel, G., & Wickens, T. D. (2004). *Design and Analysis: A Researcher's Handbook* (Fourth Edition). Englewood Cliffs.

Questions

Questions for 3.1 Purpose of Evaluation

P3.1   How is the process of evaluation related to that of understanding in the human factors design cycle?

P3.2   What are the three general purposes of evaluation?

P3.3   Would qualitative or quantitative data be more useful in diagnosing why a design is not performing as expected?

P3.4   Would qualitative or quantitative data be more useful in assessing whether a design meets safety and performance requirements?

P3.5   What is the role of quantitative and qualitative data in system design?

P3.6   Why is qualitative data an important part of usability testing given its role in the design process?

P3.7   Give examples of qualitative data in evaluating a vehicle entertainment system.

P3.8   Give examples of quantitative data in evaluating a vehicle entertainment system.

P3.9   Describe the role of formative and summative evaluations in design.

P3.10   Identify a method suited to formative evaluation and a method suited to summative evaluation.

Questions for 3.2 Timing and Types of Evaluation

P3.11   Identify the evaluation method best suited to early design concepts.

P3.12   Identify the evaluation method best suited to the prototypes,

P3.13   Identify the evaluation method best suited to pre-production designs

P3.14   Identify the evaluation method best suited to designs that are in service.

P3.15   Why are evaluation methods that do not require human subjects data collection useful in design.

P3.16   Describe two evaluation methods that do not require human collection.

P3.17   What is an important limit of both cognitive walkthroughs and heuristic evaluation?

P3.18   Describe the steps of heuristic evaluation.

P3.19   What might differ when applying a heuristic evaluation to a design of a manufacturing cell and to a website.

P3.20   How many analysts should be used to assess a system with heuristic evaluation?

P3.21   What is the main difference between a cognitive walkthrough and a heuristic evaluation?

P3.22   What evaluation techniques would be particularly useful in a scrum development environment?

P3.23   Before a large system is deployed that is being developed using a Vee development cycle, what evaluation technique would you be expected to use?

P3.24   In using scrum in a high-risk domain what evaluation technique might be difficult to complete even though it might be the right thing to do?

Questions for 3.3 Study Design

P3.25   How many participants do you need for a usability study?

P3.26   How many usability tests would you recommend as part of an iterative design process?

P3.27   What is the difference between a controlled experiment and a descriptive study?

P3.28   How does a quasi experiment relate to a descriptive study and to an experiment?

P3.29   When would you use a between subjects experimental design and when would you use a within subjects design?

P3.30   How many participants do you need for a controlled experiment?

Questions for 3.4 Measurement

P3.31   In the evaluation of an entertainment system for a car, what would be dependent variables of interest?

P3.32   What is the benefit of subjective measures?

P3.33   What is a limitation of subjective measures?

P3.34   What is the relationship between a construct and a measure?

Questions for 3.5 Data Analysis

P3.35   Describe how the driving performance data in Figure 3.9 represents a two-way interaction, and what the graph would look like without the interaction.

P3.36   What is the role of descriptive statistics and how does it differ from inferential statistics?

P3.37   Describe a descriptive statistic in assessing the distraction potential of a vehicle entertainment system.

P3.38   Describe an inferential statistic in assessing the distraction potential of a vehicle entertainment system.

P3.39   What inferential statistical approach is most commonly used for multi-factor experiments?

P3.40   What inferential statistical approach is most commonly used for descriptive studies?

P3.41   Describe what is meant by experimental control and its role in designing an experiment, quasi-experiment, and a descriptive study.

P3.42   Describe an example of confounding in a field test of a vehicle entertainment system.

P3.43   What is meant by representative sampling in selecting people, tasks, and situations in designing a study?

P3.44   What is the purpose of representative sampling in selecting people, tasks, and situations in designing a study?

Questions for 3.6 Drawing Conclusions and Communicating Results

P3.45   Describe the difference between a Type I and Type II error and its implications for system evaluation.

P3.46   What is the difference between practical and statistical significance?

P3.47   Show why it is useful to consider confidence intervals and not rely on *p* values.

Questions for 3.8 Ethical Issues

P3.48   Describe four essential aspects of protecting participants in research.

Part I

Cognitive Considerations

In the next eight chapters, we consider the basic processes by which people perceive, reason, and respond to the built environment. These processes are generally grouped under the label of cognition. We discussion the implications of cognition for design, specifically displays, controls, interfaces, and automation.

The environment: To organize this discussion, we provide a general framework shown in the figure below. This framework has the environment at its base, which includes physical features, such as light and sound levels, but also features that support cognition, such as the post-it notes and stacks of papers that help us remember. Each chapter begins with a discussion of the environment.

People: With the environment as a background, we consider characteristics of people (at the top), technology (in the middle) and the controlled system and people's goals and activities associated with that system (at the bottom). Chapter 4 and 5 discuss perception. Chapter 6 then considers how memory and attention influence how people interpret information and develop expectations. Expectations guide perception and action: we see what we expect to see and respond quickly to what we expect. Above the perception-interpretation=action-expectation cycle rests macrocognition at the top of the figure, which we discuss in Chapter 7. Macrocognition considers the cognitive processes associated with decision making, mental models, and situation awareness.

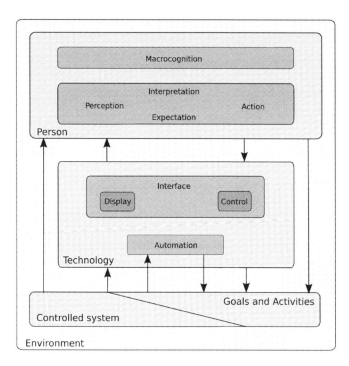

The figure shows cognition as a perception-action cycle in the context of technology and the environment.

Technology: The arrows in the figure show that information flows from the controlled system to the person. Because information often passes through a display, in Chapter 8 we discuss the im-

Theory, Principles, Guidelines

Chapters 8-11 begin with a discussion of the relevant technology, task, and system characteristics. We then present theory, principles, and guidelines to support consideration of people in design.

Theory represents basic findings of sensory and cognitive process. Theories are broadly applicable and highlight general considerations, but they are often challenging to apply to design.

Design principles provide general guidance that is independent of specific technology, but require translation to a specific application.

Design guidelines offer specific suggestions for how designers and engineers should consider people. We present guidelines in the context of specific applications, such as websites or wearable computers.

Table 3.4 Theory, principles, and guidelines .

plications of perception and cognition for display design. Displays can make complex systems easier for people to understand. The arrows in the figure also show that control flows from the person to the controlled process. In Chapter 9, we discuss the implications for designing controls, and for the cycle of perception and action. Action on the controlled system generates new information to be sensed and interpreted. This interpretation guides subsequent actions. Providing timely and understandable feedback is critical for correcting errors and controlling most systems.

Chapter 10 considers the integration of controls and displays in the design of interfaces and interactions–the realm of human-computer interaction (HCI). Chapter 11 considers automation, which extends aspects of HCI to consider technology that takes an increasingly active role in doing things for people. The arrows between the automation and the controlled system show that automation can sense and response to the system with only intermittent supervision from the person. Chapter 11 considers how to design this relationship.

The controlled system, goals, and activities: People act on the systems to achieve goals. The bottom of the figure shows the controlled system, such as a car, and the goal and activity of the person, such as navigating to a destination at a high speed. Performance and safety depends on characteristics of the controlled system as well as the goals people choose to pursue. The mass and inertia of cars at high speed contribute to challenge of controlling them and the consequences of a crash. Driving a car at high speed along a winding road requires quick responses. An oil tanker has very different dynamics and responds so slowly that "driving" it requires the person to predict the future course of the ship.

Response time of people and technology: System dynamics highlight the importance of time. The general cycle of acting on a system and perceiving the effect of these actions shown in the figure does not specify a time period. For the situations considered in this book, the relevant time periods vary enormously. People learn and adapt to the world over years as they become experts (shown at the top of the table). At the other extreme, they are able to perceive a gap in a continuous sound as short as one millisecond.

This table shows that people cannot respond instantaneously. Similarly, technology takes time to respond. Delays undermine performance and safety, and also represent one of the most critical influences of system satisfaction [67, 68]. Good design harmonizes the delays of people and technology. Some activities, such as clicking on an icon, require the system to respond within 100 ms, but other activities, such as rendering a page of a website can take several seconds without disrupting a person's task. Unexpected delays annoy people. People can generally notice a delay that is 8% longer than an expected delay of 2-4 seconds, and will notice a delay that is 13% larger than an expected delays of 6 and 30 seconds. A delay twice as long as expected will generate frustration [69]. The following table and subsequent chapters describe how to harmonize delays and other characteristics of people and systems.

Timescale	Cognitive process	Consequence
Years	Expertise (10,000 hours)	Expertise requires 10 years of deliberative practice and shifts decision processes [70] (Chapter 7) .
Months	Habits (2 months)	Consistently performing an activity over approximately two months creates an automatic routine that requires effort to suppress [71] (Chapter 7).
Days	Deliberation for big decision (1-10 days)	Large purchase decisions, such as a car, require an effort and time [67] (Chapter 7).
Hours	Circadian rhythm (24 hours)	Circadian cycles govern sleep and influences cognitive performance when awake [72] (Chapter 6) .
Minutes	Vigilance decrement (5-20 minutes)	Monitoring is effortful and stressful and can only be sustained for a limited time [73] (Chapter 6).
	Attention to video scenes (1-3 minutes)	Mean duration of online videos is 2.7 minutes [74]. Duration of scenes in films is 1.5-3 minutes [75].
Seconds	Working memory decay (15-30 seconds)	Items in working memory decay without rehearsal and interruptions prevent rehearsal [76] (Chapter 6).
	Unbroken attention to a task (6-30 seconds)	The limits of sustained attention defines a natural task duration [77, 67].
	Psychological present (2.5-3.5 seconds)	Maximum time between events to be perceived as part of a whole [78]. Longer delays lead people to start another activity [79] (Chapter 6). Easy to read sentences have 14 words, which take 3.3 seconds to read, assuming a reading speed of 250 wpm [80].
	Conversation continuity (0.5-2.0 seconds)	Maximum gap between "turns" in a conversation [81].
	Reaction to unexpected event (1.0-2.0 seconds)	Expectations strongly influence response time and unexpected events require more time [82, 83] (Chapter 6).
Sub-second	Event perception (100 milliseconds)	A system must respond within 100 ms of a mouse click to avoid a noticeable delay [67] (Chapter 5).
	Sensory integration (10 milliseconds)	Perceptible delay in drawing with e-ink with stylus; lowest update rate rate for e-ink to feel like real ink [67]. Haptic feedback response time for a virtual button on a tablet computer (Chapter 5).
	Perceptible sound gap (1 milisecond)	Maximum dropout duration in an auditory signal [67] (Chapter 5).

The table shows timescales and temporal requirements of human interaction. (Adapted from multiple sources: [67, 84, 77, 78].)

Chapter 4

Visual Sensory System

At the end of this chapter you will be able to...

1. identify required illumination in a work environment

2. choose colors to enhance perception, communication, and aesthetics

3. specify image size to ensure visibility and legibility

4. design to enhance top-down and bottom-up processing

5. design to accommodate the limits of absolute judgment, and capitalize on the capacity of relative judgment

The 50-year-old traveler, arriving in an unfamiliar city on a dark, rainy night, is picking up a rental car. The rental agency bus driver points to "the red sedan over there" and drives off, but in the dim light of the parking lot, our traveler cannot easily tell which car is red and which is brown. He climbs into the wrong car, realizes his mistake, and moves to the correct vehicle. He pulls out a city map to figure out the way to his destination, but in the dim illumination of the dome light, the printed street names on the map are just a black haze. Giving up on the map, he remains confident that he will see the road sign for his intended route (Route 60) for his destination so he starts the motor to pull out of the lot. The darkness and heavy rain has him fumbling to turn on the headlights and wipers, which are both in a different position from his own vehicle. A little fumbling, however, and both are on, and he emerges from the lot onto the highway. The rapid traffic closing behind him and bright glare of headlights in his rearview mirror force him to accelerate to an uncomfortably high speed. He cannot read the first sign to his right as he speeds by. Did that sign say Route 60 or Route 66? He drives on, assuming that the turnoff will appear in the next sign; he peers ahead, watching for the sign. Suddenly, there it is on the left side of the highway, not the right where he had expected it, and he passes it before he can change lanes. Frustrated, he turns on the dome light to glance at the map again, but in the second his head is down, the sound of gravel on the undercarriage signals that his car has slid off the highway. As he drives along the gravel, waiting to pull back on the road, he fails to see the huge pothole that unkindly brings his car to an abrupt halt.

Our unfortunate traveler is in a situation that is quite common. Night driving in unfamiliar locations is one of the more hazardous endeavors that humans undertake, especially as they become older. One reason why the dangers are so great relates to the pronounced limits of the visual sensory system. Many of these limits reside within the peripheral features of the eyeball itself and the neural pathways that send visual information to the brain. Others reside in the more central cognitive processing we discuss in Chapter 6. In this chapter we discuss the nature of light stimulus and the anatomy of the eye that processes this light. We then discuss several important characteristics of human visual performance as it is affected by this interaction between characteristics of the stimulus and cognitive processes of the human perceiver.

4.1   Visual Environment

The visual environment defines what a person can see. At the most fundamental level the properties of light entering the eye define what can be seen. The physics of light defines color in terms of wavelength and light intensity in terms of radiant energy. The structure of the light array reflected off features of the environment indicates surfaces and motion through the environment.

4.1.1 Wavelength and Color

Visual stimuli perceived by humans can be described as a wave of electromagnetic energy, that can be represented as a point along the *visual spectrum*. As shown in Figure 4.1, light has a *wavelength*, typically expressed in nanometers along the horizontal axis, and an amplitude on the vertical axis. The wavelength determines the *hue* of the stimulus that is perceived, and the amplitude determines its *brightness*. As the figure shows, the range of wavelengths typically visible to the eye runs from short wavelengths of around 400 nm (typically observed as blue-violet) to long wavelengths of around 700 nm (typically observed as red). In fact, the eye rarely encounters "pure" wavelengths, but instead mixtures of different wavelengths. For example, Figure 4.1 depicts the entire electromagnetic spectrum and, within that, the spectrum of visible light. On the other hand, the pure wavelengths, characterizing a hue, like blue or yellow, may be "diluted" by mixture with varying amounts of gray or white (called *achromatic light*). This is light with no dominant hue and therefore not represented on the spectrum. When wavelengths are not diluted by gray, like pure red, they are said to be *saturated*. Diluted wavelengths, like pink, are unsaturated. Hence, a given light stimulus can be characterized by its hue (spectral values), saturation, and brightness.

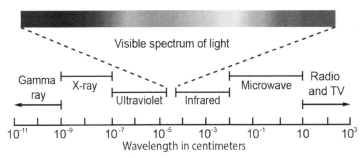

Figure 4.1 The visible spectrum of electromagnetic energy (light). Very short (ultraviolet) and very long (infrared) wavelengths falling just outside of this spectrum are shown.

The actual hue of a light is typically specified by the combination of the three primary colors—red, green, and blue—that combine to produce it. This specification follows a procedure developed by the Commission Internationel de L'Elairage and hence is called the CIE color system. Figure 4.2 shows that the CIE color system represents all colors in terms of two primary colors of long and medium wavelengths specified by the x and y axes respectively [85]. Those colors on the rim of the curved lines defining the space are pure, saturated colors. A monochrome light is represented by point in the middle of the space. The figure does not represent brightness, but this could be shown as a third dimension running above and below the color space of Figure 4.2. Use of this standard coordinate system allows common specification of colors across different users. For example, a "lipstick red" color would be established as having 0.50 units of long wavelength and 0.33

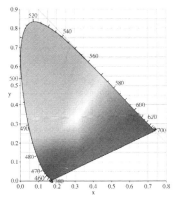

Figure 4.2 The CIE color space, showing some typical colors created by levels of x and y specifications. The x and y values precisely specify hue, but z values are needed to specify brightness. (Source: By User:PAR (Own work) [Public domain], via Wikimedia Commons, https://commons.wikimedia.org/wiki/File%3ACIExy1931.png).

units of medium wavelength (Widdel and Post provide a detailed discussion of color standardization issues for electronic displays [86]).

4.1.2 Light Intensity

We can measure or specify the hue of a stimulus reaching the eyeball by its wavelength. However, the measurement of brightness is more complex because there are several different meanings of light intensity [87]. This complexity is shown in Figure 4.3, where we see a source of light, like the sun or, in this case, the headlight of our driver's car. This source may be characterized by its luminous intensity, or *luminous flux*, which is the actual light energy of the source. It is measured in units of candela. But the amount of this energy that actually strikes the surface of an object to be seen—the road sign, for example—is a very different measure, described as the *illuminance* and measured in units of *lux* or *foot candles*. The amount of illuminance an object receives depends on the distance of the object from the light source. In Figure 4.3 this is illustrated by the values under the three signs at increasing intervals of two units, four units, and six units away from the headlight—the illuminance declines with the square of the distance from the source. *Brightness* is the subjective experience of the perceiver that depends on luminance.

Hence, the term illumination characterizes the lighting quality of a given working environment.

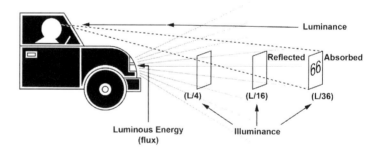

Figure 4.3 Concepts behind the perception of visual brightness. Luminous energy (flux) is present at the source (the headlight), but for a given illuminated area (illuminance), this energy declines with the square of the distance from the source (road sign).

Some of the illuminance (solid rays) is absorbed by the sign, and the remainder is reflected back to the observer (dashed rays), characterizing the luminance of the viewed sign. Although we are concerned with the illumination of light sources in direct viewing such as the amount of glare produced by headlights shining from the oncoming vehicles [88], we are more often concerned about the illumination of work place. Human factors is also concerned with the amount of light reflected off of objects to be detected, discriminated, and recognized by the observer when these objects are not themselves the source of light. This may characterize, for example, the road sign in Figure 4.3. We refer to this measure as the *luminance* of a particular stimulus typically measured in foot lamberts (fl). Luminance is different from illuminance because of differences in the amount of light that surfaces either reflect or absorb. Black surfaces absorb most of the illuminance striking the surface, leaving little luminance to be seen by the observer. (A useful hint is to think of the illuminance light, leaving some of itself [the "il"] on the surface and sending back to the eye only the luminance.) White surfaces reflect most of the illuminance. In fact, we can define the *reflectance* of a surface as: luminance(fl)/illuminance (fc). Clearly luminance can be provided by non-illuminated surfaces, such as that emanating from the mobile phone or computer screen.

The *brightness* of a stimulus, then, is the actual experience of visual intensity, an intensity that often determines its visibility. From this discussion, we can see how the visibility or brightness of a given stimulus may be the same if it is a dark (poorly reflective) sign that is well illuminated or a white (highly reflective) sign that is poorly illuminated. In addition to brightness, the ability to actually see an object—its visibility—is also affected by the *contrast* between the stimulus and its surround, but that is another story that we shall describe in a few pages.

Table 4.1 summarizes these various measures of light and the units by which they are typically measured. A *photometer* is an electronic device that measures luminous in terms of foot lamberts or candela/m^2. A *luxmeter* is a device that measures illuminance.

Table 4.2 shows the requirements of various tasks for illuminance, which is the luminous flux distributed over the work area. The tasks are divided into three groups: ABC, DEF, GHI. The ABC group concerns tasks associated with walking through areas of buildings and outdoor spaces. These tasks do not require people to process details of the scene. The DEF group concerns tasks that involve visual processing of details of the work area, such as reading text or assembly. Task types in this group progress from D, consisting of simple assembly and reading well-rendered text, to F, which consists of difficult assembly and reading poorly reproduced text. The final group describes visually demanding tasks that progress from very difficult assembly (group G) to tasks at the limit of visual perception, such as intricate tasks that might occur in a hospital operating room (group I) [89]. Naturally these values must also consider reflectance (for non electronic text) and luminance (of electronic information), and, a key consideration below, the *contrast*. Nevertheless, the order in the Table 4.2, in terms of visual performance from A to I is critical.

Luminous flux:
1 candela or
12.57 lumens (lm)

Illuminance:
1 foot candle (fc) (lm/ft^2) or
10.76 lux (lm/m^2)

Luminance:
1 foot lambert (fL) $1/\pi$ candela/ft^2 or
3.425 candela/m^2

Reflectance: A ratio—
$$R = \frac{Luminance}{Illuminance}$$
For a perfectly diffuse reflecting surface:
$$Luminance = \pi \times Illuminance/R$$
Luminance is often shown as L
Illuminance is often shown as E

Table 4.1 Physical quantities of light and their units.

IESNA Cate-gory	Space and task type	Example	Illuminance (lux)
Orientation and simple visual tasks			
A	Public space with dark surroundings	Parking lots	20–30–50
B	Simple orientation for short visits	Storage spaces	50–75–100
C	Simple, occasional visual tasks	Hallways, stairways, restrooms, elevators	100–150–200
Common visual tasks			
D	Tasks with high contrast and large size	Simple assembly, rough machining, reading	200–300–500
E	Tasks with high contrast and small size or low contrast and large size	Office, library, supermarket, kitchen	500–750–1000
F	Tasks with low contrast or very small size	Difficult assembly, poorly reproduced text, painting, polishing, operating room	1000–1500–2000
High-demand visual tasks			
G	Tasks with low contrast or very small size over prolonged period	Very difficult assembly	2000–3000–5000
H	Exacting tasks over a very longed period	Very precise assembly	5,000–7,500–10,000
I	Tasks near perceptual threshold—very small or very low contrast	Paint inspection, operating table	10,000–15,000–20,000

Table 4.2 Illuminance required for various tasks. (Abstracted from the *Lighting Handbook* [90].)

$$E = \frac{F \times N \times UF \times MF}{A}$$

Illuminance at work surface

E = Illuminance
F = Luminous flux of each lamp (lm)
N = Number of lamps
UF = Utilization factor to reflect light reflection on room surfaces
UF ranges from 0 to 1, with 1 being perfect delivery of light from the lamp to work surface, and 0.85 is representative
MF = Maintenance factor to reflect age-related decline in efficiency, as well as dirt and dust accumulation
Ranges from 0 to 1, with 1 being a new unit, and .85 is representative
A = Area of work surface (m^2)

(4.1)

The lighting requirements for these tasks ranges from 20 to 20,000 lux. For comparison, moonlight generates 1 lux, sunrise and sunset 400 lux, shaded area at midday 20,000 lux and bright sunlight 100,000 lux. There is also a range of lighting requirements for each task in the table. The range associated with each task reflects the need to adjust for age—older people tend to need more intense lighting. Also, room surfaces that are highly reflective would indicate the need for less intense lighting, but tasks that require a particularly high degree of speed and accuracy might merit more intense lighting.

The lumen method is a common way of identifying what lamps are needed to replace or supplant natural light to preform the tasks in Table 4.2. The central idea with the lumen method is that the luminous produced by a lamp distributed over the work area is the illuminance available to support perception. Most simply, the luminance flux of the lamp divided by the size of the work area. A slightly more sophisticated calculation is shown in Equation 4.1. This equation considers the effect of combining multiple lamps (N), and the effectiveness of these lamps and the properties of the room in directing light to the work surface (UF). The equation also considers the loss of effectiveness of the lamp as it ages and gets dirty. The lumen method and the associated values in Table 4.2 depends on many assumptions and more detailed analyses, such as those outlined in the *Lighting Handbook* [90], which should be consulted for more precise estimates of lighting requirements.

4.1.3 Light Sources

The sun and moon are natural light sources, but workplace and home environments require artificial light sources. Selecting the appropriate light source and configuration to light a space is a complex task and a whole field of lighting design addresses its details. At a basic level, four important properties guide the selection of lighting: efficiency, rendering index, color temperature, and control. Efficiency refers to the luminous flux produced by the lamp as a function of power. Greater efficiency means the lamp costs less to operate, longer operation from battery powered devices, and smaller environmental consequences.

One drawback of some highly efficient light sources, such as low pressure sodium lamps, is that they fail to render color accurately. The color rendering index describes the ability of a lamp to render color accurately with 100 being a perfect score, rendering colors in a manner similar to sunlight. Don't take a selfie under a fluorescent light or sodium light if you want to reveal your true colors.

The color temperature of a light source describes the distribution of wavelengths it produces, and is referenced to the temperature of a black body radiator. Higher temperatures correspond to shorter wavelengths. Counterintuitively, lower color temperature corresponds to warmer feeling light. High color temperatures tend to emphasize the blues and greens of a scene and low color temperatures tend to emphasize the orange and reds. Higher color temperatures are best suited to the workplace and lower temperatures to homes and restaurants. Generally higher color temperatures—cooler light—should be used for applications requiring higher intensity lighting [91].

Table 4.3 shows efficiency, rendering score, and color tempera-

Lighting specifications confusingly specify "cool" light as corresponding to a higher color temperature.

Lamp type	Efficiency (lumens/watt)	Color rendering index (CRI)	Color temperature (degrees Kelvin)
Incandescent	12	97	2500
Compact fluorescent	25-70	82	3000 (warm)
			5500 (cool)
Fluorescent	50-100	52 (warm white)	3800-4000
		62 (cool white)	
LED	60-100	80	6000 (cool white)
High pressure sodium	70-120	25	2000-2700
Low pressure sodium	100-200	0	1800
Natural light			
Noon (100,000 lux)		100	6000
An hour before sunset (400 lux)		100	3500
Full moon (0.25 lux)		100	4100

Table 4.3 Efficiency and color rendering capability of common light sources.

ture for typical light sources. Manufactures indicate efficiency, in lumens/watt and color rendering index (CRI) on their products.

One way to meet the requirements of particular tasks that require more light, such as reading, without wasting light when performing less light-demanding tasks, is to make the lights adjustable either automatically or by individuals. This can increase the satisfaction with the work environment, particularly for those with increasing age or visual impairment. Most mobile phones, tablets, and laptop computers automatically adjust the intensity of their screens according to the ambient light levels. A smartphone screen will be more intense during the day than at night. Similarly, rooms can be instrumented to dim lamps when they are not occupied or when the sun provides sufficient natural light.

4.1.4 Optic Flow and Ecological Optics

The physical properties of the wavelength and radiant energy of light help us understand what people see, but do not fully describe how the basic properties of light entering the eye affect behavior. One such property is optic flow, which is defined by the motion of objects, edges, and surfaces caused by the relative motion of the eye and the scene. Optic flow plays a critical role in guiding movement through the world. Examples include pilots landing aircraft, drivers avoiding collisions, and baseball players catching balls.

The optic flow, light intensity, and color define the basic visual stimuli that reach the eye. How those stimuli influence perception and behavior depend in part on the characteristics of receptor system—the eye—which we now discuss.

4.2 The Receptor System: The Eye

Figure 4.4 shows the receptor system for human vision, which is the eye. The eye takes light energy and transforms it into what we see. This is a complex process that is based on certain key features of its anatomy. This section describes how characteristics of these features influence perception. Designing to account for these influences is an extremely important role for human factors specialists.

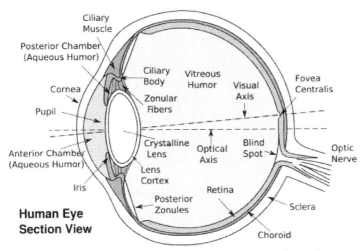

Figure 4.4 Anatomy of the eye: Transversal view of the right eye from above. (Source: Based on Eyesection.gif, by en:User_talk:Sathiyam2k. Vectorization and some modifications by user:ZStardust (Self-work based on Eyesection.gif) [Public domain], via Wikimedia Commons, https://upload.wikimedia.org/wikipedia/commons/f/f5/Eyesection.svg)

4.2.1 The Lens and Accommodation

As we see in Figure 4.4, the light rays first pass through the *cornea*, which is a protective surface that absorbs some of the light energy (and does so progressively more as we age so that 80 year olds require approximately 30 percent more light than 25 year olds). Light rays then pass through the *pupil*, which opens or dilates (in darkness) and closes or constricts (in brightness) to allow more light to enter when illumination is low and less when illumination is high. The lens of the eye also adjusts its shape, or *accommodates*, to bring the image to a precise focus on the back surface of the eyeball, the *retina*. This accommodation is accomplished by a set of ciliary muscles surrounding the lens. Sensory receptors located within the ciliary muscles send information regarding accommodation to the higher perceptual centers of the brain to aid in depth perception. When we view images up close, the light rays emanating from the images converge as they approach the eye, and the muscles must accommodate by changing the lens to a rounder shape, as reflected in Figure 4.4. When the image is far away and the light rays reach the eye in essentially parallel fashion, the muscles accommodate by creating a flatter lens. Somewhere in between is a point where the lens comes to a natural "resting" point, at which the muscles are doing very little work. This is referred to as the *resting state* of accommodation.

The amount of accommodation can be described in terms of the distance of a focused object from the eye. Formally, the amount of accommodation required is measured in *diopters*, which equal 1/viewing distance in meters—the reciprocal of the focal length. Thus, 1 diopter is the accommodation required to view an object at 1 meter, and zero diopters is optical infinity. Resting state accommodation is about 1.5 diopters, which corresponds to 1/1.5 meter or 67 cm.

As the driver in our story discovered when he struggled to read the fine print of the map, our eyeballs do not always accommodate easily. It takes time to change its shape, and sometimes there are factors that limit the amount of shape change that is possible. *Myopia*, or nearsightedness, results when the lens cannot flatten and hence distant objects cannot be brought into focus. *Presbyopia*, or farsightedness, results when the lens cannot accommodate to very near stimuli. As we grow older, the lens becomes less flexible in general, but farsightedness in particular becomes more evident. Young children have a degree of accommodation of 15- 20 diopters, which declines on average to approximate 10 diopters by the age of 25, and to only 1 by the age of 50. A 25-year-old driver could accommodate an object 1/10 meters away (approximately 4 inches), whereas a 50-year-old driver might be able to focus on the map only if was 1/1 meter away. Hence, we see that the older driver, when not using corrective lenses, must hold the map farther away from the eyes to try to focus, and it takes longer for that focus to be achieved.

People over the age of 50 can't focus on objects nearer than 1 meter without glasses.

Aging undermines accommodation by reducing the flexibility of the lens, which can be compensated by corrective lenses or by holding things further away. In the case of the driver holding the map further away only partially solves the problem because it reduces the effective size of the text on the map. As we will see small text, as defined by the visual angle, is hard to read.

4.2.2 The Receptors: Rods and Cones

An image, whether focused or not, eventually reaches the retina at the back of the eyeball. The image may be characterized by its intensity (luminance), its wavelengths, and its size. The image size is typically expressed by its *visual angle*, which is depicted by the two-headed arrows in front of the eyes in Figure 4.4 and can be calculated by Equation 4.2.

$$VA = 2 \times arctan\left(\frac{H}{2 \times D}\right)$$

H = Height of object   Visual angle of objects
D = Distance to object in same units as height

$$(4.2)$$

Example 4.1 Calculate the horizontal visual angle of the width of a 12-inch computer screen viewed at a distance of 24 inches (remembering to convert radians to degrees).

Solution:

Using Equation 4.2 with the width of the computer screen as H and the distance to the screen as D.

$$VA = 2 \times arctan\left(\frac{12}{2 \times 24}\right)$$

VA = .49 radians = 28.1 degrees =

$$.49 \; radians \times \frac{360 \; degrees}{2 \times \pi \; radians}$$

This is the same visual angle as a 24-inch screen viewed at 48 inches, or a 6-inch screen at 12 inches.

As a rough rule of thumb, the width of your thumb at arms length has a visual angle of about 2 degrees, your thumbnail is 1.5 degrees, and width of your index finger is about 1 degree. The rule works because people who have longer arms tend to have wider thumbs [92]. Visual angle rather than absolute size defines the image that falls on the retina and so it is visual angle that matters when we want to understand what people can see.

Importantly, the image can also be characterized by where it falls on the back of the retina because this location determines the types of visual receptor cells that are responsible for transforming electromagnetic light energy into the electrical impulses of neural energy to be relayed through the optic nerve to the brain.

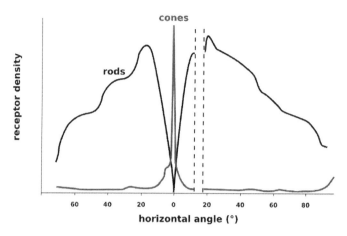

Figure 4.5 Distribution of rods and cones across the retina. (Source: Johannes Ahlmann (based on Osterberg, 1935), Available on Flickr Commons: https://www.flickr.com/photos/entirelysubjective/6146852918)

Figure 4.6 Letters of equal legibility if the center point is fixated. Because legibility depends on size the letters are equally legible for a range of distances. (Source: Reprinted from http://anstislab.ucsd.edu/ 2012/11/20/peripheral-acuity/, with permission from Anstis, adapted from Anstis (1971). A chart demonstrating variations in acuity with retinal position, *Vision Research*, 14, 7, pp. 589-592.)

There are two types of receptor cells, *rods* and *cones*, each with six distinctly different properties. Figure 4.5 shows the distribution of rods and cones across the retina at the back of the eye, with cones being concentrated at the center of the visual field. Collectively, these different properties have numerous implications for visual sensory processing.

Location. The middle region of the retina, the fovea, consisting of an area of around 2 degrees of visual angle, is inhabited exclusively by cones (Figure 4.5). Outside of the fovea, the *periphery* is inhabited by rods as well as cones, but the concentration of cones declines rapidly moving farther away from the fovea (i.e., with greater *eccentricity*.) This declining concentration means the objects in the periphery need to be larger to be seen as clearly as those in the center.

Acuity. The amount of fine detail that can be resolved is far greater when the image falls on the closely spaced cones than on the more sparsely spaced rods. We refer to this ability to resolve detail as the *acuity*, often expressed as the inverse of the smallest visual angle (in minutes of arc) that can be detected . Thus, acuity of 1.0 represents the ability to resolve a visual angle of 1 minute of arc (1/60 of 1 degree). Standard measures of visual acuity include the Landolt C or the Snellen E, both of which subtend 5 minutes of arc at 20 ft and require people to resolve a gap of 1 minute of visual arc to demonstrate 20/20 vision and the nominal limit of acuity. Because acuity is higher with cones than rods, it is not surprising that our best ability to resolve detail is in the fovea, where the cone density is greatest. Hence, we "look at" objects that require high acuity, meaning that we orient the eyeball to bring the image into focus on the fovea. However, a functional level of detail extends to approximately 10 degrees. Figure 4.6 demonstrates the decline in acuity as images extend into the periphery.

While visual acuity drops rapidly toward the periphery, the

sensitivity to motion declines at a far less rapid rate. We often use the relatively high sensitivity to motion in the periphery as a cue for something important on which we later fixate. That is, we notice motion in the periphery and move our eyes to focus on the moving object.

Sensitivity. Although the cones have an advantage over the rods in acuity, the rods have an advantage in terms of *sensitivity*, characterizing the minimum amount of light that can be detected, which is the *threshold*. Sensitivity and threshold are reciprocally related: As one increases, the other decreases. Since there are no rods in the fovea, it is not surprising that our fovea is very poor at picking up dim illumination (i.e., it has a high threshold). To illustrate this, note that if you try to look directly at a faint star, it will appear to vanish because it is now in the foveal vision where only the cones are located. *Scotopic vision* refers to vision at night when only rods are operating. *Photopic vision* refers to vision when the illumination is sufficient to activate both rods and cones—illuminance greater than 30 lux. Even though photopic vision involves both rods and cones, most of our photopic visual experience is due to cones.

Color sensitivity. Rods cannot discriminate different wavelengths of light (unless they also differ in intensity). Rods are "color blind." Only cones enable color vision, and so the extent to which hues can be resolved declines both in peripheral vision (where fewer cones are present) and at night (when only rods are operating). Hence, we can understand how our driver, trying to locate his car at night, was unable to discriminate the poorly illuminated red car from its surrounding neighbors.

Adaptation. When stimulated by light, rods rapidly lose their sensitivity, and it takes a long time for them to regain it (up to a half hour) once they are returned to the darkness that is characteristic of the rods' "optimal viewing environment." This phenomenon describes the temporary "blindness" we experience when we enter a darkened movie theater on a bright afternoon. Operators who are periodically exposed to bright light, but often need to use their scotopic vision in dimly lit workplaces, will have a hard time seeing. In contrast to rods, the sensitivity of the cones is little affected by light stimulation. However, cones may become *hypersensitive* when they have received little stimulation. This is the source of *glare* from bright lights, particularly at night.

It takes 30 minutes to adapt to the dark and 30 seconds to adapt to bright environments.

Differential wavelength sensitivity. Whereas cones are generally sensitive to all wavelengths, rods are particularly insensitive to long (i.e., red) lengths. Hence, red objects and surfaces look very black at night hence amplifying our driver's confusion between the red and brown car. More important, illuminating objects in red light in an otherwise dark environment will not destroy the rods' dark adaptation. For example, on the bridge of a ship, the navigator may use a red lamp to stimulate cones in order to read the fine detail of a chart, but this stimulation will not destroy the rods' dark adaptation and hence will not disrupt the ability of personnel to scan the horizon for faint lights or dark forms.

Using red light at night helps to preserve night vision.

Collectively, these pronounced differences between rods and

cones are responsible for a wide range of visual phenomena. We consider the implications of these phenomena to human factors issues related to several sensory processing characteristics: visual acuity, contrast sensitivity, night vision, and color vision.

4.3 Sensory Processing Characteristics

Visual acuity, contrast sensitivity, color vision, and night vision all depend on basic features of the rods and cones and their distribution over the retina. These features have important implications for designs that include text on maps and road signs as well as instrumentation in cockpits and cars.

4.3.1 Visual Acuity

The limits of visual acuity follow from the distribution of cones in the fovea and the optical properties of the eye. The approximately 100,000 cones in the central fovea, means there are approximately 2 cones per arcminute of visual angle, making it possible to resolve 1 minute of detail in a scene, such as the separate of lines in the capital letter E or the separation of two slightly nearby points [93]. Limits of visual acuity have many important consequences ranging from reading to guiding action. A visual angle of one minute corresponds to 1.05 inches at 100 yards and the accuracy of a high-quality rifle, suggesting the need for a scope to benefit from rifles of greater accuracy.

The retina's resolution also explains the naming and performance of Apple's "retina" display on its iPhones and computers. The pixels in these devices are packed so tightly that people are not able to see individual pixels—the separation of pixels is approximately 1 arc minute. An iPhone 6 has approximately 330 pixels per inch and if it is held at 10 inches, and based on Equation 4.2, this corresponds to approximately 60 pixels per degree of visual angle. Any display this density of pixels per degree would be a "retina" display.

Resolving 1 arcminute nears the limit of visual performance and is useful to specify the required resolution for computer screens and home theaters, but it does not indicate how large symbols and letters should be to ensure for fluent reading of highway signs, cockpit instrumentation, website content, and printed material. A simple rule of thumb for specifying the size of display details, such as letters, so they can be easily seen is the Bond rule, where the height of the letter is set to .007 times the viewing distance and both the viewing distance and letter height are in the same units.

A more precise answer requires the size of text to be described in terms of visual angle. Text or font size is often measured in points, where a point equals 1/72 of an inch. For, reading the critical measure of font is the x-height, which corresponds to the height of the lower case letter x, and the font size is roughly twice the value of the x-height. The x-height is relevant because people must resolve the detail in lower case letters to read. People can read

Task situation	Visual angle (arc minutes)	Physical size	Distance
Limit of simple visual acuity	1 for details 5 for letters	0.87 cm	6.1 m
750 pixel width of "Retina" display	14.2 degrees (57.6 pixels/degree)	5.79 cm	34 cm
General text—Bond Rule	24	.007×Distance	
Fluent reading (e.g., book)[94]	12 (smallest) 180 (largest)	0.14 cm (4 points) 14.0 cm (40 points) specified as x-height, font size approximately double	40 cm
Cockpit and vehicle displays[95]	24	0.40 cm	58 cm
Critical markings [96]	15-25	0.25-0.42 cm	58 cm
Street name signs	9.5	15.2 cm	54.9 m

Table 4.4 Text size required for various task situations

fluently with a relatively wide range of text sizes. For typical reading conditions, such as reading a newspaper, text size should range between 0.2° and 3°. Reading speed declines sharply when the font size drops below the critical print size (CPS) of 0.2°. For a normal reading distance of 40 cm (16 inches), this corresponds to Times New Roman type with an x-height of 0.2°, which corresponds to a 9 point font size [94]. The critical print size is at least twice the limit of visual acuity of 0.08°.

In less typical reading situations, such as reading displays while driving, vibration and the requirement to quickly glance back to the road suggests the need for text that is larger than the critical print size. Guidelines for automotive displays (e.g., warning messages, navigation systems, speedometer labels) suggest a visual angle for display text that is roughly double that of the critical print size and consistent with that recommended for cockpit displays. Importantly, these values are double the visual angle and physical font size as shown in Table 4.4.

Figure 4.7 Basic typology terminology showing x-height and font size. (Source: Max Naylor (Own work) [Public domain], via Wikimedia Commons.)

Road signs use a legibility index to define how large the lettering should be. A legibility index of 30, which is the current standard, indicates the lettering should be one inch for every 10 feet for the sign to be readable. At 300 feet the letters should be 10 inches high.

The spacing of cones in the retina limit simple acuity associated with resolving the details of objects, however; the visual system pools information across these individual receptor cells to provide a more precise sense of relative position, which is termed *hyperacuity*. Hyperaccuity can be 5 to 10 times more sensitive than simple acuity, meaning that it is possible to detect offset between two parallel lines as small as 0.17 arcminutes (10 arcseconds). Vernier scales, such as those on high-precision calipers take advantage of this ability. Unfortunately, this ability also makes rendering artifacts on computer screens, such as aliasing of lines, highly visible. As a consequence, a true retina display might need to have a resolution of 300 pixels per degree of visual angle. It also explains why a scratch on the screen of your phone is so noticeable even though it might be much smaller than the pixels that are not perceivable.

4.3.2 Contrast Sensitivity

Our unfortunate driver could not discern the wiper control label, the map detail, or the pothole for a variety of reasons, all are related to the vitally important concept of *contrast sensitivity*. Contrast sensitivity may be defined as the reciprocal of the minimum contrast between a lighter and darker spatial area that can just be detected ($CS = 1/C_{min}$); that is, with a level of contrast below this minimum, the two areas appear homogeneous. For situations where there is not a clear background and foreground, the Michelson contrast of visual pattern is expressed as the ratio of the difference between the luminance of light and dark areas as shown in Equation 4.3

Michelson contrast $$C = \frac{L - D}{L + D}$$

C = Michelson contrast
L = Luminance of light area
D = Luminance of light area

(4.3)

Contrast sensitivity influences the ability to detect and recognize shapes, whether the discriminating shape of a letter or the blob of a pothole. The higher the contrast sensitivity (CS) that an observer possesses, the smaller the minimum contrast that can be detected, CM, a quantity that describes the *contrast threshold*.

One obvious factor that influences our contrast sensitivity is that *lower* contrasts are *less easily* discerned. Hence, we can understand the difficulty our driver had in trying to read the label against the gray dashboard. Had the label been printed against a white background, it would have been far easier to read. Many people are frustrated by the black-on-black raised printing instructions (Figure 4.8). A minimum Michelson contrast of 30% is recommended [97]. Color contrast does not necessarily produce good luminance contrast. Thus, for example, PowerPoint slides that produce black text against a blue background may be very hard for an audience to read.

Figure 4.8 Low-contrast, black-on-black lettering is difficult to read. (Photograph by author: J. D. Lee.)

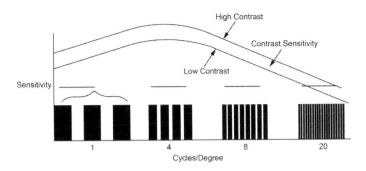

Figure 4.9 Spatial frequency grating used to measure contrast sensitivity and associated sensitivity for high and low contrast conditions.

A second factor that influences contrast sensitivity is spatial frequency. As shown in Figure 4.9, *spatial frequency* may be expressed as the number of dark-light pairs that occupy 1 degree of visual angle (cycles/degrees or c/d). If you hold this book approximately 1 foot away, then the spatial frequency of the left grating is 0.6 c/d and the grating on the right is 2.0 c/d. If the image appears like a smooth bar, as on the far right of the figure (if it is viewed from a distance), then the contrast is below the viewer's CS threshold. The human eye is most sensitive to spatial frequencies of around 3 c/d, as shown by the two CS functions drawn as curved lines across the axis of Figure 4.9. Greater contrast (between light and dark bars) makes it easier see the bars across all spatial frequencies.

The high spatial frequencies on the right side of Figure 4.9 characterize our sensitivity to small visual angles and fine detail, such as that involved in reading fine print. For text, characters subtending a visual angle of 0.5° will have a spatial frequency of approximately 4 cycles/degree, which is why your reading slows for smaller text and for lower contrast text. Much lower frequencies characterize the recognition of shapes in blurred or degraded conditions, like the road sign sought by our lost driver or the unseen pothole that terminated his trip. Contrast sensitivity declines for spatial frequency above and below 3 c/d.

A third influence on contrast sensitivity is the level of illumination of the stimulus (L + D, the denominator of Equation 4.3). Not surprisingly, lower illumination reduces the sensitivity and does so more severely for sensing high spatial frequencies (which depend on cones) than for low frequencies. This explains the obvious difficulty we have reading fine print under low illumination. However, low illumination can also disrupt vision at low spatial frequencies: Note the loss of visibility that our driver suffered for the low spatial frequency pothole.

Two final influences on contrast detection is the eye itself and the dynamic characteristics of the viewing conditions. Increasing age reduces the amount of light passing through the cornea and greatly reduces the sensitivity. This, coupled with the loss of visual accommodation ability at close viewing, produces a severe deficit for older readers in poor illumination. Contrast sensitivity also

Spatial frequency defines the size of the details the eye can resolve.

declines when the stimulus is moving relative to the viewer, as our driver found when trying to read the highway sign.

Specific considerations for making text more readable also follow from the mechanisms underling contrast sensitivity. Because of certain asymmetries in the visual processing system, dark text on lighter background ("negative contrast") also offers higher contrast sensitivity than light on dark ("positive contrast"). The disruptive tendency for white letters to spread out or "bleed" over a black background is called *irradiation*. Similarly, one should maximize contrast by employing black letters on white background rather than, less readable hued backgrounds (e.g., black on blue). Black on red is particularly dangerous with low illumination, since red is not seen by rods.

The actual font matters too. Fonts that adhere to "typical" letter shapes like the text of this book are easier to read because of their greater familiarity than those that create block letters or other nonstandard shapes. Another effect on readability is the case of the print. For single, isolated words, UPPERCASE may be better than lowercase print, as, for example, the label of an "on" switch. This advantage results in part because of the wider visual angle and lower spatial frequency. However, for multiword text, UPPERCASE PRINT IS MORE DIFFICULT TO READ than lowercase or mixed-case text. This is because lowercase text typically offers a greater variety of word shapes. This variety conveys sensory information at lower spatial frequencies that can be used to discern some aspects of word meaning in parallel with the high spatial frequency analysis of the individual letters [98, 99]. BLOCKED WORDS IN ALL CAPITALS will eliminate the contributions of this lower spatial frequency channel.

4.3.3 Color Vision

Color vision depends on a well-illuminated environment, with illuminance greater than 30 lux. Our driver had trouble judging the color of his red sedan because of the poor illumination in the parking lot. A second characteristic that limits the effectiveness of color is that approximately 7 percent of the male population is *color deficient*; that is, they are unable to discriminate certain hues from each other. Most prevalent is red-green "color blindness" (*protanopia*) in which the wavelengths of these two hues create identical sensations if they have the same luminance intensity. Many computer graphics packages use color to discriminate lines. If this is the only discriminating feature between lines, the graph may be useless for the color-deficient reader. Several web and smartphone applications show what the world looks like to those with color deficient vision (e.g., the chormatic vision simulator http://asada.tukusi.ne.jp/cvsimulator/e/)

Relying on red and green as a signal means your design will fail with 7% of men.

Because of these two important sensory limitations of color processing, a most important human factors guideline is to design for monochrome first and use color only as a redundant backup to signal important information [100]. Thus, for example, a traffic signal uses the location of the illuminated lamp (top, middle, bottom) redundantly with color.

Two additional characteristics of the sensory processing of color have some effect on its use. First, *simultaneous contrast* is the tendency of some hues to appear different when viewed adjacent to other hues (e.g., green will look deeper when viewed next to red than when viewed next to a neutral gray). Second, The *negative afterimage* is a similar phenomenon to simultaneous contrast but describes the greater intensity of certain colors when viewed after prolonged viewing of other colors.

4.3.4 Night Vision

The loss of contrast sensitivity at all spatial frequencies can inhibit the perception of print as well as the detection and recognition of objects by their shape or color in poorly illuminated viewing conditions. Coupled with the loss of contrast sensitivity due to age, it is apparent that night driving for the older population is a hazardous undertaking, particularly in unfamiliar territory [101, 102].

Added to these hazards of night vision are those associated with *glare*, which may be defined as irrelevant light of high intensity. Beyond its annoyance and distraction properties, glare temporarily destroying the rod's sensitivity to low spatial frequencies. Hence, the glare-subjected driver is less able to spot the dimly illuminated road hazard, such as the pothole or the darkly dressed pedestrian [88]. Glare can also be an important problem in offices when reflections from bright point sources of light reflect of computer screens.

4.4 Cognitive Influence on Visual Perception

Up to now, we have discussed primarily the factors of the human visual system that affect the quality of the sensory information that arrives at the brain to be perceived. As shown in Figure 4.10, we may represent these influences as those that affect processing from the *bottom* (lower levels of stimulus processing) *upward* (toward the higher centers of the brain involved with perception and understanding). Examples include the loss of acuity degrading bottom-up processing and high-contrast stimuli enhancing bottom-up processing.

In contrast, an equally important influence on processing visual information operates from the *top downward*. This is perception based on our knowledge (and desire) of what should be there. Thus, if I read the instructions, "After the procedure is completed, turn the system off," I need not worry as much if the last word happens to be printed in very small letters because I can pretty much guess what it will say. Much of our processing of perceptual information depends on the delicate interplay between *top-down processing, signaling what should be there*, and *bottom-up processing, signaling what is there*. Deficiencies in one (e.g., small, barely legible text) can often be compensated by the other other (e.g., expectations of what the text should say). Our initial introduction to the interplay between these two modes of processing is in a discussion of depth perception, and the distinction between the two modes is amplified further in our treatment of signal detection, and in subsequent chapters.

Top-down processing causes us, to some extent, to see what we expect to see.

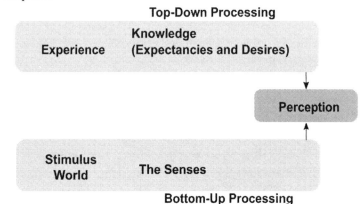

Figure 4.10 Bottom-up and top-down processing combine to define perception.

4.4.1 Depth Perception

Humans navigate and manipulate in a three-dimensional (3-D) world, and we usually do so quite accurately and automatically [103]. Yet there are times when our ability to perceive where we and other things are in 3-D space breaks down. Airplane pilots flying without using their instruments are also very susceptible to dangerous illusions of where they are in 3-D space and how fast they are moving [104, 105].

In order to judge our distance from objects (and the distance between objects) in 3-D space, we rely on a host of *depth cues* to inform us of how far away things are. The first three cues we discuss—accommodation, binocular convergence, and binocular disparity—are all inherent in the physiological structure and wiring of the visual sensory system. Hence, they may be said to operate on *bottom-up* processing.

Accommodation, as we have seen, is when an out-of-focus image triggers a change in lens shape to accommodate, or bring the image into focus on the retina. Figure 4.4 shows sensory receptors, within the ciliary muscles that accomplish this change, send signals to the higher perceptual centers of the brain that inform those centers how much accommodation was accomplished and hence the extent to which objects are close or far (within a range of about 3 m). (As we discuss in Chapter 5, these signals from the muscles to the brain are called *proprioceptive input.*)

Convergence is a corresponding cue based on the amount of inward rotation ("cross-eyedness") that the muscles in the eyeball must accomplish to bring an image to rest on corresponding parts of the retina on the two eyes. The closer the distance at which the image is viewed, the greater the amount of proprioceptive "convergence signal" sent to the higher brain centers by the sensory receptors within the muscles that control convergence.

Binocular disparity sometimes called *stereopsis*, is a depth cue that results because the closer an object is to the observer, the greater the amount of disparity there is between the view of the object received by each eyeball. Hence, the brain can use this disparity measure, computed at a location where the visual signals from the two eyes combine in the brain, to estimate how far away the object is. Most virtual reality systems and 3-D movies rely on stereopsis to convey the compelling sense of depth.

All three of these bottom-up cues are only effective for judging distance, slant, and speed for objects that are within a few meters from the viewer [106]. (However, stereopsis can be created in stereoscopic displays to simulate depth information at much greater distances, as we discuss in Chapter 8.) Judgment of depth and distance for more distant objects and surfaces depends on a host of what are sometimes called *"pictorial" cues* because they are the kinds of cues that artists put into pictures to convey a sense of depth. Because the effectiveness of most pictorial cues is based on past experience, they are subject to top-down influences. Some of the important pictorial cues to depth, as shown in Figure 4.11 include:

Linear perspective is the converging of parallel lines (i.e., the road) toward the more distant points.

Relative size is a cue based on the knowledge that if two objects are the same true size (e.g., the two trucks in the figure), then the object that occupies a smaller visual angle (the more distant vehicle in the figure) is farther away.

Interposition describes how nearer objects obscure the contours of objects that are farther away (see the two buildings).

Light and shading associated with how three-dimensional objects describes how they cast shadows and reveal reflections from illuminating light. These effects of lighting provide evidence of their location and their form and distance [107].

Textural gradients Any textured surface, viewed from an oblique angle, will show a gradient or change in texture density (spatial frequency) across the visual field (see the Illinois cornfield in the figure). The finer texture signals the more distant region, and the amount of texture change per unit of visual angle signals the angle of slant relative to the line of sight.

Relative motion, or motion parallax: More distant objects show relatively smaller movement across the visual field as the observer moves. Thus, we often move our head back and forth to judge the relative distance of objects. Relative motion also accounts for the accelerating growth in the retinal image size of things as we approach them in space, a cue sometimes called *looming* [108].

Collectively, these cues provide us with a very rich sense of our position and motion in 3-D space as long as the world through which we move is well illuminated and contains rich visual texture. Gibson [103] clearly described how the richness of these cues in our natural environment support very accurate space and motion

Figure 4.11 Some pictorial depth cues: linear perspective, relative size, interposition, light and shading, as well as texture gradients. (Photo by Brazzit available on Wikipedia/CC Attribution-SA 3.0 License.)

Most bottom-up cues and all top-down cues for depth perception do not depend on having two eyes.

perception. However, when cues are degraded, impoverished, or eliminated by darkness or other unusual viewing circumstances, depth perception can be distorted. This sometimes leads to dangerous circumstances. For example, a pilot flying at night or over an untextured snow cover has very poor visual cues to help determine where he or she is relative to the ground [109], so pilots must rely on precision flight instruments (see Chapter 8 on displays). Correspondingly, the implementation of both edge markers and high-angle lighting on highways greatly enriches the cues available for speed (changing position in depth) for judging distance hazards and allows for safer driving. In Chapter 8 we discuss how this information is useful for the design of 3-D displays.

Just as we may predict poorer performance in tasks that demand depth judgments when the quality of depth cues is impoverished, we can also predict that certain distortions of perception will occur when features of the world violate our expectations, and top-down processing takes over to give us an inappropriate perception. For example, Eberts and MacMillan [110] established that the cue of relative size contributes to higher-than-average rate at which small cars are hit from behind. A small car is perceived as more distant than it really is from the observer approaching it from the rear. Hence, a small car is approached faster (and braking begins later) than is appropriate, sometimes leading to the unfortunate collision.

Of course, clever design can sometimes turn these distortions to advantage, as in the case of the redesign of a dangerous traffic circle in Scotland [111]. Drivers tended to overspeed when coming into the traffic circle with a high accident rate as a consequence. One solution is to trick the driver's perceptual system by drawing lines across the roadway of diminishing separation, as the circle was approached. Approaching the circle at a constant (and excessive) speed, the driver experiences the optic flow of texture past the vehicle, which signals an increase in speed (i.e., accelerating). Because of the nearly automatic way in which many aspects of perception are carried out, the driver should instinctively brake in response to the perceived acceleration, bringing the speed closer to the desired safe value. This is exactly the effect that was observed in relation to driving behavior after the marked pavement was introduced, resulting in a substantial reduction in fatal accidents [112].

Designing the visual environment can directly influence people's behavior— pavement markings can reduce drivers' speed.

4.4.2   Visual Search and Detection

A critical aspect of human performance in many systems concerns the closely linked processes of *visual search* and *object detection*. Our driver at the beginning of the chapter was searching for several things: the appropriate control for the wipers, the needed road sign, and of course any number of possible hazards or obstacles that could appear on the road (the pothole was one that he missed). The goal of these searches was to *detect* the object or event in question. These tasks are analogous to the kind of processes we go through when we search the index of this book for a needed topic, search a cluttered graph for a data point, or when the quality control inspector searches a product (say, a microchip board) for a flaw. In all cases, the search may or may not successfully end in a detection.

Despite the close link between visual search and detection, it is important to separate our treatment of these topics, both because different factors affect each and because human factors specialists are sometimes interested in detection when there is no search (e.g., the detection of a fire alarm). We consider the process of search itself, but to understand visual search, we must first consider the nature of eye movements, which are heavily involved in searching large areas of space. Then we consider the process of detection.

Eye movements are necessary to search the visual field [113, 114]. Eye movements can be divided into two classes. *Pursuit* movements are those of constant velocity that are designed to follow moving targets, for example, following the rapid flight of an aircraft across the sky. More related to visual search are *saccadic eye movements*, which are abrupt, discrete movements from one location to the next designed to bring a visual item into foveal vision. You are using saccadic movements as you read. Each saccadic movement can be characterized by a set of three critical features: an *initiation latency*, a *movement time* (or speed), and a *destination*. Each destination, or *dwell*, can be characterized by both its *dwell duration* and a *useful field of view* (UFOV) . In continuous search, the initiation latency and the dwell duration cannot be distinguished.

The actual movement time is generally quite fast (typically less than 50 msec) and is not much greater for longer than for shorter movements. Most time is spent during dwells and initiations. These time limits are such that even in rapid search there are no more than about 3 to 4 dwells per second [115], and this frequency is usually lower because of variables that prolong the dwell. The destination of a scan is usually driven by top-down processes (i.e., expectancy), although on occasion a saccade may be drawn by salient bottom-up processes (e.g., a flashing light). Considerably more detailed discussion of these factors is presented in Chapter 6. The dwell duration is governed jointly by two factors: (1) the *information content* of the item fixated (e.g., when reading, long words require longer dwells than short ones), and (2) the ease of *information extraction*, which is often influenced by stimulus quality (e.g., in target search, longer dwells on a small target or one with poor contrast). Finally, once the eyes have landed a saccade on a particular location, the useful field of view defines how large an area, surrounding the center of fixation, is available for information extraction [116]. The useful field of view defines the diameter of the region within which a target might be detected if it is present, and is important because drivers with reduced UFOV are more likely to crash [117].

The *useful field of view* should be distinguished from the area of *foveal vision*, defined earlier in the chapter. Useful field of view describes the area around the fixation that people can extract information in a brief glance. Foveal vision defines a specific area of approximately 2 degrees of visual angle surrounding the center of fixation, which provides high visual acuity and low sensitivity. The diameter of the useful field of view, in contrast, is task-dependent. It may be quite small (2.5 degrees) if the operator is searching for very subtle targets demanding high visual acuity but may be much larger than the fovea if the targets are conspicuous and can be easily detected in peripheral vision (10 or more degrees).

Serial and Parallel Search. In describing a person searching any visual field for something, we distinguish between targets and nontargets (nontargets are sometimes called distractors). The latter may be thought of as "visual noise" that must be inspected in order to determine that it is not in fact the desired target. Many searches are serial in that each item is inspected in turn to determine whether it is or is not a target. If each inspection takes a relatively constant time, I, and the expected location of the target is unknown beforehand, then it is possible to predict the average time it will take to find the target as where I is the average inspection time for each item, and N is the total number of items in the search field [118, 119]. Equation 4.4 shows that on average, the target will be encountered after half of the targets have been inspected (sometimes earlier, sometimes later). This serial search model has been applied to predicting performance in numerous environments in which people search through maps or lists or computer menus [120, 121].

$$T = \frac{N \times I}{2} \text{ When there is no target present: } T = N \times I$$

Time to find a target in a set of
many possible targets

Time = Time to detect a target   ~~should be~~ $T =$
N = Number of targets
I = Time to inspect each target

(4.4)

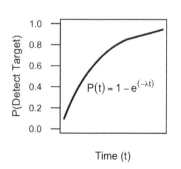

$$P(t) = 1 - e^{(-\lambda t)}$$

Figure 4.12 Predicted search success probability as a function of the time spent searching is an exponential distribution. Stopping time in search was developed by Drury [123]. (Created in R.)

If the visual search space is organized, people tend to search from top to bottom and left to right. In natural scenes there is a bias to focus on the center. If the space does not benefit from such organization (e.g., searching a map for a target or searching the ground below the aircraft for a downed airplane [122]), then people's searches tend to be considerably more random in structure and do not "exhaustively" examine all locations[122]. If targets are not readily visible, this non-exhaustive characteristic leads to an exponential search-time function as depicted in Figure 4.12.

The figure suggests that there are diminishing returns associated with giving people too long to search a given area if time is at a premium such as may be the case in industrial or baggage inspection tasks. Such a model was initially defined to examine the optimum inspection time that people should be allowed to examine each images in a quality-control inspection task [123].

Search models can be extremely important for predicting search time in time-critical environments; for example, how long will a driver keep eyes off the highway to search for a road sign? Unfortunately, however, there are two important circumstances that can render the strict serial search model inappropriate, one related to bottom-up processing and the other to top-down processing. Both factors force models of visual search to become more complex and less precise.

Conspicuity. The first circumstance that makes the serial search model inappropriate is where conspicuity of the target guides search. Certain targets are so conspicuous that they may "pop out" no matter where they are in the visual field, and so nontarget items need not be inspected [124, 125]. Psychologists describe the search for such targets as parallel because, in essence, all items are examined at once (i.e., in parallel), and in contrast to the Equation 4.4, search time does not increase with the total number of items. Such is normally the case with "attention grabbers," such as a flashing warning signal, a moving target, or a uniquely colored, highlighted item on a checklist, a computer screen, or in a phone book.

Conspicuity is a desirable property if the target is conspicuous, but an undesirable one if the conspicuous item is not relevant to the task at hand. Thus, if I am designing a checklist that highlights emergency items in red, this may help the operator in responding to emergencies, but will be a distraction if the operator is using the list to guide normal operating instructions; that is, it will be more difficult to focus attention on the normal instructions. We see this all the time with flashing, moving and annoying advertisements on web pages. As a result of these dual consequences of conspicuity, the choice of highlighting (and the effectiveness of its implementation) must be guided by a careful analysis of the likelihood the highlighted item will be the target [126]. Table 4.5 lists some key variables that can influence the conspicuity of targets and, therefore, the likelihood that the field in which they are embedded will be searched in parallel.

Expectancies. The second influence on visual search that leads to departures from the serial model has to do with the top-down implications of *searcher expectancies* of where the target might be likely to lie. Expectancies, like all top-down processes, are based upon prior knowledge. Our driver did not expect to see the road sign on the left of the highway and, as a result, only found it after it was too late. As another example, when searching a map we do not usually blanket the entire page with fixations, but our *knowledge* of the locations allows us to start the search near the target name. Similarly, when searching an index, we often have an idea what the topic is likely to be called, which guides our starting point.

It is important to realize that these expectancies, like all knowledge, come only with experience. Hence, we might predict that the skilled operator will have more top-down processes driving visual search than the unskilled one and as a result will be more in the efficient, a conclusion born out by research [127]. Novice drivers can be trained to search for hazards to bring their safety closer to that of more experienced drivers [128]. These top-down influences also provide guidance for designers who develop search tools, such as

Discriminability: Difference from background elements

1. In color (particularly if non-target items are uniformly colored)
2. In size (particularly if the target is larger)
3. In brightness (particularly if the target is brighter)
4. In motion (particularly if background is stationary)

Simplicity: Can the target be defined only by one dimension (i.e., "red") and not several (i.e., "red and small")

Automaticity: A target that is highly familiar (e.g., one's name)

Not shapes: Unique shapes (e.g., letters, numbers) do not generally support parallel search [124].

Table 4.5 Target properties inducing parallel search.

book indexes and menu pages, to understand the expected order-ings and groupings of items that users have. This topic is addressed again in Chapter 10 in the context of human-computer interaction.

In conclusion, research on visual search has four general impli-cations, all of which are important in system design:

1. Conspicuity effects can enhance the visibility of target items (reflective jogging suits [129, 130] or highlighting critical menu items). In dynamic displays, automation can high-light critical targets [131, 132], which we will discuss further in Chapter 11.

2. Serial visual search processes quantify the costs of *cluttered* displays (or search environments). When too much informa-tion is present, many maps present an extraordinary amount of clutter. For electronic displays, this fact should lead to con-sideration of *decluttering* options in which certain categories of information can be electronically turned off or deinten-sified [131]. However, careful use of color and intensity as discriminating cues between different classes of information can make decluttering unnecessary [131].

3. The role of top-down processing in visual search should lead designers to make the *structure* of the search field as appar-ent to the user as possible and consistent with the user's knowledge (i.e., past experience). For verbal information, this may involve an alphabetical organization or one based on the semantic similarity of items. In positioning road signs, this involves the use of *consistent* placement (see Chapter 17).

4. Models of visual search that combine all of these factors can predict how long it will take to find particular targets, such as the flaw in a piece of sheet metal [123], an item on a com-puter menu [126], or a traffic sign by a highway [133]. For visual search, however, the major challenge of such models is that search appears to be guided much more by top-down than by bottom-up processes [134], and developing mathe-matical terms to characterize how expertise affects top-down processing is a major challenge.

4.4.3 Detection

Once a possible target is located in visual search, it becomes nec-essary to *confirm* that it really is the item of interest (i.e., *detect* it). This process may be trivial if the target is well known and reason-ably visible (e.g., a stop sign), but it is far from trivial if the target is degraded, like a faint flaw in a piece of sheet metal, a small crack in an x-rayed bone, the faint glimmer of the lighthouse on the horizon at sea, or the missing period in a proofread manuscript. In these cases, we must describe the operator's ability to detect signals. Sig-nal detection is often critical even when there is no visual search at all. For example, the quality-control inspector may have only one place to look to examine the product for a defect. Similarly, human factors specialists are also concerned with detection of auditory signals, like the warning sound in a noisy industrial plant, when search is not at all relevant.

Signal Detection Theory. In any of a variety of tasks, the process of signal detection can be modeled by *signal detection theory (SDT)* [135, 136, 137], which is represented schematically in Figure 4.13. SDT assumes that "the world" (as it is relevant to the operator's task) can be modeled as either one in which the "signal" to be detected is present or absent, as shown across the top of the matrix in Figure 4.13. Whether the signal is present or absent, the world is assumed to contain noise: Thus, the luggage inspected by the airport security screener may contain a weapon (signal) in addition to a number of things that might look like weapons (i.e., the noise of hair drier, calculators, carabiners, etc.), or it may contain the noise alone, with no signal (i.e., items that do not resemble weapons).

The goal of the operator in detecting signals is to discriminate signals from noise. Thus, we may describe the relevant behavior of the observer as that represented by the two rows of Figure 4.13— saying, "Yes (I detect a signal)" or "No (there is only noise)." This combination of two states of the world and two responses yields four events shown in the four cells of the figure labeled *hits, false alarms, misses,* and *correct rejections* . Two of these cells (hits and correct rejections) clearly represent "good" outcomes and ideally *should* characterize much of the performance, while two are "bad" (misses and false alarms) and ideally should never occur. If several encounters with the state of the world (signal detection trials) are aggregated, some involving signals and some involving noise alone, we may then express the numbers within each cell as the *probability* of a hit [#hits/#signals = $p(H)$]; the probability of a miss [$1-p(H)$]; the probability of a false alarm [#FA/#no-signal encounters = $p(FA)$] and the probability of a correct rejection [$1-p(FA)$]. As you can see from these equations, if the values of $p(H)$ and $p(FA)$ are measured, then they determine the other two cells by subtracting these values from 1.0.

Thus, the data from a signal detection environment (e.g., the performance of an airport security inspector) may easily be represented in the form of the matrix shown in Figure 4.13, if a large number of trials are observed so that the probabilities can be reliably estimated. SDT relates these numbers in terms of two fundamentally different influences on human detection performance: *sensitivity* and *response bias* . As we describe in the next section, we can think of these two as reflecting bottom-up and top-down processes respectively.

Sensitivity and Response Bias. As Figure 4.13 shows at the bottom, the measure of sensitivity, often expressed by the measure d' (d prime) expresses how good an operator is at discriminating the signal from the noise, reflecting essentially the number of good outcomes (hits and correct rejections) relative to the total number of both good and bad outcomes. Sensitivity is higher if there are more correct responses and fewer errors. It is influenced both by the keenness of the senses and by the strength of the signal

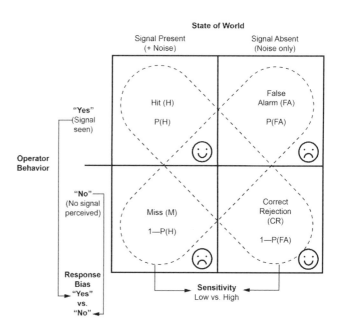

Figure 4.13 Representation of the outcomes in signal detection theory. The figure shows how changes in the four joint events influence the primary performance measures of response bias and sensitivity, shown at the bottom.

Signal detection performance depends on considering not just signals detected but also false alarms.

relative to the noise (i.e., the *signal-to-noise ratio*) . For example, sensitivity usually improves with experience on the job up to a point; it is degraded by poor viewing conditions (including poor eyesight). An alert inspector has a higher sensitivity than a drowsy one. The formal calculation of sensitivity is not discussed in this book, and there are other related measures that are sometimes used to capture sensitivity [137]. However the simple measure of percent correct often provides a good approximation.

The measure of response bias, or *response criterion*, shown in the left of Figure 4.13, reflects the bias of the operator to respond "yes, signal" versus "no, noise." Although formal signal detection theory characterizes response bias by the term *beta*, which has a technical measurement [137, 10], one can more simply express response bias as the probability that the operator will respond yes [(#yes)/(Total responses)]. Response bias is typically affected by two variables, both characteristic of top-down processing. First, increases in the operator's expectancy that a signal will be seen leads to corresponding increases in the probability of saying yes. For example, if a quality-control inspector has knowledge that a batch of products may have been manufactured on a defective machine and therefore may contain a lot of defects, this knowledge should lead to a shift in response criterion to say "signal" (defective product) more often. The consequences of this shift are to generate both more hits *and* more false alarms.

Second, changes in the *values*, or costs and benefits, of the four different kinds of events can also shift the criterion. The air traffic controller cannot afford to miss detecting a signal (a conflict between two aircraft) because of the potentially disastrous consequences of a midair collision [138]. The miss has a high cost. As a result, the controller will set the response criterion at such a level that misses are very rare, but the consequences are that the less costly false alarms are more frequent. In representing the air traffic controller as a signal detector, these false alarms are circumstances when the controller detects a potentially conflicting path and redirects one of the aircraft to change its flight course even if this was not necessary [139].

In many cases, the outcome of a signal detection analysis may be plotted in what is called a *receiver operating characteristic (ROC)* space, as shown in Figure 4.14[140]. Here p(FA) is plotted on the x axis, p(H) is plotted on the y axis, and a single point in the space (consider point A) thereby represents all of the data from one set of detection conditions. In different conditions, detection performance at B would represent improved sensitivity (higher d). Detection performance at C would represent only a shift in the response criterion relative to A (here a tendency to say yes more often, perhaps because signals occurred more frequently). More details about the ROC space can be found in [136, 137, 10].

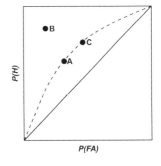

Figure 4.14 A receiver operating characteristic, or ROC curve. Each point represents the signal detection data from a single matrix.

Interventions to enhance SDT performance. The distinction between sensitivity and response criterion made by SDT is important because it allows the human factors specialist to understand the consequences of different kinds of interventions to improve detection performance in a variety of circumstances. For example, any instructions that "exhort" operators to "be more vigilant" and not miss signals will probably increase the hit rate but will also increase the false-alarm rate. This is because the instruction is a motivational one reflecting costs and values, which typically affects the setting of the response criterion, as the shift from point A to point C in the ROC of Figure 4.14. (Financially rewarding hits will have the same effect.) Correspondingly, it has been found that directing the radiologist's attention to a particular area of an x-ray plate where an abnormality is likely to be found will tend to shift the response criterion for detecting abnormalities at that location but will not increase the sensitivity [141]. Hence, the value of such interventions must consider the relative costs of misses and false alarms.

However, there are certain things that *can* be done that do have a more desirable direct influence on increasing sensitivity (that is, moving from point A to point B in Figure 4.14). As we have noted, training the operator for what a signal looks like can improve sensitivity, as can providing the inspector with a "visual template" of the potential signal that can be compared with each case that is examined [142, 143]. Several other forms of interventions to influence signal detection and their effects on sensitivity or response bias are shown in 4.6. In Chapter 5 we describe how signal detection theory is also important in the design of auditory alarms.

Providing knowledge of results usually increases sensitivity, but may calibrate response bias if it provides observer with more

Mechanism	Intervention	Effect
Top-down	Payoffs	Response bias
	Introducing "false signals" to raise signal rate artificially	Response bias
	Providing incentives and exhortations	Response bias
Top-down & Bottom-up	Providing knowledge of results	Increases sensitivity
Bottom-up	Slowing down the rate of signal presentation, such as slowing the assembly line	Increases sensitivity
	Differentially amplifying the signal more than the noise	Increases sensitivity
	Making the signal dynamic	Increases sensitivity
	Giving frequent rest breaks	Increases sensitivity
	Providing a visual (or audible) template of what the signal looks or sounds like	Increases sensitivity
	Providing experience seeing the signal	Increases sensitivity
	Providing redundant representations of the signal between auditory and visual channels	Increases sensitivity

Table 4.6 Factors influencing signal detection performance.

accurate perception of probability of signal.

In Chapter 13 we describe signal detection theory's role in characterizing the loss of vigilance of operators in low arousal monitoring tasks, like the security guard at night. For inspectors on an assembly line, the long-term decrement in performance may be substantial, sometimes leading to miss rates as high as 30 to 40 percent. The guidance offered in Table 4.6 suggests some of the ways in which these deficiencies might be addressed. To emphasize the point made above, however, it is important for the human factors practitioner to realize that any intervention that shifts the response criterion to increase hits will have a consequent increase in false alarms. Hence, it should be accepted that the costs of these false alarms are less severe than the costs of misses (i.e., are outweighed by the benefits of more hits). The air traffic control situation is a good example. When it comes to detecting possible collisions, a false alarm is less costly than a miss (a potential collision is not detected), so interventions that increase false alarm rate can be tolerated even if they also decrease miss rate.

4.4.4 Discrimination

Very often, issues in human visual sensory performance are based on the ability to *discriminate* between one of two signals rather than to *detect* the existence of a signal. Our driver was able to see the road sign (detect it) but, in the brief view with dim illumination, failed to discriminate whether the road number was 60 or 66 (or in another case, perhaps, whether the exit arrow pointed left or right). He was also clearly confused over whether the car color was red or brown. Confusion, the failure to discriminate, results whenever stimuli are similar. Even fairly different stimuli, when viewed under degraded conditions, can produce confusion.

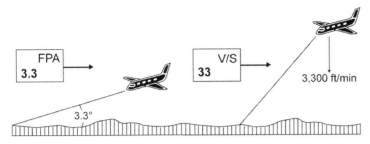

Figure 4.15 Confusion induced by the labels of the automation setting contributed to a commercial airline crash.

Figure 4.15 shows how a display can confuse. It is believed that one cause of the crash of a commercial jet liner in Europe was that the automated setting that controlled its flight path angle with the ground (3.3 degrees) looked so similar to the automated setting that controlled its vertical speed (3,300 feet/minute)[144]. The pilots believed the top condition to exist, when in fact the bottom existed. The display represents the two conditions in a very similar manner, and hence the two were quite confusable. As a result, pilots could easily have confused the two, thinking that they had "dialed in" the 3.3-degree angle when in fact they had set the 3,300 ft/min vertical speed (which is a much more rapid decent rate than that given by the 3.3-degree angle).

The extreme visual similarity of very different drug names leads to medication error [145]. Consider such names as capastat and cepastat, or mesantoin and metinon, or Norflox and Norflex; each has different health implications when administered, yet the names are quite similar in terms of visual appearance. Such possible confusions are likely to be amplified when the prescription is filtered through the physician's (often illegible) handwriting. The greater proportion of matching or similar features between the two stimuli, the harder the discriminate and the more likely errors will occur. "Tall man" labeling (see sidebar) can reduce label confusion [146, 147]

It is important for the designer to consider alternative controls, especially if they must be reached and manipulated for activation or to consider alternative displays that must be interpreted and perceived. Can they be adequately discriminated? Are they far enough apart in space, or can they be distinguished by other features, such as color, shape, or other labels, to minimize confusion? It is important to remember, however, that if only verbal labels are used to discriminate the displays or controls from each other, then attention must be given to the visibility and readability issues discussed earlier. We discuss the important and often overlooked issues of discrimination and confusion further as we address the issues of working memory in Chapter 6 and displays in Chapter 8.

buPROPion

busPIRone

Tall Man Labels (1991) One promising approach to enhancing discrimination in labels is the *"tall man" method*. This approach uses capital letters in a drug name to highlight differences in potentially confusable names [TODO JEP Reference]. Start on the left of the word and capitalize all the characters to the right once two or more dissimilar letters are encountered (e.g., acetaZOLAMIDE and acetoHEXAMIDE). If there are no common letters on the left, start on the right and capitalize all after the similar letters (e.g., DOPamine and DOBUTamine). When there are no common letters on the left or right side of the word, capitalize only the central part of the word (e.g., buPROPion and busPIRone) (https://www.ismp.org/tools/tallmanletters.pdf).

4.4.5 Absolute Judgment

Discrimination refers to judgment of differences between two sources of information that are actually (or potentially) present, and generally people are good at this task as long as the differences are not small and the viewing conditions are favorable. In contrast, *absolute judgment* refers to the limited human capability to judge the absolute value of a variable signaled by a coded stimulus. For example, estimating the height of a bar graph to the nearest digit is an absolute judgment task with 10 levels. Judging the color of a traffic signal (ignoring its spatial position) is an absolute judgment task with only three levels of stimulus value. People are not generally very good at these absolute value judgments of attaching "labels to levels" [10]. It appears that they can be certain to do so accurately only if fewer than around five levels of any sensory continuum are used and that people are even less accurate when making absolute value judgments in some sensory continua like pitch or sound loudness; that is, even with five levels they may be likely to make a mistake, such as confusing level three with level four [148].

The lessons of these absolute judgment limitations for the designer are that the number of levels that should be judged based on an absolute coding scheme, like position on a line or color of a light, should be chosen conservatively. It is recommended, for example, that no more than six colors be used if precise accuracy in judgment is required (and an adjacent color scale for comparison is not available). The availability of such a scale would turn the absolute judgment task into a relative judgment task, and hence a much easier one). This can affect performance of multi-color displays, like maps, as the number of colors grows, an issue we treat further in our discussion of absolute judgment in Chapter 8. Furthermore, even this guideline should be made more stringent under potentially adverse viewing conditions (e.g., a map that is read in poor illumination). It is quite difficult to match colors based on memory or even when they are compared side-by-side (for a demonstration see this color matching task: http://color.method.ac).

> People are poor at absolute judgment, but good at relative judgment.

4.5 Visual Influence on Cognition

The eyes and vision are often considered as a source of for navigating the world and supporting higher cognition. However, eyes can have a more direct effect on mental state through their influence on the body's internal clock—circadian rhythm— and the moods color can evoke.

4.5.1 Light and Circadian Desynchronization

Earlier in this chapter we described the eye's photoreceptors as rods and cones, but the eyes also have a third type of photoreceptors— awkwardly named intrinsically photosensitive retinal ganglion cells (*ip*RGCs)—enable the body to "see" what time it is. These cells don't help you see in the traditional sense, but they influence the Pineal gland and the regulation of melatonin levels. Melatonin levels govern sleep cycles. Good quality sleep depends on exposing (*ip*RGCs) to light during the day and not at night. These photoreceptors are particularly sensitive to blue light (wavelengths between 446 and 477 nm [149]).

These results have several practical implications. Exposure to displays of computers, tablets and smartphones at night can disrupt sleep [150]. Solutions include minimizing exposure and tuning the displays to reduce the light output between 446 and 477 nm, as Apple has done in its Night Shift mode. More generally, using lower color temperature light sources (warmer light) and indirect lighting can also help. Exposing yourself to light during the day can reduce jet-lag, which is disrupted sleep that occurs when sleep cycles don't match your location. Chapter 15 discusses other factors affecting sleep and its consequence for performance.

> Exposing your eyes to blue light from computer and phone screens at night undermines your sleep.

4.5.2 Meaning and Emotional Influence of Color

Colors have intrinsic and learned associations that influence mood and performance. As an example, response times to words printed in red were faster for negative words and words associated with failure, whereas response times to words printed in green were faster for words associated with success, but not positive words [151]. More generally, red is also associated with avoidance motivation, which is an emotional response to the situation [152]. This emotional response can affect cognitive performance. Those taking an IQ test with a red cover answered 6.5 questions correctly and those with a gray or green color answered 7.8 questions correctly [152]. Beyond mood and performance, colors strongly influence the aesthetic appeal of products and many professional designers focus on picking appealing color combinations. Recently statistical models have been created that can select appealing color combinations, such as the one used on the cover of this book [153].

Color not only affects mental state, but it can affect other senses, such as taste [154]. Coloring a cherry-flavored drink green and a lime-flavored drink red led 40% of people to report the cherry-flavored drink as tasting like lemon, lime, or lemon-lime. When the lime-flavored drink was colored red 18% report it as tasting of cherry [155]. Colors that fail to match the expected taste also led to less intense and less acceptable taste [156]. We will return to these interactions between senses at the end of Chapter 5.

4.6 Summary

We have seen in this chapter how limits of the visual system influence the nature of the visual information that arrives at the brain for more elaborate perceptual interpretation. We have also begun to consider some aspects of this interpretation, as we considered top-down influences like expectancy, learning, and values. In Chapter 5, we consider similar issues regarding the processing of auditory and other sensory information. Together, these chapters describe the sensory processing of the "raw" ingredients for the more elaborative perceptual and cognitive aspects of understanding the world. Once we have addressed these issues of higher processing in Chapter 6, we can consider how all of this knowledge—of bottom-up sensory processing, perception, and understanding—can guide the design of displays that support tasks confronting people. This is the focus of Chapter 8.

Additional Resources

Several useful resources that expand on the content touched on in this chapter include:

1. **Color matching game:** This interactive website can help sharpen your perception of color and appreciate the limits of absolute judgment (http://color.method.ac)

2. **Chromatic vision simulator:** This free smartphone app allows you to see the world as those with various color vision deficiencies do (http://asada.tukusi.ne.jp/cvsimulator/e/)

3. **Web content accessibility guidelines:** This document outlines considerations for making web page content accessible (http://www.w3.org/TR/WCAG20/#perceivable)

4. **Lighting handbook:** This resource covers a broad range of considerations in the design of lighting for various tasks and activities.

   DiLaura, D., Houser, K. W., Misrtrick, R. G., and Steffy, R. G. (2011). *The Lighting Handbook 10th Edition: Reference and Application.* Illuminating Engineering Society of North America

Questions

Questions for 4.1 Visual Environment

P4.1   Does the wavelength of light primarily affect our perception of hue or brightness?

P4.2   For display designs, what is the purpose of the CIE color space?

P4.3   How is reflectance defined?

P4.4   In a perfectly reflective surface, would luminance be equal to illuminance? Explain.

P4.5   How do orientation and simple visual tasks differ from complex visual task in terms of illuminance?

P4.6   How does a brightly lit operating room compare to the illuminance outside at noon.

P4.7   Calculate the number of highly efficient 0.5 watt LED lamps needed to light a 1 m x 1.5 m desk from a height of 2 m. Assume the person will be reading poor photocopies to digitize their contents.

P4.8   You approach an intersection at the same time a car approaches on a perpendicular street. The image of the other car does not move relative to a spot on your windshield. Explain what happening in terms of optic flow.

Questions for 4.2 The Receptor System: The Eye

P4.9   What information do the ciliary muscles send the brain?

P4.10   Calculate the font size (in points) needed for comfortable reading at the resting state of accommodation.

P4.11   Why do older people need to hold a book farther from them to read it if the do not use reading glasses? Explain in terms of age-related changes to the optics of the eye.

P4.12   Why do you use the visual angle to measure the size of text and symbols in degrees and minutes rather than a ruler to measure their size in meters?

P4.13   What happens to visual acuity as you move from the center of the fovea to the edge?

P4.14   What is responsible for the change in acuity from the center of the fovea to the edge?

P4.15   Explain where an object falls on your fovea when you look directly at it.

P4.16   Would the light level of a dimly lit parking lot be sufficient of engage photopic vision and if not what would be consequences for searching for your car?

P4.17   How could the characteristics of the visual receptor system distort or disrupt our ability to see objects during a nighttime drive?

Questions for 4.3 Sensory Processing Characteristics

P4.18   Identify which depth cues are considered to be more bottom-up processing and which are considered to be more top-down. Explain.

P4.19   Calculate the visual angle (θ) subtended by a 2 foot high letter on a highway sign that is 100 feet away.

P4.20   At night, a light of 60 foot candelas shines on a road sign and it reflects back 45 foot lamberts. What is the reflectance of the sign?

P4.21   What color lights should you use in lighting a dashboard of a car if you want to preserve drivers night vision?

P4.22   How many pixels per degree of visual angle is required for a "retina" as defined by a display where you cannot see individual pixels?

P4.23   Apply the Bond rule to estimate the size of lettering needed on a road sign that is intended to be seen from 100 feet away.

P4.24   How does the critical print size relate to the limits of visual acuity?

P4.25   What happens when text is smaller than the critical print size?

P4.26   How much more sensitive is vision as measured by hyperacuity compared to simple acuity?

P4.27   Does negative contrast enhance or undermine the readability of text?

P4.28   Explain the neural mechanism underlying hyperacuity.

P4.29   Describe an application of hyperacuity in the workplace.

P4.30   Describe the shape of the curve that describes contrast sensitivity as a function of spatial frequency.

P4.31   Describe an example of how particularly low and high and high spatial frequencies affect performance in the workplace.

P4.32   Can all capital letters be an effective way to present a single word? Or several sentences? Explain.

P4.33   Why is sentence case lettering often more effective than all caps lettering?

P4.34   What is the most important human factors guideline in designing with color?

P4.35   What is glare and how does it manifest in driving and office work?

Questions for 4.4 Cognitive Influence on Visual Perception

P4.36 How do the concepts of top-down and bottom-up processing characteristics influence your approach to improving visual performance?

P4.37 Describe a bottom-up and a top-down component of depth perception.

P4.38 What would you do to ensure a pilot could land successfully on a field of new snow?

P4.39 Using the concepts of optic flow, describe how you would paint a bike path to encourage cyclists to slow before an intersection.

P4.40 What are some key features of a target in a parallel search?

P4.41 Calculate the average time to find the TRUE Waldo (in a "Search for Waldo" game) if there were 19 Waldo look-alikes in the search field, and it takes on average 10 seconds to inspect each potential Waldo.

P4.42 There are 100 pints of assorted ice cream in your supermarket's freezer. How long will it take you to search for your favorite brand assuming an unstructured search of 2 seconds for each pint?

P4.43 What is the difference between conspicuity and expectancy?

P4.44 Describe a top-down intervention to improve TSA agents' performance in detecting explosives at an airport checkpoint.

P4.45 Describe a bottom-up intervention to improve TSA agents' performance in detecting explosives at an airport checkpoint.

P4.46 Would a top-down intervention to improve TSA agents' performance have a greater effect on the agents' sensitivity or their bias?

P4.47 Would a bottom-up intervention to improve TSA agents' performance have a greater effect on the agents' sensitivity or their bias?

P4.48 Explain how increasing the response bias in signal detection performance could be a good thing.

P4.49 Describe how the "tall man" strategy reduces confusion between drug labels.

P4.50 Considering the limits of absolute judgment, what are the consequences for selecting colors for a graph, where the colors are meant to identify the quarterly revenue of 12 different products?

Questions for 4.5 Visual Influence on Cognition

P4.51 Describe how reading from a tablet computer before you go to bed can disrupt your sleep?

P4.52 How would you design a house to help people make nighttime trips to the restroom in a way that balances the need to provide light (to minimize chance of falling) and minimize exposure to short-wavelength light that can disrupt sleeping?

Chapter 5

Auditory, Tactile, and Vestibular Systems

At the end of this chapter you will be able to...

1. identify how the physical properties of sound contribute to its perception

2. calculate permissible noise levels in a work environment and implement hazard management to address them

3. design effective auditory alarms

4. use properties of top-down and bottom-up processing to enhance speech-based communication

5. understand what makes noise annoying and annoyance can be mitigated

6. explain the basic elements of other sensory modalities and sensory integration

A worker at the small manufacturing company was becoming increasingly frustrated by the noise at her workplace. It was unpleasant and stressful, and she came home each day with a ringing in her ears and a headache. What particularly concerned her was an incident the day before when she could not hear the emergency alarm go off on her own equipment, a failure of hearing that nearly led to an injury. Asked by her husband why she did not wear earplugs to muffle the noise, she said, "They're uncomfortable. I'd be even less likely to hear the alarm, and besides, it would be harder to talk with the worker on the next machine, and that's one of the few pleasures I have on the job." She was relieved that an inspector from Occupational Safety and Health Administration (OSHA) would be visiting the plant in the next few days to evaluate her complaints.

The worker's concerns illustrate the effects of three different types of sound: the undesirable noise of the workplace, the critical warning of the alarm, and the important communications through speech. Our ability to process these three sources of acoustic information, whether we want to (alarms and speech) or not (noise), and the influence of this processing on performance, health, and comfort are the focus of the first part of this chapter. We conclude by discussing three other sensory channels: tactile, proprioceptive-kinesthetic, and vestibular, as well as their integration. These senses play a smaller, but significant, role in the design of human-machine systems.

5.1   Auditory Environment

The stimulus for hearing is sound, a compression and rarefaction of the air molecules, which is a wave with amplitude and frequency. This is similar to the fundamental characteristics of light discussed in Chapter 4, but with sound, the waves are acoustic rather than electromagnetic. The amplitude of sound waves contributes to the perception of the loudness of the sound and its potential to damage hearing, and the frequency contributes to the perception of its pitch. Before we discuss the subjective experience of loudness and pitch, we need to understand the physics of sound and its potential to damage hearing.

5.1.1   Amplitude, Frequency, Envelope, and Location

Sound can be represented as a sum of many sine waves, each with a different amplitude and frequency. Figure 5.1a shows the variation of sound pressure over time, and 5.1b shows three of the many sine waves that make up this complex signal. Each of these sine waves has a different frequency and amplitude. The line at the top of Figure 5.1b shows a high frequency signal and the line at the bottom shows a low frequency signal. These are typically plotted as a power spectrum, as shown in 5.1c. The horizontal position of each bar represents frequency of the wave, expressed in cycles/second or *Hertz* (Hz). The height of each bar reflects the amplitude of the wave and is typically plotted as the square of the amplitude, or the power. 5.1d shows the *power spectrum* from the full range of the many sine waves that make up a more complex sound signal.

The power spectrum is important because it shows the range of frequencies in a given sound. Similar to light, people can only perceive a limited range of sound frequencies. The lowest perceptual frequency is approximately 20 Hz and the highest 20,000 Hz. Above this range is ultrasound and below this range, sound is felt more than heard. People are most sensitive to sounds in the range of 2,000-5,000 Hz.

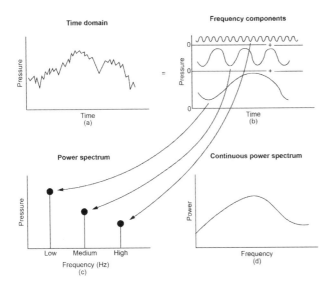

Figure 5.1 Different schematic representations of a speech sound: (a) timeline of a speech sound; (b) three frequency components of (a); (c) the power spectrum of (b); (d) a continuous power spectrum of the full range of frequencies in (a).

In addition to amplitude and frequency, two other critical dimensions of the sound stimulus are associated with temporal characteristics, sometimes referred to as the *envelope* in which a sound occurs, and its *location*. The temporal characteristics are what may distinguish the wailing of a siren from the steady blast of a car horn, and the location (relative to the hearer) is, of course, what might distinguish the siren of a fire truck pulling up behind from that of a fire truck about to cross the intersection ahead [157]. The envelope of sound is particularly critical in describing the sound of speech.

Figure 5.2(top) shows the timeline of sound waves of someone saying the /d/ sound as in "day". Such signals are more coherently presented by the power spectrum, as shown in Figure 5.2(middle). However, for speech, unlike noise or tones, many of the key properties are captured in the time-dependent changes in the *power spectrum*—the sound envelope. To represent this information graphically, speech is typically described in a *spectrogram*, shown in Figure 5.2(bottom). One can think of each vertical slice of the spectrogram as the momentary power spectrum, existing at the time labeled on the horizontal axis. Darker areas indicate more power. The spectral content of the signal changes as the time axis moves from left to right. The particular speech signal shown at the bottom of Figure 5.2c begins with a short burst of relatively high frequency sound and finishes with a longer lower frequency component. Collectively this pattern characterizes the sound of a human voice saying the /d/ sound. As a comparison, Figure 5.3 shows the timeline, power spectrum and spectrogram of an artificial sound, that of a collision warning alert.

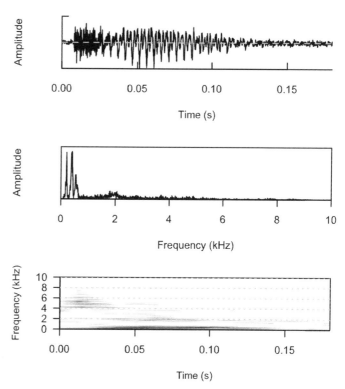

Figure 5.2 Graphs of the speech sounds associated with saying the /d/ sound in the word "day": (a) timeline; (b) power spectrum; (c) spectrogram. (Created in R.)

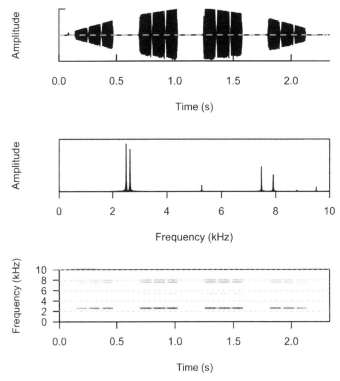

Figure 5.3 Timeline, power spectrum, and spectrogram for a collision warning sound. (Created in R.)

5.1.2 Sound Intensity

Similar to light reaching the eye, sound reaching the ear begins at a source, spreads through space, and reflects off surfaces. Sound energy at the source is defined in terms of watts (W) and the pressure of the sound waves (P) squared is proportional to this energy. In an open space, sound energy spreads across a sphere as it radiates, and so the intensity of the sound energy per square meter decreases with the square of the distance. The ear transforms sound pressure variation (Figure 5.3) into the sensation of hearing and it is this sound pressure variation that can be measured. When describing the effects on hearing, the amplitude is typically expressed as a *ratio* of sound pressure, P, to a reference P_0. Table 5.1 summarizes these measures and their units.

As a ratio, the decibel scale can be used in two ways: as a measure of absolute intensity relative to a standard reference (P and P_0) and as a ratio of two sounds (P_1 and P_2). As a measure of absolute intensity, P is the sound pressure being measured, and Po is a reference value near the threshold of hearing (i.e., the faintest sound that can be heard under optimal conditions). This reference value is a pure tone of 1,000 Hz at 20 micro Pascals or Newtons/m^2. Decibels represent the ratio of a given sound to the threshold of hearing. Most commonly, when people use dBs to refer to sounds

Quantity (Units)	Definition
Energy intensity I(dB)	$$10 \times log\left(\frac{I}{I_0}\right) \sim log\left(\frac{P^2}{P_0^2}\right)$$ $$I_0 = 10^{-12}\, watts/\, m^2$$
Sound pressure level L(dB)	$$20 \times log\left(\frac{P}{P_0}\right) = log\left(\frac{P^2}{P_0^2}\right)$$ $$P_0 = 2 \times 10^{-5}\, Newton/\, m^2$$

Table 5.1 Physical characteristics governing sound intensity and their units.

they mean dB relative to this threshold. Table 5.2 shows examples of the sound pressure levels of everyday sounds along the decibel scale, as well as their sound pressure level. Because it is a ratio measure, the decibel scale can also characterize the ratio of two audible sounds; for example, the OSHA inspector at the plant may wish to determine how much louder the alarm is than the ambient background noise. Using the ratio of the sound pressure levels, we might say it is 15 dB more intense. As another example, we might characterize a set of earplugs as reducing the noise level by 20 dB.

Because the ear is sensitive to such a range of sound levels we use a logarithmic scale— dB.

Because sound is measured in dB, which is a ratio on a logarithmic scale, the combined sound pressure level produced by multiple sound sources cannot be determined by simply adding the dB values. For example, the combination of a passing bus (90 dB) and a passing car (70 dB) is not 160 dB. Instead, Equation 5.1 must be used, where L_n is the sound pressure level in dB of each of N sources.

Sound intensity increases with the number and intensity of sources

$$L_{sum} = 10 \times log_{10} = \sum_{n=1}^{N} 10^{L_n} \tag{5.1}$$

Using this equation, the total sound pressure level associated with the 90 dB bus and the 70 dB car is:

$$L_{sum} = 10 \times \left(log_{10}10^{90/10} + log_{10}10^{70/10}\right) = 90.04 dB$$

Example 5.1 The sound pressure level of one person talking is 60 dB, the total sound pressure level of two people talking

Sound Pressure Level (dB)	Absolute Sound Pressure Level (Newton/m)	Example
140	200	Jet at take-off
130	63	Pain threshold
120	20	Rock concert
110	6.3	Lawn mower
100	2.0	Subway train
90	0.63	Shouting
80	0.20	Busy road
70	6.3×10^{-2}	Average car
60	2.0×10^{-2}	Normal conversation
50	6.3×10^{-3}	Quiet restaurant
40	2.0×10^{-3}	Quiet office
30	6.3×10^{-4}	Whisper
20	2.0×10^{-4}	Rustling leaves
10	6.3×10^{-5}	Normal breathing
0	2.0×10^{-5}	Threshold of hearing

Table 5.2 The Decibel scale with examples.

is not 60+60=120 dB. Calculate the combined sound pressure level.

$$L_{sum} = 10 \times \left(log_{10}10^{60/10} + log_{10}10^{60/10} \right) = 63.01 dB$$

Solution: Instead of the 120 db, the combined sound pressure level is 63.01 dB. Adding two sound pressure levels of equal intensity always leads to a combined sound pressure level of 3.01 dB more than the individual sounds.

Exactly how sound propagates and combines from different sources depends on the environment in which the sound occurs. A lawnmower creates a different distribution of sound in a garage than in the middle of a large yard.

5.1.3 Sound Field

Similar to light, sound propagates from its source and reflects off the surfaces it hits. Figure 5.4 and Equation 5.2 shows how the intensity of sound changes with distance. Because the power of the sound is spread over an area that is proportional to the square of the distance, each doubling of the distance leads to a 6 dB decline in the intensity of the sound.

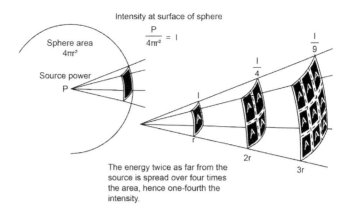

Figure 5.4 Propagation of sound energy from a source. (Figure reprinted from HyperPhysics (http://hyperphysics.phy-astr.gsu.edu/hbase/Ph4060/p406i.html) with permission from Carl Rod Nave.)

Sound intensity declines with distance from source

$$L_2 = L_1 - 20 \times log_{10}\left(\frac{d_2}{d_1}\right)$$

L_2 = Sound pressure at second distance
L_1 = Sound pressure at first distance
d_2 = Distance to second distance
d_1 = Distance to first distance

(5.2)

Equation 5.2 relies on the assumption that sound propagates uniformly from a point source. Such a situation is called a *free field*, but when substantial sound energy reflects off nearby surfaces it is a *reverberant field*. Near field is where the sound source is so close that it does not act as a uniformly radiating source. The near field concept is important because measuring sound in this range will be imprecise. The wavelength of the relevant sound waves and the dimensions of the source define the near field. The near field is the largest of either the length of the relevant wavelength (e.g., 2 meters for 161.5 Hz assuming speed of sound at sea level of 343 m/sec) or twice the longest dimension of the sound source (e.g., approximately 0.5 meters for a lawn mower).

Reflection of sound off walls or other surfaces is termed *reverberation* and has three important effects. First, it provides a sense of space and affects the feel of a room. The reverberations of a small room provide a sense of intimacy. Second, reverberation can interfere with speech communication if the reflected sound returning to a person is greater than 50 msec. Reflected sound less than 50 msec tends to improve the intelligibility of speech. Third, reverberations can increase the sound pressure level beyond that predicted by Equation 5.2. For that reason, measuring sound in the workplace is an important aspect of estimating whether the level of noise poses a risk to a worker's hearing, a topic we discuss in the next section.

5.1.4 Sound Sources and Noise Mitigation

Sound pressure levels can be measured by a sound intensity meter. This meter has a series of scales that can be selected, which weight frequency ranges differently. In particular, the A scale deferentially weights frequencies of the sound to reflect the characteristics of human hearing, providing greatest weighting at those frequencies where we are most sensitive. The C scale weights all frequencies nearly equally and therefore is less closely correlated with the characteristics of human hearing and used for very high sound pressure levels. As you might expect, the B scale weighs frequencies in a manner between the A and C weightings, and the Z scale does not weight the frequencies differently. Only the A scale is the commonly used to assess noise levels in the workplace, and is typically indicated as dBA.

During the last few decades in the United States, the Occupational Safety and Health Administration (OSHA) has taken steps to try and protect workers from the hazardous effects of prolonged noise in the workplace by establishing standards that can be used to trigger remediating action (CFR 29 1910.95; [158]). Even brief exposure to sounds over 100 dB can cause permanent hearing damage, but 85 dB sounds can lead to hearing damage with prolonged exposure. Intense sounds can damage hearing and the longer people are exposed to intense sound the greater damage. Of course, many workers do not experience continuous noise of these levels but may be exposed to bursts of intense noise followed by periods of greater quiet. How would you combine these exposures to estimate the risk to a worker?

Sustained exposure to noise above 85 dB–a lawn mower– can cause permanent hearing damage.

OSHA provide means of converting the varied time histories of noise exposures into the single equivalent standard—the time weighted average. The time weighted average (TWA) of noise experienced in the workplace, which trades off the intensity of noise exposure against the duration of the exposure. If the TWA exceeds 85 dBA (a weighted measure of noise intensity), the action level, employers are required to implement a hearing protection plan in which ear protection devices are made available, instruction is given to workers regarding potential damage to hearing and steps that can be taken to avoid that damage, and regular hearing testing is implemented. If the TWA is above 90 dBA, the permissible exposure level, then the employer is required to takes steps toward noise reduction through procedures that we will discuss. Beyond the workplace, the popularity of portable music players has introduced a new source of sound that can damage hearing: 58% of one sample of college students found that they expose themselves to a TWA greater than 85 dBA [159].

$$T = 8 \times 2^{(90-L)/5}$$

T = Permissible time in hours
L = A-weighted sound pressure level in dBA

Permissible noise exposure declines with noise intensity

(5.3)

Dose of noise exposure depends on sum of exposures

$$D = 100 \times \sum \frac{C_n}{T_n}$$

D = Percent of permissible dose
C_n = Actual exposure in hours
T_n = Permissible exposure in hours
n = Number of time periods with different noise levels

(5.4)

Time-weighted average of noise based on dose in Equation 5.4

$$TWA = 90 + \frac{5}{log_{10}(2)} \times log_{10}\left(\frac{D}{(100)}\right)$$

TWA = Time weighted average (dB)
D = Percent of permissible dose

(5.5)

Equation 5.3 shows the permissible time a worker can be exposed to a given level of noise, and Equation 5.4 shows how the percent of a permissible dose depends on each of the time spent exposed to different noise levels. All noise exposure over 80 dBA must be integrated into this calculation. Equation 5.5 converts this dose into the TWA. A dose greater than 100 exceeds the permissible exposure level and is equivalent to a TWA of 90 dBA. All measurements should use the A scale of the sound intensity meter.

The noise level at a facility cannot always be expressed by a single value but may vary from worker to worker, depending on his or her location relative to the source of noise. For this reason, TWAs might be best estimated using noise *dosemeters*, which are worn by individual workers and collect the data necessary to compute the TWA over the course of the day.

Example 5.2

A worker spends 4 hours at a lathe where the sound intensity meter shows 90 dBA. She also spends 2 hours in an office area where the sound level is 70 dBA, and another 2 hours in a packaging area where the sound level is 95 dBA. Calculate the time-weighted average exposure.

Solution: Because the time in the office is less than 80 dBA it is not considered in the calculation. The table below shows how these data can be combined to estimate the TWA. A dose of 100 is converted to a TWA of 90 dBA using Equation 5.5.

Source	Duration (hours)	Intensity (dBA)	Time allowed (hours)	Dose (percent)
Lathe	4	90	8	$100 \times 4/8 = 50$
Office	2	70	—	
Packaging	2	95	4	$100 \times 2/4 = 50$
Total	8			100

The steps that should be taken to remediate the effects of noise might be very different, depending on the particular nature of the noise-related problem and the level of noise that exists before remediation. On the one hand, if noise problems relate to communication difficulties when the noise level is below 85 dBA (e.g., an idling truck), then signal enhancement procedures may be appropriate, such as increasing the volume of alarms. On the other hand, if noise is above the action levels (a characteristic of many industrial workplaces), then noise reduction procedures must be adopted because enhancing the signal intensity (e.g., louder alarms) will do little to alleviate the possible health and safety problems. Finally, if noise is a source of irritation and stress in the environment (e.g., residential noise from an airport or nearby freeway), then many of the sorts of solutions that might be appropriate in the workplace, like wearing earplugs, are obviously not applicable. We may choose to reduce noise in the workplace by focusing on the source, the path or environment, or the listener. The first is the most preferred method; the last is the least.

The Source: Equipment and Tool Selection can often reduce the noise to the appropriate and careful choice of tools or sound-producing equipment. Crocker [160] provides some good case studies where this has been done. Ventilation or fans, or handtools, for example, vary in the sounds they produce, and appropriate choices in purchasing such items can be made. The noise of vibrating metal, the source of loud sounds in many industrial settings, can be attenuated by using damping material, such as rubber. One should consider also that the irritation of noise is considerably greater in the high-frequency region (the shrill pierced whine) than in the mid- or low-frequency region (the low rumble). Hence, to some extent the choice of tool can reduce the irritating quality of its noise.

The Environment: Sound Path or path from the sound source to the human can also be altered in several ways. Changing the environment near the source, for example, is illustrated in Figure 5.5, which shows the attenuation in noise achieved by surrounding a piece of equipment with a plexiglass shield. Sound absorbing walls, ceilings, and floors can also be very effective in reducing the noise coming from reverberations. Finally, there are many circumstances when repositioning workers relative to the source of noise can be effective. The effectiveness of such relocation is considerably enhanced when the noise emanates from only a single source. This is more likely to be the case if the source is present in a more sound-absorbent environment (less reverberating).

The Listener: Ear Protection is a possible solution if noise cannot be reduced to acceptable levels at the source or path. Ear protection devices that must be made available when noise levels exceed the action level are of two generic types: earplugs, which fit inside the ear, and ear muffs, which fit over the top of the ear. As commercially available products, each is provided with a certified noise reduction ratio (NRR), expressed in decibels, and each may also have very different spectral characteristics (i.e., different decibel reduction across the spectrum). For both kinds of devices, it appears that the manufacturer's specified NRR is typically greater (more optimistic) than is the actual noise reduction experienced by users in the workplace [161]. This is because the manufacturer's NRR value is typically computed under ideal laboratory conditions, whereas users in the workplace may not always wear the device properly.

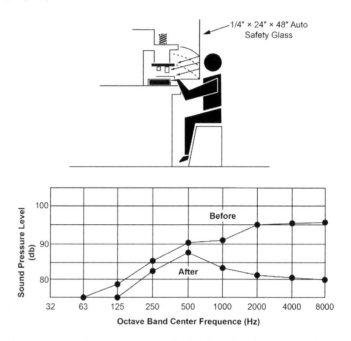

Figure 5.5 Use of a 1/4 in (6 mm)-thick safety glass barrier to reduce high-frequency noise from a punch press. (Source: American Industrial Hygiene Association, 1975, Figure 11.73. Reprinted with permission by the American Industrial Hygiene Association.)

Of the two devices, earplugs can offer a greater overall protection if properly worn. However, this qualification is extremely important because earplugs are more likely than ear muffs to be worn improperly. Hence, without proper training (and adherence to that training), muffs may be more effective than plugs. A second advantage of muffs is that they can readily double as headphones through which critical signals can be delivered, simultaneously achieving signal enhancement and noise reduction.

Comfort must be considered in assessing hearing protection in the workplace. Devices that are annoying and uncomfortable may go unused in spite of their safety effectiveness (see Chapter 14). Interestingly, concerns such as that voiced by the worker at the beginning of this chapter that hearing protection will not allow her to hear conversations are not always well grounded. The ability to hear conversation is based on the signal-to-noise ratio. Depending on the precise spectral characteristics and amplitude of the noise and the signal and the noise-reduction function, wearing such devices may actually enhance rather than reduce the signal-to-noise ratio, even as both signal and noise intensity are reduced. The benefit of earplugs to increasing the signal-to-noise ratio is greatest with louder noises, above about 80 to 85 dB [162, 163].

Finally, it is important to note that the adaptive characteristics of the human speaker may themselves produce some unexpected consequences on speech comprehension. We automatically adjust our voice level, in part, on the basis of the intensity of sound that we hear, talking louder when we are in a noisy environment [164, 165] or when we are listening to loud stereo music through headphones. Hence, it is not surprising that speakers in a noisy environment talk about 2 to 4 dB softer (and also somewhat faster) when they are wearing ear protectors than when they are not. This means that hearing conversations can be more difficult in environments in which all participants wear protective devices, unless speakers are trained to avoid this automatic reduction in the loudness of their voice [166].

5.2   The Receptor System: The Ear

The preceding discussion focused on the physical properties of sound: frequency and intensity. We now turn to the properties of the receptor system (the ear), and how these properties explain how noise can damage hearing and interfere with communication.

5.2.1   Anatomy of the Ear

The ear has three primary components responsible for differences in our hearing experience. As shown in Figure 5.6, the *pinna* both collects sound and, because of its asymmetrical shape, provides some information regarding where the sound is coming from (i.e., behind or in front). Mechanisms of the *outer* and *middle* ear (the ear drum or tympanic membrane, and the hammer, anvil, and stirrup bones) conduct and amplify the sound waves into the inner ear and are potential sources of breakdown or deafness (e.g., from a rupture of the eardrum or buildup of wax). The muscles of the middle ear respond to loud noises and reflexively contract to attenuate the amplitude of intense sound waves before it is conveyed to the inner ear. This aural reflex thus offers some ion to the inner ear.

The *inner ear*, consisting of the *cochlea*, within which lies the basilar membrane, is where the physical movement of sound energy is transduced to electrical nerve energy that is then passed through the auditory nerve to the brain. This transduction is accomplished by displacement of tiny hair cells along the basilar membrane as the membrane moves differently to sounds of different frequency. Intense sound experience can lead to selective hearing loss at particular frequencies as a result of damage to the hair cells at particular locations along the basilar membrane. Finally, the neural signals are compared between the two ears to determine the delay and amplitude differences between them. These differences provide another cue for sound localization, because these features are identical only if a sound is presented directly along the midplane of the listener.

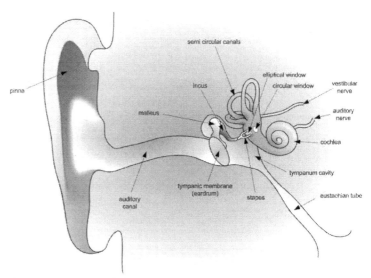

Figure 5.6 Anatomy of the ear. (Source: Dan Pickard [Public domain], via Wikimedia Commons, https://commons.wikimedia.org/wiki/File: HumanEar.jpg)

5.2.2   Masking, Temporary Threshold Shift, and Permanent Threshold Shift

The worker in our story was concerned about the effect of noise on her ability to hear at her workplace. When we examine the effects of noise, we consider three components of the potential hearing loss: *masking*, a loss of sensitivity to a signal while the noise is present; *temporary threshold shift*, transient loss of sensitivity due to exposure to loud sounds; *permanent threshold shift*, permanent loss of hearing due to aging or repeated exposure to loud sounds.

Masking of one sound by other sounds depends on both the intensity (power) and frequency of that signal [166]. These two variables are influenced by the speaker's gender and by the nature of the sound. First, since the female voice typically has a higher base frequency than the male, it is not surprising that the female voice is more vulnerable to masking of noise. Likewise consonant sounds, like s and ch, have distinguishing features at very high frequencies, and high frequencies are more vulnerable to masking by low frequencies than the converse. Hence, it is not surprising that consonants are much more susceptible to masking and other disruptions than are vowels. This characteristic is particularly disconcerting because consonants typically transmit more information in speech than do vowels. One need only think of the likely possibility of confusing "fly to" with "fly through" in an aviation setting to realize the danger of such consonant confusion [167]. Miller and Nicely [168] provide a good analysis of the confusability between different consonant sounds. We return to the issue of sound confusion in Chapter 6.

As our worker at the beginning of the chapter discovered, sounds can be masked by other sounds. The nature of masking is actually quite complex [169] , but a few of the most important principles for design are the following:

1. The minimum intensity difference necessary to ensure that a sound can be heard is around 15 dB (intensity above the mask).

2. Sounds tend to be masked more by sounds in a critical frequency band surrounding the sound that is masked.

3. Low-frequency sounds mask high-frequency sounds more than the converse. Thus, a woman's voice is more likely to be masked by other male voices than a man's voice would be masked by other female voices even if both voices are speaking at the same intensity level.

Temporary Threshold Shift (TTS) is the second form of noise-induced hearing loss [166], which occurs after exposure to intense sounds. If our worker steps away from the machine to a quieter place to answer the telephone, she may still have some difficulty hearing because of the "carryover" effect of the previous noise exposure. This temporary threshold shift is large immediately after the noise is terminated but declines over the following minutes as hearing is "recovered" (Figure 5.7). The TTS is typically expressed as the amount of loss in hearing (shift in threshold in dB) that is present two minutes after the source of noise has terminated. The TTS increases with a longer and greater noise exposure. The TTS can be quite large. For example, the TTS after being exposed to 100 dB noise for 100 minutes is 60 dB, meaning you might not be able to hear a normal conversation after a loud concert.

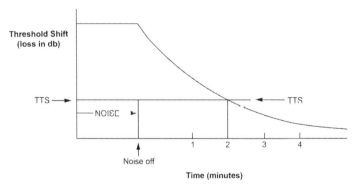

Figure 5.7 TTS following the termination of noise. Note that sensitivity is recovered (the threshold shift is reduced over time). Its level at two minutes is arbitrarily defined as the TTS.

Permanent Threshold Shift (PTS) is the third form of noise-induced hearing loss experienced by our worker, and it has the most serious implications for worker health. This measure describes the "occupational deafness" that may set in after workers have been exposed to months or years of high-intensity noise at the workplace. Also, PTS tends to be more pronounced at higher frequencies, usually greatest at around 4,000 Hz [166]. Workplace noise that is concentrated in a certain frequency range has a particularly strong effect on hearing while in that frequency range. Note in Figure 5.2 that the consonant /d/ is located in that range, as are many other consonants. Consonants are most critical for discrimination of speech sounds. Hence a PTS can be particularly devastating for conversational understanding. Like the TTS, the PTS is greater with both louder and longer prior exposure to noise [170]. Age contributes to a large portion of hearing loss, particularly in the high-frequency regions, a factor that should be considered in the design of alarm systems, particularly in nursing homes.

5.3   Auditory Sensory Processing Characteristics

To amplify our previous discussion of the sound stimulus, the three dimensions of the raw stimulus map onto psychological experience of sound: Loudness maps to intensity, pitch maps to frequency, and frequency distribution maps on to sound quality. In particular, the timbre of a sound stimulus—what makes the trumpet sound different from the flute—is determined by the set of higher harmonic frequencies that lie above the fundamental frequency (which determines the pitch of the note). Various other temporal characteristics, including the envelope and the rhythm of successive sounds, also determine the sound quality. As we shall see, differences in the envelope are critically important in distinguishing speech sounds.

5.3.1   Loudness and Pitch

Loudness is a psychological experience that correlates with, but is not identical to, the physical measurement of sound intensity. There are two important reasons why loudness and intensity do not directly correspond: the psychophysical scale of loudness and the modifying effect of pitch.

Psychophysical Scaling relates the physical stimulus to what people perceive. A simple form of discrimination characterizes the ability of people to notice the change or difference in simple dimensional values, for example, a small change in the height of a bar graph, the brightness of an indicator, or the intensity of a sound. In the classic study of psychophysics (the relation between the psychological sensations and physical stimulation), such difference thresholds are called just noticeable difference, or JND. Designers should assume that people cannot reliably detect differences that are less than a JND. For example, if a person is meant to detect fluctuations in a sound, those fluctuations should be scaled so that those fluctuations are greater than a JND.

$$JND = \frac{K \times \Delta I}{I}$$

JND = Just noticeable difference
K = Constant for particular sensory stimulus
ΔI = Change in intensity
I = Absolute level of intensity

(5.6)

Weber's law relates perceived change in the stimulus to the actual change in the stimulus

Along many sensory continua, the JND for judging intensity differences increases in proportion to the absolute amount of intensity, a simple relationship described by Weber's law (Equation 5.6). Here ΔI is the change in intensity, I is the absolute level of intensity, and K is a constant, defined separately for different sensory continua (such as the brightness of lights, the loudness of sounds, or the length of lines). Importantly, Weber's law also describes the psychological reaction to changes in other non-sensory quantities. For example, how much a change in the cost of an item means to you (i.e., whether the cost difference is above or below a JND) depends on the cost of the item. You may stop riding the bus if the bus fare is increased by $1.00, from $0.50 to $1.50; the increase was clearly greater than a JND of cost. However, if an air fare increased by the same $1.00 amount (from $432 to $433), this would probably have little influence on your choice of whether or not to buy the plane ticket. The $1.00 increase is less than a JND compared to the $432 cost. We will discuss the such influences on decision making in Chapter 7.

Consistent with Equation 5.6, equal increases in sound intensity (on the decibel scale) do not create equal increases in loudness; for example, an 80 dB sound does not sound twice as loud as a 40 dB sound. Instead, the scale that relates physical intensity to the psychological experience of loudness, expressed in units called sones, is that shown in Figure 5.8. One sone is established arbitrarily as the loudness of a 40 dB tone of 1,000 Hz. A tone twice as loud will be two sones. As an approximation, we can say that loudness doubles with each 10 dB increase in sound intensity. For example, an increase in 20 dB would be associated with approximately a sound four times as loud. However, the loudness of the intensity levels are also influenced by the frequency (pitch) of the sound, and so we must now consider that influence.

Frequency Influence describes how loudness depends on frequency. Figure 10 shows a series of equal-loudness curves, where all points lying on a single line are perceived as equally loud. Thus, a 1,000-Hz tone of 40 dB sounds about the same loudness (40 phons) as an 8,000-Hz tone of around 60 dB. That is, every point along a line sounds just as loud as any other point along the same line. The l equal loudness curves are described in units of phons. One phon = 1 dB of loudness of a 1,000-Hz tone, the standard for calibration. Thus, all tones lying along the 40 phon line have the same loudness—1 sone—as a 1,000-Hz tone of 40 dB. The equal

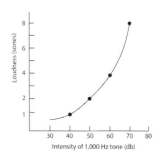

Figure 5.8 Relation between sound intensity and loudness.

loudness contours follow more or less parallel tracks. Thus as shown in the figure, the frequency of a sound, plotted on the x axis, influences all three of the critical levels of the sound experience: threshold, loudness, and danger levels.

5.4 Cognitive Influence on Auditory Perception

Just as cognition influences visual perception, it also influences auditory perception. Top-down influences associated with expectations influence what we hear, particularly as we localize sounds, interpret alarms, and understand what others are saying. Such cognitive influences also influence how annoying a particular sound might be.

5.4.1 Detection and Localization

In Chapter 4 we described the role of the visual system in searching worlds as guided by eye movements. The auditory system is not as well suited for precise spatial localization but nevertheless has some very useful capabilities in this regard, given the differences in the acoustic patterns of a single sound, processed by the two ears [171, 172]. The ability to identify location of sounds is better in azimuth (e.g., left-right) than it is in elevation, and front-back confusions are also prominent. Overall, sound localization is less precise than visual localization.

Despite the limited ability to localize sounds, in environments where the eyes are heavily involved with other tasks or where signals could occur in a 360-degree range around the head (whereas the eyes can cover only about a 130-degree range with a given head fixation), sound localization can provide considerable value. For example, auditory warnings can provide pilots with guidance as to the possible location of a midair conflict [172]. In particular, a redundant display of visual and auditory location can be extremely useful in searching for targets in a 3-D 360-degree volume. The sound can guide the head and eyes very efficiently to the general direction of the target, allowing the eyes then to provide more precise localization [173].

Unlike visual stimuli that require people to direct their eyes to the source of information, the auditory system is omnidirectional; that is, unlike visual signals, we can sense auditory signals no matter how we are oriented. Furthermore, it is much more difficult to "close our ears" than it is to close our eyes [174]. For these and other reasons, auditory warnings induce a greater level of compliance than do visual warnings [175], but can also be more annoying as we will discuss in the next section.

5.4.2 Alarms

The design of effective alarms, like the one that was nearly missed by the worker in our opening story, depends very much on matching the modality of the alarm (e.g., visual or auditory) to the requirements of the task. If a task analysis indicates that an alarm signal must be sensed, like a fire alarm, it should be given an auditory form, although redundancy in the visual or tactile channel may be worthwhile in when there is a high level of background noise or for people who do not hear well.

While the choice of modality seems straightforward, the issue of how auditory alarms should be designed is more complicated. Consider the following quotation from a British pilot, taken from an incident report, which illustrates many of the problems with auditory alarms.

> I was flying in a jetstream at night when my peaceful revelry was shattered by the stall audio warning, the stick shaker, and several warning lights. The effect was exactly what was not intended; I was frightened numb for several seconds and drawn off instruments trying to work out how to cancel the audio/visual assault, rather than taking what should be instinctive actions. The combined assault is so loud and bright that it is impossible to talk to the other crew member and action is invariably taken to cancel the cacophony before getting on with the actual problem. [176]

Designing alarms well can avoid, or at least minimize, the potential costs described above. First, as we have noted, *environmental and task analysis* can identify the quality and intensity of other sounds (noise or communications) that might characterize the environment in which the alarm is presented to guarantee detectability and minimize disruption of other essential tasks.

Second, to guarantee informativeness and to minimize confusability, designers should try to make alarm sounds as different from each other as possible by capitalizing on the various dimensions along which sounds differ. These dimensions include: pitch (fundamental pitch or frequency band), envelope (e.g., rising, *woop woop*, constant *beep beep*), rhythm (e.g., synchronous *da da d*a versus asynchronous da **da** da **da**), and timbre (e.g., a horn versus a flute). Two alarms will be most discriminable (and least confusable) if they are constructed at points on opposite ends of all four of the above dimensions, similar to selecting colors from distant points in the color space.

Third, combine the elements of sound to create the overall alarm system. Patterson [176] recommends the procedure outlined in Figure 5.9. The top of Figure 5.9 shows the smallest components of the sound—pulses—that occur over 100 to 500 msec. These show an acoustic wave with rounded onsets and offsets. The middle row shows bursts of pulses that play out over 1 to 2 seconds, with a distinctive rhythm and pitch contour. The bottom row shows how these bursts of varying intensity combine into the overall alarm, which might play over 10 to 40 seconds.

At the top of the figure, each individual pulse in the alarm is configured with an envelop rise time that is not too abrupt (i.e., at least 20 msec) to avoid the "startle" created by more abrupt rises [177]. The set of pulses in the alarm sequence, shown in the middle of the figure, are configured with two goals in mind: (1) The pauses between each pulse can be used to create a unique rhythm that can help minimize confusion; and (2) the increase then decrease in intensity gives the perception of an approaching then receding sound, which creates a psychological sense of urgency.

Finally, the bottom row of Figure 5.9 shows how repeated presentations of the bursts can be implemented. The first two presentations may be at high intensity to guarantee their initial detection (first sequence) and identification (first or second sequence). Under the assumption that the operator has probably been alerted, the third and fourth sequences can be less intense to minimize annoyance and possible masking of other sounds (e.g., the voice communications that may be initiated by the alarming condition). An intelligent alarm system may infer, after a few sequences, that no action has been taken and then repeat the sequence at a higher intensity.

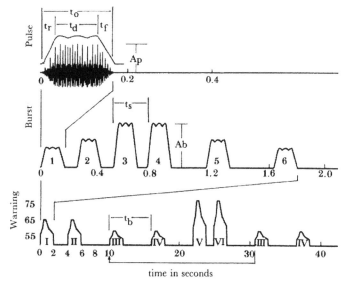

Figure 5.9 The modules of a prototype warning sound: The sound pulse at the top; the burst shown in the middle row is a set of pulses; the complete sequence at the bottom. (Source: Patterson, R. D., 1990. Auditory warning sounds in the work environment, *Philosophical Transactions B, 327, p. 490, Figure 3*, by permission of The Royal Society through the Copyright Clearance Center.)

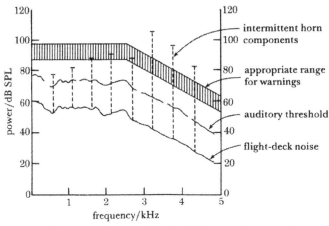

Figure 5.10 The range of appropriate levels for warning sound components on the flight deck of the Boeing 737 (vertical line shading). (Source: Patterson, R. D., 1990. Auditory warning sounds in the work environment, *Philosophical Transactions B, 327, p. 487, Figure 1*, by permission of The Royal Society through the Copyright Clearance Center.)

Specific Design Criteria for Alarms include those shown as shown in Figure 5.9. Generally alarm design avoid two opposing problems of detection, experienced by our factory worker at the beginning of the chapter, the "overkill" and annoyance experienced by the pilot.

1. Most critically, the alarm must be heard above the ambient background noise. This means that the noise spectrum must be carefully measured at the location where everyone must respond to the alarm. Then, the alarm should be set at least 15 dB above the noise level, and to guarantee detection, set at 30 dB above the noise level. It is also wise to include components of the alarm at several different frequencies, well distributed across the spectrum, in case the particular malfunction that triggered the alarm creates its own noise (e.g., the whine of a malfunctioning engine), which exceeds the ambient noise level.

2. The alarm should not exceed the danger level for hearing, whenever this condition can be avoided. (Obviously, if the ambient noise level is close to the danger level, one has no choice but to make the alarm louder by criterion 1, which is most important.) This danger level is around 85 to 90 dB. Careful selection of frequencies of the alarm can often be used to meet both of the above criteria. For example, if the ambient noise level is very intense (90 dB), but only in the high frequency range, it would be counterproductive to try to impose a 120-dB alarm in that same frequency range when several less intense components in a lower frequency range could be heard.

3. Ideally, the alarm should not startle. As noted in Figure 5.9,

this can be addressed by tuning the rise time of the alarm pulse so that the rise time is at least 20 ms.

4. In contrast to the experience of the British pilot, the alarm should not interfere with other signals (e.g., other simultaneous alarms) or any background speech communications that may be essential to deal with the alarm. This criterion implies that a careful task analysis should be performed of the conditions under which the alarm might sound and of the necessary communications tasks to be undertaken as a consequence of that alarm.

5. The alarm should be informative, signaling to the listener the nature of the emergency and, ideally, some indication of the appropriate action to take. The criticality of this informativeness criterion can be seen in one alarm system that was found in an intensive care unit of a hospital (an environment often in need of alarm remediation [176, 178]. The unit contained six patients, each monitored by a device with 10 different possible alarms: 60 potential signals that the staff may have had to rapidly identify. Some aircraft have been known to contain at least 16 different auditory alerts, each of which, when heard, is supposed to trigger identification of the alarming condition in the pilot's mind. Such alarms are often found to be wanting in this regard.

6. In addition to being informative, the alarm must not be confusable with other alarms that may be heard in the same context. As you will recall from our discussion of vision in Chapter 4, this means that the alarm should not impose on the human's restrictive limits of absolute judgment. Just four different alarms may be the maximum allowable to meet this criterion if these alarms differ from each other on only a single physical dimension, such as pitch.

Voice Alarms and Meaningful Sounds, such as alarms composed of synthetic voice, provide one answer to the problems of discriminability and confusion. Unlike "symbolic" sounds, the hearer does not need to depend on an arbitrary learned connection to associate sound with meaning. The loud sounds *Engine fire!* or *Stall!* in the cockpit mean exactly what they seem to mean. Voice alarms are employed in several circumstances (the two aircraft warnings are an example). But voice alarms themselves have limitations that must be considered. First, they are likely to be more confusable with (and less discriminable from) a background of other voice communications, whether this is the ambient speech background at the time the alarm sounds, the task-related communications of dealing with the emergency, or concurrent voice alarms. Second, unless care is taken, they may be more susceptible to frequency-specific masking noise. Third, care must be taken if the meaning of such alarms is to be interpreted by listeners in a multilingual environment who are less familiar with the language of the voice.

The preceding concerns with voice alarm suggest the advisability of using a redundant system that combines the alerting, distinctive features of the (nonspeech) alarm sound with the more informative features of synthetic voice [179]. Combining stimuili from multiple modalities often promotes more reliable performance although not necessarily a faster response. Such *redundancy gain* is a fundamental principle of human performance that can be usefully employed in alarm system design.

Another possible design that can address some of the problems associated with comprehension and masking is to synthesize alarm sounds that sound like the condition they represent, called auditory icons or *earcons* [180, 181]. Belz, Robinson, and Casali [182], for example, found that representing hazard alarms to automobile drivers in the form of earcons (e.g., the sound of squealing tires representing a potential forward collision) significantly shortened driver response time relative to conventional auditory tones. In particular, to the extent that such signals sound like their action meanings, like the crumpling paper signaling delete or squealing tires signaling braking, auditory icons can be quite effective in signaling actions.

False Alarms, such as those discussed in terms of human signal detector in Chapter 4, also plague warning systems because warnings do not always indicate an actual hazard. When sensing low-intensity signals from the environment (a small increase in temperature, a wisp of smoke), the system sometimes makes mistakes, inferring that nothing has happened when it has (the miss) or inferring that something has happened when it has not (the false alarm [183]).

Most alarm designers and users set the alarm's criterion as low as possible to minimize the miss rate for obvious safety reasons. But as we learned in our discussion of signal detection in chapter 3, when the low-intensity signals on which the alarm decision is made, are themselves noisy, the consequence of setting a miss-free criterion might be an unacceptable false alarm rate: To paraphrase from the old fable, the system "cries wolf" too often [184, 185]. Such was the experience with the initial introduction of the ground proximity warning system in aircraft, designed to alert pilots that they might be flying dangerously close to the ground. Unfortunately, when the conditions that trigger the alarm occur very rarely, an alarm system that guarantees detection will, almost of necessity, produce a fair number of false alarms, or "nuisance alarms" [186].

From a human performance perspective, the obvious concern is that users may come to distrust the alarm system and perhaps ignore it even when it provides valid information [187, 188]. More serious yet, users may attempt to disable the annoying alarms. Many of these concerns are related to the issue of *trust* in automation, discussed in Chapter 16 [189, 190].

Five steps can help mitigate the problems of false alarms. First, it is possible that the alarm criterion itself has been set to such an extremely sensitive value that readjustment to allow fewer false alarms will still not appreciably increase the miss rate. Second, more sophisticated decision algorithms within the system may be developed to improve the sensitivity of the alarm system, a step that was taken to address the problems with the aircraft ground proximity warning system. Third, users can be trained about the inevitable tradeoff between misses and false alarms and therefore can be taught to accept the false alarm rates as an inevitable consequence of automated protection in an uncertain probabilistic world rather than as a system failure. (This acceptance will be more likely if care is taken to make the alarms noticeable by means other than shear loudness [191, 192]). Fourth, designers should try to provide the user with the "raw data" or conditions that triggered the alarm, at least by making available the tools that can verify the alarm's accuracy.

Finally, a *graded* or *likelihood alarm* systems in which more than a single level of alert can be provided. Hence, two (or more) levels can signal to the human the system's own confidence that the alarming conditions are present. That evidence in the fuzzy middle ground (e.g., the odor from a slightly burnt piece of toast), which previously might have signaled the full fire alarm, now triggers a signal of noticeable but reduced intensity [193]. This mid-level signal might be liked to a *caution*, with the more certain alert likened to a *warning*.

An important facet of alarms is that experienced users often employ them for a wide range of uses beyond those that may have been originally intended by the designer (i.e., to alert to a dangerous condition of which the user is not aware[194]). For example, in one study of alarm use in hospitals by anesthesiologists noted how anesthesiologists use alarms as a means of verifying the results of their decisions or as simple reminders of the time at which a certain procedure must be performed[178]. One can imagine using an automobile headway monitoring alert of "too close" as simply a means of establishing the minimum safe headway to "just keep the alert silent".

5.4.3 Speech Communication

Our example at the beginning of the chapter illustrated the worker's concern with her ability to communicate with her neighbor in the workplace. A more tragic illustration of communications breakdown contributed to the 1979 collision between two jumbo jets on the runway at Tenerife airport in the Canary Islands, in which over 500 lives were lost [167]. One of the jets, a KLM 747, was poised at the end of the runway, engines primed, and the pilot was in a hurry to take off while it was still possible before the already poor visibility got worse and the airport closed operations. Meanwhile, the other jet, a Pan American airplane that had just landed, was still on the same runway, trying to find its way off in the fog. The air traffic controller instructed the pilot of the KLM: "Okay, stand by for takeoff and I will call." Unfortunately, because of a less than perfect radio channel and because of the KLM pilot's extreme desire to proceed with the takeoff, he apparently *heard* just the words "Okay . . . take off." The takeoff proceeded until the aircraft collided with the Pan Am 747, which had not yet cleared the runway.

In Chapter 4, we discussed the influences of both bottom-up (sensory quality) and top-down (expectations and desires) processing on perception. The Canary Island accident tragically illustrates the breakdown of both processes. The communications signal from ATC was degraded (loss of bottom-up quality), and the KLM pilot used his own expectations and desires to "hear what he wanted to hear" (inappropriate top-down processing) and to interpret the message as authorization to take off. In this section we consider in more detail the role of both of these processes in what is arguably the most important kind of auditory communications, the processing of human speech. We have already discussed the communication of warning information. Now we consider speech communication and ways to measure and improve its effectiveness.

There are two different approaches to measuring speech communications, based on bottom-up and top-down processing respectively. The bottom-up approach derives some objective measure of speech quality. It is most appropriate in measuring the potential degrading effects of noise. Thus, the *speech intelligibility index (SII)*, similar to *articulation index (AI)|seealsospeech*, represents the signal-to-noise ratio (dB of speech sound minus dB of background noise) across a range of the frequency spectrum where useful speech information is located. Figure 5.11 shows how to calculate AI with four different frequency bands. This measure can be weighted by the different frequency bands, providing greater weight to the ratios within bands that contribute relatively more heavily to the speech signal. Remember that the higher frequency is home of the most important consonants.

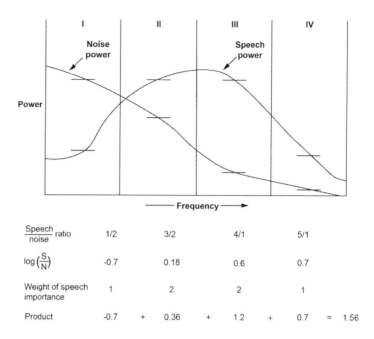

Figure 5.11 Schematic representation of the calculation of an AI. The speech spectrum has been divided into four bands, weighted in importance by the relative power that each contributes to the speech signal. The calculations are shown in the rows below the figure. (Source: Wickens, C. D. Engineering Psychology and Human Performance, 2nd ed.))

SII differs from the Articulation Index (AI) because it considers a broader range of factors affecting speech intelligibility, such as upward spreading of masking to higher frequencies, level distortion, and how the importance of noise frequencies depend on the types of speech (e.g., isolated words compared to sentences). SII also reflects the benefit of visual information associated with face-to-face communication. These additional considerations lead to a much more complicated analysis procedure that has been implemented in readily available software tools (https://cran.r-project.org/web/packages/SII/index.html)

This calculation makes it possible to predict how the background noise will interfere with speech communication and how much amplification or noise mitigation might avoid these problems. When the SII is below 0.2, communication is severely impaired and few words are understood and above 0.5 the communication is typically good with most words being heard correctly [195].

SII does not consider reverberation and the detrimental effects of elements of the sound signal that arrive after the direct sound, where the direct sound is the sound wave that first hits the ear. Sound that arrives after the direct sound comes from reflections from the surroundings, such as walls or the ceiling. Sounds arriving 50 ms after the direct sound interfere with intelligibility and are quantified in terms of C50, which is sometimes termed sound clarity. C50 is the ratio of the signal (sound in the first 50 ms) and noise (sound following the initial 50 ms). A signal to noise ratio of 4 dB is considered very good.

While the merits of the bottom-up approach are clear, its limits in predicting the understandability of speech should become apparent when one considers the contributions of top-down processing to speech perception. For example, two letter strings, *abcdefghij* and *wcignspexl*, might both be heard at intensities with the same SII. But it is clear that more letters of the first string would be correctly understood [196]. Why? Because the listener's knowledge of the predictable sequence of letters in the alphabet allows perception to "fill in the gaps" and essentially guess the contents of a letter whose sensory clarity may be missing. This, of course, is the role of top-down processing.

A measure that takes top-down processing into account is the *speech intelligibility level (SIL)*. This index directly measures the percentage items correctly heard. At any given SII level, this percentage will vary as a function of the listener's expectation of and knowledge about the message communicated. This complementarity relationship between bottom-up (as measured by SII) and top-down processing (as influenced by expectations).Sentences that are known to listeners can be recognized with just as much accuracy as random isolated words, even though the latter are presented with nearly twice the bottom-up sensory quality. Combining the information describing the top-down influences on hearing with the bottom-up influences described by AI or SII makes it possible to anticipate when speech communication will likely fail. Thus, for example, automated readings of a phone number should slow down, and perhaps increase loudness slightly of the critical and often random four digits of the suffix.

Both bottom-up and top-down influences need to be considered to know if people will understand voice communication.

Speech Distortion. While the AI can objectively characterize the damaging effect of noise on bottom-up processing of speech, it cannot do the same thing with regard to distortions. Distortions may result from a variety of causes, for example, clipping of the beginning and ends of words, reduced bandwidth of high-demand communications channels, echoes and reverberations, and even the low quality of some digitized synthetic speech signals [197].

While the bottom-up influences of these effects cannot be as accurately quantified as the effects of noise, there are nevertheless important human factors guidelines that can be employed to minimize their negative impact on voice recognition. One issue that has received particular attention from acoustic engineers is how to minimize the distortions resulting when the high-information speech signal must be somehow "filtered" to be conveyed over a channel of lower bandwidth (e.g., through digitized speech).

For example, a raw speech waveform may contain over 59,000 bits of information per second [198]. Transmitting the raw waveform over a single communications channel might overly restrict that channel, which perhaps must also be shared with several other signals at the same time. There are, however, a variety of ways to reduce the information content of a speech signal. One may filter out the high frequencies, digitize the signal to discrete levels, clip out bits of the signal, or reduce the range of amplitudes by clipping out the middle range. Human factors studies have been able to inform the engineer which way works best by preserving the maximum amount of speech intelligibility for a given resolution in information content. For example, amplitude reduction seems to preserve more speech quality and intelligibility than does frequency filtering, and frequency filtering is much better if only very low and high frequencies are eliminated [198].

Of course, with the increasing availability of digital communications and voice synthesizers, the issue of transmitting voice quality with minimum bandwidth is lessened in its importance. Instead, one may simply transmit the symbolic contents of the message (e.g., the letters of the words) and then allow a speech synthesizer at the other end to reproduce the necessary sounds. (This eliminates the uniquely human, nonverbal aspects of communications—a result that may not be desirable when talking on the telephone.) Then, the issue of importance becomes the level of fidelity of the voice synthesizer necessary to (1) produce recognizable speech, (2) produce recognizable speech that can be heard in noise, and (3) support "easy listening." The third issue is particularly important, as Pisoni [197] has found that listening to synthetic speech takes more mental resources than does listening to natural speech. Thus, listening to synthetic speech can produce greater interference with other ongoing tasks that must be accomplished concurrently with the listening task (see Chapter 6) which, in turn, will be more disrupted by the mental demands of those concurrent tasks.

The voice, unlike the printed word, is transient. Once a word is spoken, it is gone and cannot be referred back to. The human information-processing system is designed to prolong the duration of the spoken word for a few seconds through what is called *echoic memory*. However, beyond this time, spoken information must be actively rehearsed, a demand that competes for resources with other tasks. Hence, when displayed messages are more than a few words, they should be delivered visually or at least backed up with a redundant and more permanent "visual echo".

Besides obvious solutions of "turning up the volume" (which may not work if this amplifies the noise level as well and so does not change the signal-to-noise ratio) or talking louder, there may be other more effective solutions for enhancing the amplitude of speech or warning sound signals relative to the background noise. First, careful consideration of the *spectral content* of the masking noise may allow one to use signal spectra that have less overlap with the noise content. For example, the spectral content of synthetic voice messages or alarms can be carefully chosen to lie in frequency regions where noise levels of the ambient environment are lower. Since lower frequency noise masks higher frequency signals, more than the other way around, this relation can also be exploited by trying to use lower frequency signals. Also, synthetic speech devices or earphones can often be used to bring the source of signal closer to the operator's ear than if the source is at a more centralized location where it must compete more with ambient noise.

There are also signal-enhancement techniques that emphasize more the *redundancy* associated with top-down processing. As one example, it has been shown that voice communications is far more effective in a face-to-face mode than it is when the listener cannot see the speaker [199]. This is because of the contributions made by many of the redundant cues provided by the lips [200], cues of which we are normally unaware unless they are gone or distorted. (To illustrate the important and automatic way we typically integrate sound and lip reading, recall, if you can, the difficulty you may have in understanding the speech of poorly dubbed foreign films when speech and lip movement are not synchronized in a natural way.)

Another form of redundancy is involved in the use of the phonetic alphabet ("alpha, bravo, charlie, . . ."). In this case, more than a single sound is used to convey the content of each letter, so if one sound is destroyed (e.g., the consonant *b*), other sounds can unambiguously "fill in the gap" (*ravo*).

In the context of communications measurement, improved top-down processing can also be achieved through the choice of vocabulary. Restricted vocabulary, common words, and standardization of communications procedures, such as that adopted in air traffic control (and further emphasized following the Tenerife disaster), will greatly restrict the number of possible utterances that could be heard at any given moment and hence will better allow perception to "make an educated guess" as to the meaning of a sound if the noise level is high.

5.5 Auditory Influence on Cognition: Noise and Annoyance

We have discussed the potential of noise as a health hazard in the workplace, a factor disrupting the transmission of information. Here we consider its potential as an annoyance in the environment. In Chapter 13, we also consider noise as a stressor that has degrading effects on performance other than the communications masking effect discussed here. In Chapter 14 we consider broader issues of workplace safety. We conclude by offering various possible remediations to the degrading effects of noise in all three areas: communications, health, and environment.

Noise in residential or city environments, while presenting less of a health hazard than at the workplace, is still an important human factors concern, and even the health hazard is not entirely absent. Meecham [201], for example, reported that the death rate from heart attacks of elderly residents near the Los Angeles Airport was significantly higher than the rate recorded in a demographically equivalent nearby area that did not receive the excessive noise of aircraft landings and takeoffs.

Measurement of the irritating qualities of environmental noise levels follows somewhat different procedures from the measurement of workplace dangers. In particular, in addition to the key component of intensity level, there are a number of other "irritant" factors that increase annoyance. For example, high frequencies are more irritating than low frequencies. Airplane noise is more irritating than traffic noise of the same level. Nighttime noise is more irritating than daytime noise. Noise in the summer is more irritating than in the winter (when windows are likely to be closed). While these and other considerations cannot be precisely factored into an equation to predict "irritability," it is nevertheless possible to estimate their contributions in predicting the effects of environmental noise on resident complaints. One study found that the percentage of people "highly annoyed" by residential noise follows a logistic function of the mean day and night sound intensity, see Equation (5.7), and for noise levels above 70 dB it is roughly linear, see Equation (Equation 5.8)[202]. A noise level of 80 dB would lead approximately 52% of people ($20 + 3.2 \times 10$) to be highly annoyed [203, 204]

Annoyance increases with noise intensity

$$Annoyed = \frac{100}{1 + e^{11.13 - 0.14 \times L_{dn}}}$$

$Annoyed$ = Percent of people highly annoyed
L_{dn} = Mean day and night noise intensity

(5.7)

A linear approximation of equation 5.7

$$Annoyed = 20 + 3.2 \times L_{over70dB}$$

Annoyed = Percent of people highly annoyed
$L_{over\ 70dB}$ = Noise level over 70 dB

(5.8)

Noise concentrated at a single frequency is more noticeable and annoying than when distributed more broadly. Apple capitalized on this phenomenon when it created asymmetric fan blades for the cooling fans for its laptops, helping the computer maintain an illusion of quite operation and avoiding annoyance [205].

Is All Noise Bad? Before we leave our discussion of noise, it is important to identify certain circumstances in which softer noise may actually be helpful. For example, low levels of continuous noise (the hum of a fan) can mask the more disruptive and startling effects of discontinuous or distracting noise (the loud ticking of the clock at night or the conversation in the next room). Soft background music may accomplish the same objective. These effects also depend on the individual, with some people much more prone to being annoyed by noise [206]. Under certain circumstances, noise can perform an alerting function that can maintain a higher level of vigilance [207] (see also Chapter 6). For this reason, many seek out coffee shops for their engaging level of noise. More generally, the background *soundscape* of a design studio, hotel lobby, restaurant, or home can have broad implications for productivity and positive feelings [208].

This last point brings us back to reemphasize one final issue that we have touched on repeatedly: the importance of *task analysis*. The full impact of adjusting sound frequency and intensity levels on human performance can never be adequately predicted without a clear understanding of what sounds will be present when, who will listen to them, who *must* listen to them, and what the costs will be to task performance, listener health, and listener comfort if hearing is degraded. Furthermore, one person's noise may be another person's "signal" (as is often the case with conversation).

5.6   Other Senses

Vision and hearing have held the stage during these last two chapters for the important reason that the visual and auditory senses are of greatest implications for the design of human-machine systems. The "other" senses, critically important in human experience, have played considerably less of a role in system design. Hence, we do not discuss the senses of smell and taste, important as both of these are to the pleasures of eating (although smell can provide an important safety function as an advanced warning of fires and advanced detection of carbon monoxide). We discuss briefly, however, three other categories of sensory experiences that have some direct relevance to design: touch and feel (the tactile and haptic sense), limb position and motion (proprioception and kinesthesis), and whole-body orientation and motion (the vestibular senses). All of these offer important channels of information that help coordinate human interaction with many physical systems.

5.6.1   Touch: Tactile and Haptic Senses

Lying just under the skin are sensory receptors that respond to pressure on the skin and relay their information to the brain regarding the subtle changes in force applied by the hands and fingers (or other parts of the body) as they interact with physical things in the environment. Along with the sensation of pressure, these senses, tightly coupled with the proprioceptive sense of finger position, also provide *haptic*/indexhaptic information regarding the shape of manipulated objects and things [209].The combination of tactile with auditory and visual is often referred to as "multi-modal". We see the importance of these sensory channels in the following examples:

1. A problem with the membrane keyboards sometimes found on calculators is that they do not offer the same "feel" (tactile feedback) when the fingers are positioned on the button as do mechanical keys (see Chapter 9).

2. Gloves, to be worn in cold weather (or in hazardous operations) should be designed to maintain some tactile feedback if manipulation is required [210].

3. Early concern about the confusion that pilots experienced between two very different controls—the landing gear and

the flaps—was addressed by redesigning the control handles to feel quite distinct. The landing gear felt like a wheel—the plane's tire—while the flap control felt like a rectangular flap. Incidentally this design also made the controls feel and look somewhat like the system that they activate; see Chapter 9 where we discus control design.

4. The tactile sense is well structured as an alternative channel to convey both spatial and symbolic information for the blind through the braille alphabet.

5. Designers of virtual environments, which we discuss in Chapter 10, attempt to provide artificial sensations of touch and feel via electrical stimulation to the fingers, as the hand manipulates "virtual objects" [211], or use tactile stimulation to enable people to "see" well enough to catch a ball rolled across when they are blindfolded [212].

6. In situations of high visual load, tactile displays can be used to call attention to important discrete events [213]. Such tactile alerts cannot convey as much information as more conventional auditory and visual alerts, but are found to be more noticeable than either of the others, particularly in workplace environments often characterized by a wide range of both relevant and irrelevant sights and sounds.

5.6.2   Proprioception and Kinesthesis

We briefly introduced the proprioceptive channel in the previous section in the context of the brain's knowledge of finger position. In fact, a rich set of receptor systems, located within all of the muscles and joints of the body, convey to the brain an accurate representation of muscle contraction, joint angles, and limb position in space. The *proprioceptive channel* is tightly coupled with the *kinesthetic channel*, receptors within the joints and muscles, which convey a sense of the motion of the limbs as exercised by the muscles. Collectively, the two senses of kinesthesis and proprioception provide rich feedback that is critical for our everyday interactions with things in the environment. One particular area of relevance for these senses is in the design of manipulator controls, such as the joystick or mouse with a computer system, the steering wheel on a car, the clutch on a machine tool, and the control on an aircraft (see Chapter 9). As a particular example, an isometric control is one that does not move but responds only to pressure applied upon it. Hence, the isometric control cannot benefit from any proprioceptive feedback regarding how far a control has been displaced, since the control does not move at all. Early efforts to introduce isometric side-stick controllers in aircraft were, in fact, resisted by pilots because of this elimination of the "feel" associated with control movement. More of this information characterizing control sticks will be found in chapter 9 (Controls).

Tactile stimulation on the tongue, based on output from a video camera, enables you "see" simple objects.

5.6.3 The Vestibular Senses

Located deep within the inner ear are two sets of receptors, located in the semicircular canals and in the vestibular sacs. These receptors convey information to the brain regarding the angular and linear accelerations of the body respectively. Thus, when I turn my head with my eyes shut, I "know" that I am turning, not only because kinesthetic feedback from my neck tells me so but also because there is an angular acceleration experienced by the semicircular canals. Associated with the three axes along which the head can rotate, there are three semicircular canals aligned to each axis. Correspondingly, the vestibular sacs (along with the tactile sense from the "seat of the pants") inform the passenger or driver of linear acceleration or braking in a car. These organs also provide the constant information about the accelerative force of gravity downward, and hence they are continuously used to maintain our sense of balance (knowing which way is up and correcting for departures). When gone, as in outer space, designers might create "artificial gravity", by rotating the space craft around an axis.

Not surprisingly, the vestibular senses are most important for human-system interaction when the systems either move directly (as vehicles) or simulate motion (as vehicle simulators or virtual environments). The vestibular senses play two important (and potentially negative) roles here, related to *spatial disorientation* and to *motion sickness*.

Spatial disorientation illusions of motion, occur because certain vehicles, particularly aircraft, place the occupants in situations of sustained acceleration and nonvertical orientation for which the human body is not naturally adapted. Hence, for example, when the pilot is flying in the clouds without sight of the ground or horizon, the vestibular senses may sometimes be "tricked" into thinking that up is in a different direction from where it really is. This presents a real danger that has contributed to the loss of control of the aircraft [104].

The vestibular senses also play a key role in motion sickness. Normally, our visual and vestibular senses convey compatible and redundant information to the brain regarding how we are oriented and how we are moving. However, there are certain circumstances in which these two channels become decoupled so that one sense tells the brain one thing and the other tells it something else. These are conditions that invite *motion sickness* [214, 215, 216]. One example of this decoupling results when the vestibular cues signal motion and the visual world does not. When riding in a vehicle with no view of the outside world (e.g., a toddler sitting low in the backseat of the car, a ship passenger below decks with the portholes closed, or an aircraft passenger flying in the clouds), the visual view forward, which is typically "framed" by a manufactured rectangular structure, provides no visual evidence of movement (or evidence of where the "true" horizon is). In contrast, the continuous rocking, rolling, or swaying of the vehicle provides very direct stimulation of movement to the vestibular senses to all three of these passengers. When the two senses are in conflict, motion sickness often results

Motion sickness is an unpleasant reminder that our senses work together and are not independent information channels.

(a phenomenon that was embarrassingly experienced by the second author while in the Navy at his first turn to "general quarters" with the portholes closed below decks). Automated vehicles may produce a similar effect when people turn their attention inside the vehicle rather than being focussed on the road and so designers should consider tuning vehicle dynamics and passenger feedback systems to mitigate this risk [217].

Conflict between the two senses can also result from the opposite pattern. The visual system can often experience a very compelling sense of motion in video games, driving or flight simulators, and in virtual environments, even when there is no motion of the platform [218]. Again, there is conflict and the danger of a loss of function (or wasted training experience) when the brain is distracted by the unpleasant sensations of motion sickness. We return to this topic in Chapter 13.

The effect of motion can be quantified to describe the proportion of people who will vomit—the motion sickness incidence (MSI). MSI depends on the magnitude of acceleration associated with the oscillations (e.g., waves), the frequency of these oscillations, and the duration of exposure. Figure 5.12 shows iso-emesis curves, where each curve demarks an exposure that will lead the same proportion of people to vomit [219].

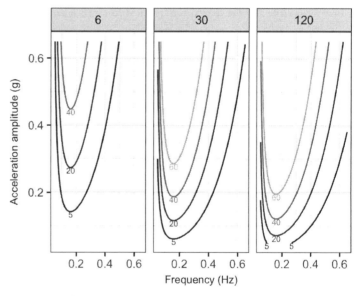

Figure 5.12 Effect of motion on motion sickness illness. Each curve describes the percent of people vomiting for 6, 30, and 120 minutes of exposure to motion of different frequencies. Created using R statistical Software package.

5.7 Summary

Hearing, when coupled with vision and the other senses, offers array of information. Each sensory modality has particular strengths and weaknesses, and collectively the ensemble nicely compensates for the weaknesses of each sensory channel alone. Clever designers can capitalize on the strengths and avoid the weaknesses in rendering the sensory information available to the higher brain centers for interpretation, decision making, and guiding action. In the following two chapters, we consider the characteristics of these higher level information-processing or cognitive operations before addressing, in Chapter 8, the implications sensory and information processing characteristics the design of displays.

Additional Resources

Several useful resources that expand on the content touched on in this chapter include several packages for analysis of sound data for the statistical language R, as well as handbooks on sound design and noise control:

1. **Calculations of power spectra and spectrograms:** Seewave (http:// rug.mnhn.fr /seewave/)

2. **Calculations for the speech intelligibility index based on ANSI standard:** SII (https:// cran.r-project.org / web/packages/ SII/index.html)

3. **Noise and vibration control handbook:** This book covers a broad range of considerations assessing and mitigating the damaging effects of noise and vibration.

   Crocker, M. J. (2007). *Handbook of Noise and Vibration Control.* (M. J. Crocker, Ed.). Hobocken, NJ: Wiley.

4. **Handbook for sound design:** This book describes how to craft sound with a focus on music and entertainment.

   Ballou, G. (Ed.). (2008). *Handbook for Sound Engineers* (Fourth). New York: Taylor & Francis.

Questions

Questions for 5.1 Auditory Environment

P5.1   What is the range of frequencies in which people can hear?

P5.2   What range of frequencies are people most sensitive to?

P5.3   How is the distribution of frequencies in a sound typically displayed?

P5.4   What does the envelope of a sound represent?

P5.5   What information does a spectrogram represent?

P5.6   How does the sound pressure level relate to the sound energy at the source of the sound?

P5.7   How can dBs describe the intensity of a particular sound given that it is a unit related to the log of a ratio?

P5.8   What is the dB level of the most intense sound you can hear, the softest sound, and the dB level for conversation?

P5.9   Calculate the combined sound pressure level of an 85 dB, 90 dB, and 95 dB sound source.

P5.10   Calculate the combined sound pressure level of three 85 dB sounds.

P5.11   How much does the sound pressure level increase when you double the number of identical sound sources?

P5.12   What does the sound field describe?

P5.13   Why would you want to avoid measuring sounds in the near field, and how far should you measure to avoid near field artifacts?

P5.14   In measuring sound pressure levels to assess risk for hearing damage, what scale on the sound meter should you select?

P5.15   If you have 10 machines each producing 90 dB of noise, how many machines would need to be removed for the combined sound level to seem half as loud as all 10

P5.16   An important characteristic of the sound field are reverberations associated with reflections of sound off walls and other obstacles. What effect do reverberations that arrive at the ear within 50ms of the initial sound wave have on speech intelligibility?

P5.17   An operator spends 4 hours at a milling machine (85 dBA), 3 hours at a press (80 dBA, and 1 hour (92 dBA) at a finishing booth. Calculate the time weighted average (TWA) and decide whether a hearing conservation program is required.

P5.18   A machine shop has a sound level of 87dB. What must be done to comply with OSHA standards to protect hearing? If the sound pressure level was 93 dB what must be done?

P5.19   Describe three strategies you might use to address the problem of noise in a machine shop.

P5.20 Hazard mitigation to address noise that might damage hearing includes addressing the source, the environment, and the listener. Which of these is the most and the least desirable strategy?

Questions for 5.2 The Receptor System: The Ear

P5.21 What does the aural reflex have to do with hearing?

P5.22 What is the mechanism for hearing loss and how does it explain why people exposed to a particular frequency of sound have hearing loss at that frequency?

P5.23 How does masking relate to glare, which was discussed in Chapter 4?

P5.24 How do the mechanisms of hearing explain masking, particularly the effect of the frequency distribution of the masking noise?

Questions for 5.3 Auditory Sensory Processing Characteristics

P5.25 What sound pressure level of the target sound, above the masking sound, is needed to ensure it is heard?

P5.26 Beyond increasing the volume of the target sound, how can the sound be adjusted to avoid being masked?

P5.27 Do people become less sensitive to high or low frequency sounds with increasing age?

P5.28 Describe why age-related hearing loss makes it particularly difficult to understand conversations.

P5.29 What is the relationship between pitch and frequency?

P5.30 What is the relationship between sound pressure level and loudness?

P5.31 What is a sone and what is a phon?

P5.32 How do sones and phons relate to the perception of loudness?

P5.33 Approximately what sound pressure level is needed if you want a tone to sound 4 times as loud as a 50 dB tone?

Questions for 5.4 Cognitive Influence on Auditory Perception

P5.34 What is the purpose of an environment and task analysis in alarm design?

P5.35 Describe how you would minimize confusability of an alarm.

P5.36 How would you avoid startling people using adjustments in specific properties of warning sounds?

P5.37 What is an important benefit of voice alarms?

P5.38 An alarm indicating that a prisoner is trying to escape frequently occurs when no prisoner is trying to escape (i.e., a false alarm), describe how you might address this problem using three of the five strategies described in this chapter.

P5.39   Describe how your alarm clock adheres to or neglects criteria of alarm design.

P5.40   Why is psychophysical scaling important to designing systems to fit human capabilities?

P5.41   Describe one top-down method and one bottom-up strategy to enhance speech communi-cation.

P5.42   How many dB above the background noise must an alarm be to reliably detected, how many to guarantee detection?

Questions for 5.5 Auditory Influence on Cognition: Noise and Annoyance

P5.43   A new train line will generate 78 db of noise. What proportion of people will be annoyed?

Questions for 5.6 Other Senses

P5.44   From the perspective of proprioceptive sensation, why might an isometric joystick be less effective than a joystick that moves when a person presses it?

P5.45   In terms of spatial disorientation, describe why pilots should ignore their feelings about what direction is up and rely on the instruments instead.

P5.46   Explain why video games and riding on a ship could make you sick.

Chapter 6

Cognition

At the end of this chapter you will be able to...

1. describe how different cognitive environments affect cognition

2. relate selective and divided attention to system design, particularly those systems that demand multi-tasking

3. use the properties of working memory to predict memory-related errors and design to avoid these errors

4. use the principles of long-term memory to guide effective learning and retention of information

5. use the features that promote habits to change long-term behavior patterns

Arriving at the airport rental lot after a delayed flight, Laura picked up a car and began to drive away, tapping the brakes to familiarize herself with the feel of the car. Laura was running late for an appointment in a large, unfamiliar city and relied on her new navigation device to guide her. She had read the somewhat confusing instructions and realized the importance of the voice display mode so that she could hear the directions to her destination without taking her eyes off the road. She had reminded herself to activate it before she got into heavy traffic, but the traffic suddenly increased, and she realized that she had forgotten to do so. Being late, however, she did not pull over but tried to remember the sequence of mode switches necessary to activate the voice mode. She couldn't get it right, but she managed to activate the electronic map. However, transposing its north-up representation to accommodate her south-bound direction of travel was too confusing. Finally lost, she pulled out her cellular phone to call her destination, glanced at the number she had written down, 303-462-8553, and dialed 303-462-8533. Getting no response, she became frustrated. She looked down to check the number and dial it carefully. Unfortunately, she did not see the car rapidly converging along the entrance ramp to her right, and only at the last moment the sound of the horn alerted her that the car was not yielding. Slamming on the brakes, heart beating fast, she pulled off to the side to carefully check her location, read the instructions, and place the phone call in the relative safety of the roadside.

Each day, we process large amounts of information from our environment to accomplish various tasks and make our way successfully through the world. The opening vignette represents a typical problem that one might experience because of a poor match between engineered equipment (or the environment) and the human information-processing system. Sometimes these mismatches cause misperceptions, and sometimes people just forget things. While the scenario described above may seem rather mundane, there are dozens of other cases where such difficulties result in injury or death [220, 10]. Some of these cases are discussed in Chapter 14 on safety. In this chapter, we provide a framework of the basic mechanisms by which people perceive, think, and remember, and processes information, which are generally grouped under the label of cognition. As we learn about the various capabilities and limitations of human cognition, we consider the implications for creating systems that are a good match.

The human information-processing system is conveniently represented by different stages at which information gets transformed: (1) *sensation*, by which the senses, described in chapters 4 and 5 transform physical into neural energy, (2) *perception* of information about the environment, (3) *central processing* or transforming and remembering that information, and (4) *responding* to that information. We highlight the second and third stages as the processes involved in cognition and most typically represented in the study of applied cognitive psychology [221, 12]. This chapter then picks up where our discussions of the more sensory aspects of auditory and visual processing left off in the previous two chapters. In Chapter 7, we describe more complex cognitive processes that form the basis of decision making, in Chapter 8 we discuss the implications of perception and cognition for display design, and in Chapter 9 we discuss the implications for control, and in Chapter 10 the implications for i human-computer interaction and interface design. Finally, our discussion of memory has many direct implications for learning and training, as discussed in Chapter 18.

6.1 Cognitive Environment

Similar to the properties of the visual and auditory environment, described in terms of the intensity, frequency, and distribution of the light and sound energy, the properties of the cognitive environment govern how people come to understand and respond to the world around them. Important dimensions of the cognitive environment include its *bandwidth* (c.g., how quickly it changes), *familiarity* (e.g., how often and how long the person has experienced the environment), and the degree of *knowledge in the world* (e.g., to what extent information that guides behavior is indicated by features in the environment). Driving, as in the opening vignette, is a high bandwidth environment that can change very quickly and can demand a response in less than a second, other environments change slowly and allow people minutes or hours to respond.

Figure 6.1 shows a cognitive environment that requi a high degree of knowledge in the world that guides cognition. It includes a broad array of information that supports long-term memory of facts and ideas. Obviously the knowledge in the world offered by the densely packed office depends on the professor's familiarity with the material. For someone who has not lived with this accumulating mass of material, the office would not offer knowledge, but simply overwhelming clutter. Unlike driving, the professor's environment shown in this figure doesn't require a response in seconds, but instead gives him hours or weeks to respond to the demands of student questions and research.

Figure 6.1 Cognitive environments with a high density of information. (Source: Photo of Professor Paul Milgram in his office at the University of Toronto. Photograph by Benjamin Rondel, reprinted with permission from photographer.

6.2   Information Processing Model of Cognition

Figure 6.2 shows a model of information processing that highlights those aspects that typically comprise cognition: perceiving, thinking about, and responding to the world. The senses, shown to the left of the figure, gather information, which is then perceived, providing a meaningful interpretation of what is sensed as aided by prior knowledge, through a mechanism that we described in Chapter 4 as *top-down processing*. This prior knowledge is stored in long-term memory.

Sometimes, perception leads directly to the selection and execution of a response, as when the driver swerved to avoid the converging car in the opening story. Quite often, however, an action is delayed, or not executed at all, as we "think about" or manipulate perceived information in *working memory*. This stage of information processing plays host to a wide variety of mental activities that are in our consciousness, such as rehearsing, planning, understanding, visualizing, decision making, and problem solving. *Working memory* is a temporary, effort-demanding store.

One of the activities for which working memory is used is to create a more permanent representation of the information in *long-term memory*, where it may be retrieved minutes, hours, days, or years later. These are the processes of learning (putting information into long-term memory) and retrieval. As we see in the figure, information from long-term memory is retrieved every time we perceive familiar information. Whether information is directly perceived, transformed in working memory, or retrieved from long-term memory we then select an action or response, and then our muscles carry out or execute that response.

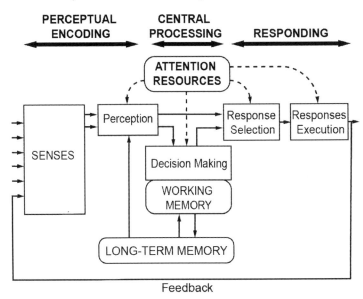

Figure 6.2 A model of human information processing.

At the top of the figure we note that many of the stages of information processing depend upon a limited pool of *attentional resources* that can be allocated to processes as required. The figure highlights a distinction that has important implications for design. On the left, we see the role of attention in selecting sensory channels for further processing, as when our eyes focus on one part of the world and not another. Attention here is a *filter*. In contrast, the other role of attention is indicated by the dashed arrows that show attention dividing between tasks and information processing stages. Attention here is a *fuel*, or mental energy that supports information processing. These two aspects of attention, selection and division, or the filter and the fuel, are treated separately in this chapter: selective attention at the start and divided attention at the end.

Finally, we note the feedback loop. Although we have discussed the sequence as starting with the senses, the sequence of information processing can start anywhere. Our actions often generate new information to be sensed and perceived. For example, sometimes we initiate an action from a decision with no perception guiding it. We then may evaluate the consequence of that decision later, through sensation and perception. Sometime we might even act to produce information to help us understand the world, as when Laura tapped the brakes to get a feel of the car. We consider the importance of this closed feedback loop in Chapter 9.

Perception guides action, but action also generates the input for perception.

6.3 Selective Attention and Perception

Many fatal accidents in commercial aviation occur when a pilot flies a perfectly good airplane into the ground. These accidents are labeled "controlled flight into terrain," and they are often caused by a failure of selective attention to those sources of information regarding the plane's altitude above the ground [222]. Similarly, inattention accounts for many car crashes in which people fail to see and respond to seemingly obvious hazards [223].

6.3.1 Mechanisms of Selective Attention

Selective attention does not guarantee perception, but it is usually necessary to achieve it. Stated in other terms, we normally look at the things we perceive and perceive the things we look at. We considered the role of visual scanning in selective attention in Chapter 4. While we do not have "earballs" that can index selective auditory attention as we have eyeballs in the visual modality, there is nevertheless a corresponding phenomenon in selecting auditory information. For example, we may tune our attention to concentrate on one conversation in a noisy workplace while filtering out the distraction of other conversations and noises. We also deploy tactile selective attention when we choose to feel a texture, or try to ignore the uncomfortable feeling of an itchy shirt.

The selection of channels to attend (and filtering of channels to ignore) is typically driven by four factors: *salience, effort, expectancy*, and *value* [224, 225]. They can be represented in the same contrasting framework of stimulus-driven bottom-up processes versus knowledge-driven top-down processes that we applied to perception in Chapters 4 and 5. Salience contributes to the bottom-up process of allocating attention, influencing *attentional capture*, which occurs when the environment directs attention [125]. The car horn, for example, clearly captured Laura's attention. Salient stimulus dimensions are chosen by designers to capture attention and signal important events via alarms and alerts [183]. Abrupt onsets [125], distinct visual stimuli and auditory stimuli [226, 174, 213], and tactile stimuli [227] are salient.

In contrast to salient features that capture attention, many events that do not have these features may not be noticed, even if they are significant, a phenomenon known as *change blindness* or *attentional blindness* [228, 229, 230]. Change blindness leads people to miss surprisingly large features of the environment even though they may look directly at them. The following link provides a demonstration: http://www.theinvisiblegorilla.com/videos.html

Selective attention also depends on effort. We prefer to scan short distances rather than long ones, and we often prefer to avoid head movements to select information sources. It is for this reason that drivers, particularly fatigued ones (who have not much "effort to give"), fail to look behind them to check their blind spot when changing lanes.

Expectancy and value together define what are characteristically called top-down or knowledge-driven factors in allocating attention. That is, we tend to look at, or "sample," the world where we expect to find valuable information. Laura looked downward because she expected to see the phone number there. She felt free

Attention leaves us surprisingly blind to much of what happens around us.

to look down because she did not expect to see traffic suddenly appear in her forward field of view. As an example in visual search, discussed in Chapter 4, a radiologist looks most closely at those areas of an x-ray plate most likely to contain an abnormality. Correspondingly, a pilot looks most frequently at the instrument that changes most rapidly because that is where the pilot expects to see change [231].

The frequency of looking at or attending to channels is also modified by how valuable it is to look at (or how costly it may be to miss an event on a channel) [115]. This is why a trained airplane pilot will continue to scan the world outside the cockpit for other airplanes. Although close encounters with other airplanes are rare [229, 232], the costs of not seeing another airplane (and colliding with it) are large. It takes years of experience or special training for drivers to scan places in the roadway environment where hazards might unexpectedly occur [128].

In addition to understanding that failures to notice often contribute to accidents [233], understanding bottom-up processes of attentional capture is important for alarm design ([194, 183, 10], Chapters 5 and 8) and automated cueing (Chapter 11). Understanding the role of effort in inhibiting attention movement is also important in both designing integrated displays (Chapter 8) and configuring the layout of workspaces (Chapter 10).

The most direct consequence of selective attention is perception, which involves the extraction of meaning from an array (visual) or sequence (auditory) of information processed by the senses, filtered and perceived. Our driver, Laura, eventually looked to the roadside (selection) and perceived the hazard of the approaching vehicle. Sometimes, meaning may be extracted (perception) without attention. In this way, our attention at a party can be "captured" in a bottom-up fashion when a nearby speaker utters our name even though we were not initially selecting that speaker. This classic phenomenon is sometimes labeled the "cocktail party effect" [234, 235]. Correspondingly, the driver may not be consciously focusing attention on the roadway, even though she is adequately perceiving roadway information enough to steer the car.

Even though attention can sometimes be directed and information extracted without effort, the more general lesson is that focus of attention is typically much more narrow than our intuition might suggest [236]—only a small slice of the world is attended and subsequently perceived. This is why presenting information in a head-up display can give a driver the illusion that because she is looking to the road she is able to see the road and see the email message projected on the display at the same time. It is also why many drivers feel entirely confident that they can engage in cell phone conversation and drive safely [237]; but just because the eyes and ears are both free from interference, perceptual and cognitive operations in the brain must share resources, undermining driving safety.

When a driver's eyes are on the road it doesn't necessarily mean the driver's mind is also on the road.

6.3.2 Mechanisms of the Perceptual Processes

Once attention is directed to an object or area of the environment, perception proceeds by three often simultaneous and concurrent processes: (1) bottom-up feature analysis, (2) top-down processing, and (3) unitization [238]. The latter two are based on long-term memory, and each has different implications for design. The distinction between bottom-up and top-down processing was discussed in Chapter 4 in the context of visual search and signal detection, and in Chapter 5 in the context of speech perception. Bottom-up processing depends on the physical make up of the stimulus. Top-down processing, based on knowledge and context, depends on expectancies stored from experience in long term memory. But the third component, unitization joins the physical stimulus and experience.

Perception proceeds by analyzing the raw features of a stimulus or event, whether it is a word (the features may be letters), a symbol on a map (the features may be the color, shape, size, and location), or a sound (the features may be the phonemes of the word or the loudness and pitch of an alarm). Every event could potentially consist of a huge combination of features. However, to the extent that past experience has exposed the perceiver to sets of features that occur together and their co-occurrence is familiar (i.e., represented in long-term memory), these sets are said to become *unitized*. The consequence of unitization is rapid and *automatic processing*, also called automaticity.

Unitization explains the difference between perceiving the printed words of a familiar and an unfamiliar language is that the former can be perceived as whole units, and their meaning is directly accessed (retrieved from long-term memory), whereas the latter may need to be analyzed letter by letter or syllable by syllable, and the meaning is more slowly and effortfully retrieved from long-term memory. This distinction between the effortful processing of feature analysis and the more automatic processing of familiar unitized feature combinations (whose combined representation is stored in long-term memory), can be applied to almost any perceptual experience, such as perceiving symbols and icons (see Chapter 8), depth cues (Chapter 4), or alarm sounds (Chapter 5).

Whether unitized or not, stimulus elements and events may be perceived in clear visual or auditory form (reading large text in a well-lighted room or hearing a clearly articulated speech) or may be perceived in a degraded form. For a visual stimulus, degradation occurs with short glances, tiny text, and poor illumination or low contrast. For an auditory stimulus, masking noise and low intensity or unfamiliar accents produce degradation. This degradation undermines bottom-up processing. The perception of such degraded stimuli is better if they are unitized and familiar. The ability of familiarity to offset degraded bottom-up processing, reflects the third aspect of perceptual processing: top-down processing .

You can think about top-down processing as the ability to correctly guess what a stimulus or event is, even in the absence of clear physical features necessary to precisely identify it using bottom-up processing. Such guesses are based upon expectations, and these expectations are based upon past experience, which is, by definition, stored in long-term memory. That is, we see or hear what we expect to see or hear. High expectations are based on events that we have encountered frequently in the past. They are also based on associations between the perceived stimulus or event, and other stimuli or events that are present in the same context and have been joined in past experience.

The concepts of frequency and context in supporting top-down processing can be illustrated by the following example of an industrial trash compactor. A status indicator for this trash compactor—a very reliable piece of equipment—can be either green, indicating normal operations, or red, indicating failure. Given our past experience of red and green in designed systems, we associate these two colors to their meaning (OK versus danger) fairly automatically. A brief glance at the light, in the glare of the sun, makes it hard to see which color it is (poor bottom-up processing). The past high reliability of the system allows us to "guess" that it is green (top-down processing based upon frequency) even if the actual color is hard to see. Hence, quick glance confirms that it is green. The sound of smooth running and good system output provides a context to amplify the "perception of greenness" (top-down processing based upon context). An abnormal sound gradually becomes evident. The context has now changed, and red becomes somewhat more expected. The same ambiguous stimulus (hard to tell the color) is now perceived to be red (changing context). Now a very close look at the light, with a hand held up to shield it from the sun (improved bottom-up processing), reveals that it in fact is red, and it turns out that it was red all along. Perception had previously been deceived by expectancy.

We now consider two other examples of the interplay between, and complementary of, bottom-up and top-down processing. As one example, in reading, bottom-up processing is degraded by speed (brief glances) as well as by legibility, factors discussed in Chapter 4. With such degradation, we can read words more easily than random digit strings (phone numbers, basketball scores, or stock prices), because each word provides an context-based expectation for the letters within, and when text is presented, the words of a sentence provide context for reading degraded words. For example, if we read the sentence "Turn the machine off when the red light indicates failure" and find the fourth word to be nearly illegible (poor bottom-up cues), the context of the surrounding words allows us to guess that the word is probably "off."

Furthermore, there are usually less serious consequences for failing to perceive the name correctly than for failing to perceive the phone number correctly. The latter will always lead to a dialing error. Like the digits in the phone number, the letters in an email address should also be larger, since the lack of standardization of email addresses (and the fact that many people don't know the middle initial of an addressee) removes context that could otherwise help support top-down processing.

Top-down processing, through frequency and context, helps us see what would otherwise be very difficult to see, and sometimes even see what isn't there.

In short,
Adam Humfac: Adamjhumfa @xxx.yyy 444-455-2995
is a better design than

Adam Humfac: adamjhumfa@xxx.yyy 444-455-2995

6.3.3 Implications of Selective Attention and Perception for Design

The proceeding examples lead us to a few simple guidelines for supporting attention and perception.

1. **Maximize bottom-up processing** (Chapters 4 and 5). This involves not only increasing visibility and legibility (or audibility of sounds), but also paying careful attention to confusion caused by similarity of message sets that could be presented in the same context.

2. **Maximize automaticity and unitization** by using familiar perceptual representations (those encountered frequently in long-term memory). Examples include the use of familiar fonts and lowercase text (Chapter 4), meaningful icons (Chapter 8), and words rather than abbreviations.

3. **Maximize top-down processing** when bottom-up processing may be poor (as revealed by analysis of the environment and the conditions under which perception may take place), and when unitization may be lacking (unfamiliar symbology or language). Improving top-down processing means providing the best opportunities for guessing. For example, putting information in a consistent location, as is done with the height of stop signs.

4. **Maximize discriminating features** to avoid confusion:

   - **Use a smaller vocabulary**. This has a double benefit of improving guess rate and allowing the creation of a vocabulary with more discriminating features. This is why in aviation, a restricted vocabulary is enforced for communications with air traffic control.

   - **Create context.** For example, the meaning of "your is low" is better perceived than that of the shorter phrase "fuel low," particularly under noisy conditions [239].

   - **Exploit redundancy.** This is quite similar to creating context, but redundancy often involves direct repetition of content in a different format. For example, simultaneous display of a visual and auditory message is more likely to guarantee correct perception in a perceptually degraded environment. The phonetic alphabet exploits redundancy by having each syllable convey a message concerning the identity of a letter (alpha = a).

- **Test symbols and icons in their context of use.** When doing usability testing of symbols or icons, make sure that the testing situation is similar to that in which they will eventually be used [240, 241]. This provides a more valid test of the effective perception of the icons, because context affects perception.

- **Consider expectations** Be wary of the "conspiracy" to invite perceptual errors when encountering unexpected situations when bottom-up processing is degraded. An example of such conditions is flying at night and encountering unusual aircraft attitudes, which can lead to illusions. Another example is driving at night and encountering unexpected roadway construction. In these cases, as top-down processing attempts to compensate for the bottom-up degradation, it encourages the perception of the expected, which will not be appropriate. Under such conditions, perception of the unusual must be supported by providing particularly salient cues.

  A special case here is the poor perception of negation in sentences. For example, "do not turn off the equipment" may be readily perceived as "turn off the equipment" if the message is badly degraded, because our perceptual system treats the positive meaning of the sentence as the "default" [242]. We return to this issue in our discussion of comprehension and working memory. If negation is to be used, highlight it to avoid misinterpretation.

One downside of the redundancy and context, which support top-down processing, is that they increase the length of perceptual messages, thereby reducing the *efficiency* of information transfer [213]. For example, "alpha" and "your fuel is low" both take longer to say than "A" and "fuel low" (although they do not necessarily take longer to understand). The printed message "failure" occupies more space than the letter "F" or a small red light. Thus, redundancy and context can improve perceptual accuracy, but at the expense of efficiency. This is a tradeoff that designers should address by analyzing the consequences of perceptual errors and the extent that the environmental and stress may degrade bottom-up processing. We consider these aspects of stress, such as time pressure and imminent danger in more detail in Chapter 13.

Perception is often relatively automatic (but becomes less so as bottom-up processing is degraded and as top-down and unitization processes become less effective). However, as the extent of the perceptual process increases, we speak less of perception and more of *comprehension*, which is less automatic. The border between perception and comprehension is a fuzzy one, although we usually think of perceiving a word, but comprehending a series of words that make up a sentence. As we shall see, comprehension, like perception, is very much driven by top-down processing, from past experience and long-term memory. However, comprehension tends to also rely heavily upon the capabilities of working memory in a way that perception does not. We address the issues of perception further in Chapter 8 on displays.

6.4   Working Memory

Everyone suffers from memory failures—and relatively frequently [243]. Sometimes, the failures are trivial, such as forgetting a new password that you just created. Other times, memory failures are more critical. For example, in 1915 a railroad switchman at a station in Scotland forgot that he had moved a train to an active track. As a result, two oncoming trains used the same track and the ensuing crash killed over 200 people [244].

The next few sections focus on the part of cognition that involves human memory systems. Substantial evidence shows that there are two very different types of memory. The first, *working memory* (sometimes termed *short-term memory*), is relatively transient and limited to holding a small amount of information that may be rehearsed or "worked on" by other cognitive processes [245, 246]. It is the temporary store that keeps information available while we are using it, until we use it, or until we store it in long-term memory.

Examples of working memory include looking up a phone number and then holding it in working memory until we have completed dialing, remembering the information in the first part of a sentence as we hear the later words and integrate them to understand the sentence meaning, "holding" subsums while we multiply two two-digit numbers, remembering numbers on one computer screen until a second screen can be accessed for comparison, and constructing an image of the way an intersection will look from a view on a map.

The other memory store, long-term memory, involves the storage of information after it is no longer active in working memory and the retrieval of the information at a later point in time. When retrieval fails from either working or long-term memory, it is termed forgetting. Conceptually, working memory is the temporary holding of information that is active, either perceived from the environment or retrieved from long-term memory, while long-term memory involves the relatively passive store of information, which is activated only when it needs to be retrieved. The limits of working memory hold major implications for system design.

6.4.1   Mechanisms of Working Memory

Working memory can be understood in the context of a model proposed by Baddeley [247, 246], consisting of four components. In this model, a *central executive* component acts as an attentional control system that coordinates information from three "storage" systems: *visuospatial sketchpad, episodic buffer, phonological loop* (Figure 6.3).

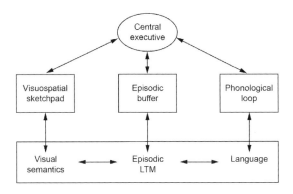

Figure 6.3 Elements of working memory. [247]

The *visuospatial sketchpad* holds information in an analog spatial form (e.g., visual imagery) [248]. These images consist of encoded information that has been brought from the senses or retrieved from long-term memory. Thus, the air traffic controller uses the visual-spatial sketchpad to retain information regarding where planes are located in the airspace. This representation is essential for the controller if the display is momentarily lost from view. This spatial working-memory component is also used when a driver tries to construct a mental map of necessary turns from a set of spoken navigational instructions. Part of the problem that Laura had in using her north-up map to drive south into the city was related to the mental rotation in spatial working memory that was necessary to bring the map into alignment with the world out her windshield.

The *phonological loop* represents verbal information in an acoustical form [246]. It is kept active, or "rehearsed," by articulating words or sounds, either vocally or sub-vocally. Thus, when we are trying to remember a phone number, we silently sound out the numbers until we no longer need them. As when we have dialed the number, or memorized it.

The *episodic buffer* orders and sequences events and communicates with long-term memory to provide meaning to the information held in the phonological loop and visuospatial sketchpad. The episodic buffer is important for design because it enables a meaningful sequence of events—a story—to be remembered much more easily than an unordered sequence [247].

Working memory holds two different types of information: verbal and spatial. The central executive then operates on this material that is temporarily and effortfully preserved, either in the phonological loop or the visual spatial scratchpad. Whether material is verbal (in the phonetic loop) or spatial (in the visuospatial sketchpad), our ability to maintain information in working memory is severely limited in four interrelated respects: how much information can be kept active (its capacity), how long it can be kept active, how similar material is to other elements of working memory and to ongoing information processing, and the availability and type of attentional resources required to keep the material active. We

The episodic buffer is one reason why people naturally remember information presented as a story.

describe each of these influences in turn.

6.4.2 Limits of Working Memory

Limits of working memory are substantial and severely limit how people can process and retain information they see and hear.

Capacity: The upper limit or the capacity of working memory is *4 chunks* [245], although it often mistakenly described as 7±2 chunks [148]. A chunk is the unit of working memory space, defined jointly by the physical and cognitive properties that bind items within the chunk together. Thus, the sequence of four unrelated letters, X F D U, consists of four chunks, as does the sequence of four digits, 8 4 7 9. However, the four letters DOOR or the four digits 2004 consist of only one chunk, because these can be coded into a single meaningful unit. As a result, each chunk occupies only one "slot" in working memory, and so our working memory could hold four words, or four familiar dates, or four unrelated letters or digits.

What then binds the units of an item together to make a single chunk? As the examples suggest, it is familiarity with the links or associations between the units, a familiarity based upon past experience and therefore related to long-term memory. The operation is analogous to the role of unitization in perception, discussed earlier. As a child learns to read, the separate letters in a word gradually become unified to form a single chunk. Correspondingly, as an expert gains familiarity with a domain, an acronym or abbreviation that was once several chunks (individual letters) now becomes a single chunk.

Chunks in working memory can be thought of as "memory units," but they also have physical counterparts in that perceptual chunks may be formed by providing spatial separation between them. For example, the social security number 123 45 6789 contains three physical chunks. Such physical chunking is helpful to memory, but physical chunking works best when it is combined with cognitive chunking. In order to demonstrate this, ask yourself which of the following would be the easiest to remember: FBI CIA USA, or FB ICIAU SA.

Chunking benefits working memory in several ways. First, and most directly, it reduces the number of items in working memory and therefore increases the capacity of working memory storage. Second, chunking makes use of meaningful associations in long-term memory, and this aids in retention of the information. Third, because of the reduced number of items in working memory, material can be more easily rehearsed and is more likely to be transferred to long-term memory (which then reduces load on working memory).

People vary in the overall capacity of working memory, and these individual differences can be readily measured [249] and can predict differences in performance of more complex tasks, such as detecting failures in automation systems [250].

Time: The capacity limits of working memory are closely related to the second limitation of working memory, the limit of how long information may remain. The strength of information in working memory decays over time unless it is periodically reactivated, or "pulsed" [245], a process called maintenance rehearsal [251]. Maintenance rehearsal for acoustic items in verbal working memory is essentially a serial process of subvocally articulating each item—repeating items in your head. Thus, for a string of items like a phone number or a personal identity number (PIN), the interval for reactivating any particular item depends on the length of time to proceed through the whole string. For a seven-digit phone number, we can serially reactivate all items in a relatively short time, short enough to keep all items active (i.e., so that the first digit in the phone number will still be active by the time we have cycled through the last item). The more chunks contained in working memory (like a seven-digit phone number plus a three-digit area code), the longer it will take to cycle through the items in maintenance rehearsal, and the more likely it will be that the early rehearsed items have decayed beyond the point where they can be reactivated.

Two specific features should be noted in the proceeding example, relevant to both time and capacity. First, with rehearsal, seven digits (when corganized into three chunks) is at the working memory limit, but 10 digits clearly exceeds it. Hence, requiring area codes to be retained in working memory, particularly unfamiliar ones, is a bad design. Second, familiar area codes create one chunk, not three, and a familiar prefix also reduces three chunks to one. Thus, a familiar combination, such as one's own phone number, will occupy 6, not 10, slots of working memory capacity.

To help predict working memory decay for differing numbers of chunks, Card, Moran, and Newell [252] combined data from several studies to determine the "half-life" of items in working memory (the delay after which recall is reduced by half). The half-life was estimated to be approximately 7 seconds for three chunks and 70 seconds for one chunk.

Confusability and Similarity: In Chapters 4 and 5 we saw that perceptual confusability is source of error. Likewise, similarity of the features of different items can lead to confusion in working memory because as their representation decays before reactivation, it is more likely that the discriminating details will be gone. For example, the ordered list of letters E G B D V C is less likely to be correctly retrieved from working memory than is the list E N W R U J because of the greater confusability of the acoustic features of the first list. (This fact, by the way, demonstrates the dominant auditory aspect of the phonetic loop, since such a difference in working memory confusion is observed no matter whether the lists are heard or seen). Thus, decay and time are more disruptive to material that is more similar, particularly when such material needs to be recalled in a particular order [245]. A particularly lethal source of errors concerns the confusability of repeated items. For example, as Laura discovered in the driving example, the digit string 8553 is particularly likely to be erroneously recalled as 8533.

Availability and Type of Attention: Working memory, whether verbal or spatial, is resource-limited. Working memory depends very much on the limited supply of attentional resources. If such resources are fully diverted to a concurrent task, rehearsal will stop, and decay will be rapid. In addition, if the activity toward which resources are diverted uses similar material, like diverting attention to listening to basketball scores while trying to retain a phone number, the added confusion (here digits with digits) may be particularly lethal to the contents of working memory. The diversion of attention need not be conscious and intentional to disrupt working memory. For example, sounds nearly automatically intrude on the working memory for serial order [174]. We return to this issue of auditory disruption at the end of the chapter, just as we highlighted its attention-capturing properties in our discussion of selective attention. In terms of Baddeley's model of working memory, the visual spatial scratchpad is more disrupted by other spatial tasks, like pointing or tracking, and the phonetic loop is more disrupted by other verbal or language-based tasks, like listening or speaking [253, 254].

6.4.3 Implications of Working Memory Limits for Design

1. **Minimize working memory load.** An overall rule of thumb is that both the time and the number of alphanumeric items that human operators have to retain in working memory during task performance should be kept to a minimum [255]. In general, designers should try to avoid long codes of arbitrary digit or numerical strings [256]. Hence, any technique that can offload information in working memory sooner is valuable. Windows in computer systems that support comparisons between side-by-side information sources that avoid the demands on working memory imposed by switching between screens. Working memory is surprisingly limited—four chunks is a rough estimate—an even lower limit should be used if material is to be retained for more than a few seconds, in a noisy context.

   Designs that require working memory for more that 3 items for more than 7 seconds or one item for more than 70 seconds invite errors.

2. **Provide visual echoes.** Wherever an auditory presentation is used to convey messages, these should, be coupled with a redundant visual display of the information to minimize the burden on working memory. For example, when automated telephone assistance "speaks" phone numbers with a synthetic voice, this visual display of the smart phone should display the same number in the form of a redundant "visual echo." The visual material can be easily rescanned. In contrast, auditory material whose memory may be uncertain cannot be reviewed without an explicit request to "repeat."

3. **Provide placeholders for sequential tasks.** Tasks that require multiple steps, whose actions may be similar in appearance or feedback, benefit from some visual reminder of what steps have been completed, so that the momentarily distracted operator will not return to the task, forgetting what was done, and needing to start from scratch [257].

4. **Exploit chunking.** We have seen how chunking can increase the amount of material held in working memory and increase its transfer to long-term memory. Thus, any way we can take advantage of chunking is beneficial, including:

   - **Physical chunk size:** For presenting arbitrary strings of letters, numbers, or both, the optimal chunk size is three to four numbers or letters per chunk [258, 256].

   - **Create meaningful sequences.** The best procedure for creating cognitive chunks out of random strings is to find or create meaningful sequences within the total string of characters. A meaningful sequence should already have an integral representation in long-term memory. This means that the sequence is retained as a single item rather than a set of the individual characters. Meaningful sequences include things such as 555, 4321, or a friend's initials.

   - **Superiority of letters over numbers:** In general, letters induce better chunking than numbers because of their greater potential for meaningfulness. Advertisers have capitalized on this principle by moving from numbers such as 1-800-663-5900, which has eight chunks, to letter-based chunking such as 1-800-GET HELP, which has three chunks ("1-800" is a sufficiently familiar string that it is just one chunk). Grouping letters into one word, and thus one chunk, can ease working memory demands.

   - **Keep numbers separate from letters:** If displays must contain a mixture of numbers and letters, it is better to keep them separated [259]. For example, a license plate containing one numeric and one alphabetic chunk, such as 458 GST, will be more easily kept in working memory than a combination such as 4G5 8ST.

5. **Minimize confusability:** Confusability in working memory can be reduced by building physical distinctions into material to be retained. We have already noted that making words and letters sound more different reduces the likelihood that they will be confused during rehearsal. This can sometimes be accommodated by deleting common elements between items that might otherwise be confused. For example, confusion between 3 and 2 is less likely than between A5433 and A5423, even though in both cases only a single digit discriminates the two strings. Spatial separation also reduces confusability [260]. A display that has four different windows for each of four different quantities will be easier to keep track of than a single window display in which the four quantities are cycled. Spatial location represents a salient, discriminating cue to reduce item confusability.

6. **Avoid unnecessary zeros in codes to be remembered:** The zeros in codes like 002385, which may be created because

of an anticipated hundredfold increase in code number, will occupy excessive slots of working memory.

7. **Ensure congruence of instructions:** Congruence reduces working memory load by aligning the order of words and actions [10]. Congruence is critical in situations where there is no tolerance for error, such as instructions designed to support emergency procedures. To understand how we comprehend sentences, it is useful to assume that most words in a sentence will need to be retained in working memory for a person to interpret the meaning of the sentence [261, 262] Thus, long sentences obviously create vulnerabilities. So too do sentences with unfamiliar words or codes. Particularly vulnerable are instructions in which information presented early must be retained until the meaning of the whole string is understood. Such an example might be procedural instructions that reads:

> Before doing X and Y, do A.
>
> Here, X and Y must be remembered until A is encountered.
>
> A better order is: Do A. Then do X and Y.

The reason the second design is better is because it maintains congruence between the order of text and the order of action [10].

8. **Avoid the negative:** Finally, reiterating a point made in the context of perception, *negation* imposes an added chunk in working memory. Even if the negation may be perceived in reading or hearing an instruction, it may be forgotten from working memory as that instruction is retained before being carried out. In such circumstances, the default memory of the positive is likely to be retained, and the user may do the opposite of what was instructed. This is another reason to advocate using positive assertions in instructions where possible [10]. More details on instructional design are given in Chapter 18.

6.5   Long-Term Memory

We maintain information in working memory for its immediate use (less than a minute), but we also need a mechanism for storing information and retrieving it later. This mechanism is termed long-term memory. Learning is the process of storing information in long-term memory, and when specific procedures are designed to facilitate learning, we refer to this as instruction or training, an issue treated in depth in Chapter 18. Our emphasis in the current chapter is on retrieval and forgetting and the factors that influence them.

The ability to retrieve key information from long-term memory is important for many tasks in daily life. We saw at the beginning of this chapter that Laura's failure to recall instructions was a major source of her subsequent problems. In many jobs, forgetting to perform even one part of a job sequence can have catastrophic consequences. In this section, we review the basic mechanisms that underlie storage and retrieval of information from long-term memory and how to design around the limitations of the long-term memory system.

Long-term memory can be distinguished by whether it involves memory for general knowledge, called *semantic memory*, memory for specific events, called *episodic memory*, and memory of how to do things *procedural memory* [263]. Semantic memory concerns what you might learn for an exam or a friend's favorite color. Episodic memory concerns memory of sequences of activities and events, such as what you might have done last weekend. People can talk about semantic and episodic memory, but procedural memory is not easily put into words: imagine explaining how to tie your shoe to someone without actually showing them. Many routine activities of daily life as well as highly practiced skills used in sports rely on procedural memory. This is one reason why interviewing experts about how they do their job, as we described in Chapter 2, is so difficult. So much of their expertise relies on procedural, non-verbal, knowledge.

> You can't put procedural memory into words.

In contrast to both procedural and declarative knowledge, which is acquired from multiple experiences, the personal knowledge or memory of a specific event or episode is, almost by definition, acquired from a single experience. This may be the first encounter with an employer or coworker, a particular incident or accident at home or the workplace, or the eyewitness view of a crime or accident. Such memories are heavily based on visual imagery. In addition, the memories themselves are not "video replays" of the events, but reconstructions that change with each recall, and so have a number of biases discussed below.

Figure 6.5 shows processes involved in the formation, storage, and retrieval of episodic memories. Here an "event" occurs, which defines what actually happened. The event is observed and some information about it is encoded, which reflects the allocation of selective attention and may reflect some of the top down biases of expectancy on perception that we described earlier. As time passes, the memory of the episode is maintained in long-term memory, where it will show some degradation (forgetting), and the memory may be distorted by influences related to both schemas—general ways of thinking about the world—and specific intervening events [264]. Finally, the memory may be retrieved in a variety of circumstances, and this retrieval changes the memory. For example, a witness picks out a suspect from a police lineup, the witness is interviewed by police as the prosecution develops its case, or the witness responds to queries during actual courtroom testimony. Each instance of retrieval slightly changes the person's memory [265].

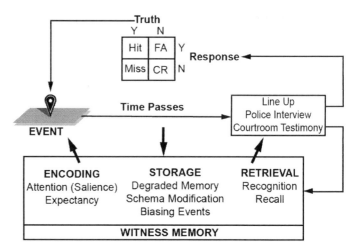

Figure 6.4 The processes involved in episodic memory and the influences of these processes. Retrieval in a courtroom starts another memory cycle that the jury encodes.

As the figure shows, eye witness recognition can often be represented by signal detection theory (Chapter 4), where the two states of the world are whether the witness did see the suspect in question or not, and the two responses are "that's the one", or "that's not the one". Extensive research on eyewitness testimony has revealed that the episodic memory process is far from perfect [266, 267, 268, 269]. In one study of police lineup recognition, Wright and McDaid [270] estimated that an innocent person was chosen (as a guilty perpetrator) approximately 20 percent of the time. The sources of such biases can occur at all three stages. For example, at encoding, a well-established bias is the strong focus of witness attention on a weapon when one is used at the scene of the crime. In light of what we know about the limits of attention, it should come as no surprise that this focus degrades the encoding of other information in the scene, particularly the physical appearance of the suspect's face relative to crimes where no weapon is employed [271].

An important application of memory research to episodic retrieval is the *cognitive interview* (CI) technique for assisting police in interviewing witnesses to maximize the information retrieved. Their approach is to avoid recognition tasks because asking witnesses a series of yes-or-no questions ("Did the suspect have red hair?") can be quite biasing and leave vast quantities of encoded information untapped. Instead, they apply a series of principles from cognitive psychology to develop effective recall procedures [272, 273]. The CI technique:

⚐ Asking leading questions can change memories and even create memories.

- Encourages the witness to reinstate the context of the original episode, thereby possibly exploiting a rich network of associations that might be connected with the episodic memory.

- Avoids time-sharing requirements where the witness must divide cognitive resources between searching episodic memory for details of the crime and listening to the interrogator ask additional questions. Ask witnesses to recall the sequence of events and details of the situation.

- Avoids time stress, allows the witness plenty of time to retrieve information about the crime and ideally allowing the witness multiple opportunities to recall.

These features of the CI take advantage of the rich network of associations. The CI technique enables witnesses to generate between 35 and 100 percent more information than standard police interview procedures and to do so while maintaining accuracy; it has been adopted by a number of police forces [268]. The CI technique has application beyond police interviews to accident investigations mentioned in Chapter 2.

One final implication for every reader is that when you witness a serious episode about which you might be later queried, it is good advice to write down everything about it as soon as the episode has occurred and indicate your degree of certainty about the events within the incident. Your written record will now be "knowledge in the world," not susceptible to forgetting or inadvertent revision.

6.5.1 Mechanisms of Long-Term Memory

Material in long-term memory has two important features that determine the ease of later retrieval: its strength and its associations.

Strength. The strength of an item in long-term memory is determined by the frequency and recency of its use. Regarding frequency, if a password is used every day (i.e., frequently) to log onto a computer, it will probably be well represented in long-term memory and rarely forgotten. Regarding recency, if a pilot spends a day practicing a particular emergency procedure, that procedure will be better recalled (and correctly executed) if the emergency is encountered in flight the very next day than if it is encountered a month later. Because emergency procedures are generally used infrequently suggests that their use should be supported by external visual checklists rather than reliance upon memory.

The best study techniques for later retrieval of material is via repeated active retrieval of the material from long term memory. Regular practice distributed over time, called spaced practice, is more effective than massed practice that is concentrated over just a few sessions [274]. Self quizzes (or instructor imposed quizzes) are much more effective than spending the same time re-reading the material [275, 276]. Each act of retrieval better solidifies the material (as long as what is retrieved is checked to assure its accuracy).

Study tip: Spaced recall practice (quizzes daily) are much more effective than massed practice (cramming for an exam).

Associations. Each item retrieved in long-term memory may be linked or associated with other items. For example, the sound of a foreign word is associated with its meaning or with its sound in the native language of the speaker. As a different example, a particular symptom observed in an abnormal system failure will, in the mind of the skilled troubleshooter, be associated with other symptoms caused by the same failure as well as with memory of the appropriate procedures to follow given the failure. Associations

Password dilemma
An interesting struggle that illustrates the many tradeoffs that lurk behind the scenes of human factors: Most of the factors that make alphanumeric strings more memorable are the very same that make passwords less secure [277].

Considering the strengths and limits of long-term memory suggests another approach: passphrases. Passphrases are more memorable than passwords because a phrase or sentence ties into a web of associations that a single word does not. Passphrases are also more secure because they are more difficult to guess due to the many more potential combinations.

The added typing is an important tradeoff with a passphrase.

https://www.smashingmagazine.com/2015/12/passphrases-more-user-friendly-passwords/

Table 6.1 Passwords and Pass phrases

between items have a strength of their own, just as individual items do. As time passes, if associations are not repeated, they become weaker. For example, at some later point a worker might recognize a piece of equipment but be unable to remember its name.

Working Memory and Long-term Memory Interaction. Ease of retrieval depends on the richness and number of associations that can be made with other items. Like strings tied to an underwater object, the more strings there are, the greater likelihood that any one (or several) can be found and pulled to retrieve the object. Thus, thinking about the material you learn in class in many different contexts, with different illustrative examples, improves your ability to later remember that material. Doing the mental work to form meaningful associations between items describes the active role of working memory in learning [262]. As we noted in the discussion of working memory, storing such relations in long-term memory forms chunks, which reduce the load on working memory. Sometimes, however, when rehearsing items through simple repetition (i.e., the pure phonetic loop) rather than actively seeking meaning through associations, our memories may be based solely on frequency and recency, which is essentially rote memory. Rote memory is more rapidly forgotten. This is a second reason that advertisers have moved from solely digit-based phone numbers to items such as 1-800-GET-RICH. Such phone numbers have both fewer items (chunks) and more associative meaning.

Forgetting. The decay of item strength and association strength takes the form of an exponential curve, where people experience a very rapid decline in memory within the first few days. This is why evaluating the effects of training immediately after an instructional unit is finished does not accurately indicate the degree of one's eventual memory. Even when material is rehearsed to avoid forgetting, if there are many associations that must be acquired within a short period of time (massed practice), they can interfere with each other or become confused, particularly if the associations pertain to similar material. New trainees may well recall the equipment they have seen and the names they have learned, but they confuse which piece of equipment is called which name as the newer associations interfere with the older ones.

Thus, memory retrieval often fails because of (1) weak strength due to low frequency or recency, (2) weak or few associations with other information, and (3) interfering associations. To increase the likelihood that information will be remembered at a later time, it should be processed in working memory frequently and in conjunction with other information in a meaningful way.

A particularly powerful way to reinforce a memory is to recall it just before it is forgotten. Known as recall learning, it requires substantial effort to bring a nearly forgotten fact to mind and this effort is essential to forging strong memory for an item. Re-reading a textbook multiple times relies on rote memory, whereas quizzes and flashcards that cover the same engage recall learning and are much more effective.

Different forms of long-term memory retrieval degrade at different rates. In particular, recall, in which one must retrieve the required item (fact, name, or appropriate action), is lost faster than recognition, in which a perceptual cue is provided in the environment, which triggers an association with the required item to be retrieved. For example, a multiple-choice test visually presents the correct item, which must be recognized and discriminated from a set of "foils." In contrast, short-answer questions require recall. In human-computer interaction, discussed in Chapter 15, command languages require recall of the appropriate commands to make something happen. In contrast, menus allow visual recognition of the appropriate command to be clicked, which make them easier to remember.

6.5.2 Effect of Repetition: Habits

The diagram showing the information processing model (Figure 6.2) includes a feedback loop. This feedback loop indicates that tasks are not performed in isolation, but are often part of a repeating pattern of activity. The discussion of memory and automaticity indicate that repeating a task many times changes it so that it comes to mind easily and can be performed without effort. Such tasks may become *habits*, which are cued by reoccurring context, involve little effort, and are not tied to a goal or intention [278, 279]. Habits guide much of our daily routine, from tooth brushing and checking email, to driving to work.

Habits form with repetition of an activity that occurs in a consistent context. For example, you likely fell into the habit of brushing your teeth in part because you have done it many times in response to the fuzzy feeling in your mouth, at the same time of day, in the same location, as part of the same larger sequence of tasks [280]. This context provides a trigger for the habit. Beyond repetition, rewards also reinforce habits. Toothpaste includes a mild irritant that gives you that tingling fresh feeling after brushing. This feeling acts as a reward that reinforces the effect of repetition. In the context of gambling, this reward is particularly powerful and can lead to pathological gambling habits. Once established, habits occur effortlessly and actually require cognitive resources and effort to avoid. This makes bad habits hard to break.

Developing habits requires repetition over time, consistent context, and benefits from an embedded reward. The cues focus on the feeling of fuzzy teeth that indicates the need to brush your teeth, but habits depend on the broader context that includes time of day and physical location, and even companions. With inconsistent context, habits develop more slowly. Disrupting the context, such as when you change jobs or go on vacation, can disrupt habits. Likewise, periods of transition offer opportunities to insert new habits: people who have just moved into a city are more likely to make public transit a habit than those who have lived in the city for many years [278].

Habits take time to develop. By one estimate the average time to develop a habit was 66 days, but this varied by person and seems to be longer for more complex activities [71]. Rewards reinforce the tendency to repeat an activity, such as the periodic reward of receiving an email from a good friend that reinforces the habit of checking email. Rewards are most potent when they occur soon after the behavior occurs. Such rewards can lead to a craving for the activity and can quickly establish a habit. Some of the most successful ways to break bad habits are to avoid the context that triggers the habit or to substitute a different routine for the one to be avoided. Expecting to resist a bad habit through willpower tends to fail. Because habits govern so much of our daily life, they merit careful attention in design.

6.5.3 Organization of Information in Long-Term Memory

It is apparent from the description of working memory and long-term memory that we do not put isolated pieces of information in long-term memory the way we would put papers in a filing cabinet. Instead, we store items in connection with related information. The information in long-term memory is stored in associative networks where each piece of information (or image or sound) is associated with other related information. Much of our knowledge that we use for daily activities is semantic knowledge, that is, the basic meaning of things. Long-term memory is organized in four ways: Semantic networks, schemas, mental models, and cognitive maps.

Semantic Network. Our knowledge seems to be organized into semantic networks where sections of the network contain related pieces of information [281]. Thus, you probably have a section of your semantic network that relates your knowledge about college professors, both general information and specific instances, based on previous experience. These semantic networks are then linked to other associated information, such as images, sounds, and so on. A semantic network has many features in common with the network structure that may underlie a database or file structure, such as that used in an index, maintenance manual, or computer menu structure. It is important that the designer create the structure of the database to be compatible or congruent with the organization of the user's semantic network [282, 283]. In this way, items that are close together, sharing the same node in the semantic network, will be close together in the database representation of the information. For example, if the user of a human factors database represents perception and displays as closely associated, the database should also contain links between these two concepts. We see in Chapter 8 how this process can be aided by good displays.

Constructing a semantic network, or concept map, can also be an effective way of learning new material and enhancing long-term memory. Writing out concepts and connecting the related concepts with lines to form a large network helps to identify and reinforce associations in the material to be learned. Concept maps enhance retention of information at all grade levels and across a wide variety of course types [284]. Students who create concept maps increase their performance by one half a standard deviation, approximately the difference between a B+ and an A.

Schemas and Scripts. The information we have in long-term memory is sometimes organized around central concepts or topics. The knowledge structure about a particular topic is often termed a schema. People have schemas about many aspects of their world, including equipment, activities, and systems that they use. Examples of the wide variety of schemas include: college courses, kitchens, and vacations. Schemas that describe a typical sequence of activities, are called scripts like logging into a computer system, shutting down a piece of industrial equipment, or dealing with a crisis at work [285]. Schemas and scripts are important for design because they help people develop appropriate expectations and process information efficiently. The use of scenarios, user journeys, and design patterns, described in Chapter 2, will be most effective if they are consistent with existing schema and scripts.

> ⚡Like the episodic component of working memory, schemas are another reason why people are naturally inclined to remember information presented as a story.

Mental Models. People also have schemas about equipment or systems. The schemas of dynamic systems are often called mental models [286, 287, 288]. Mental models typically include our understanding of system components, how the system works, and how to use it. In particular, mental models generate a set of expectancies about how the equipment or system will behave. Mental models may vary on their degree of completeness and correctness. For example, a correct mental model of aerodynamics posits that an aircraft stays aloft because of the vacuum created over the wings. An incorrect model assumes that it stays aloft because of its speed.

Mental models may also differ in terms of whether they are personal (possessed by a single individual) or are similar across large groups of people. In the latter case the mental model defines a population stereotype [289]. Designs that are consistent with the population stereotype are said to be compatible with the stereotype (such as turning a knob clockwise should move a radio dial to the right). Later chapters on displays (Chapter 8), controls (Chapter 9), and human-computer interaction (Chapter 10) show the importance of knowing the user's mental model so that the interface can be designed to avoid a mismatch. Mental models develop expectations of how a system will respond to user inputs. Violating those expectations invites errors.

Cognitive Maps. Mental representations of spatial information, like the layout of a city, a room, or a workplace, are referred to as cognitive maps. They represent the long-term memory analogy to the visual-spatial scratchpad in working memory. Such maps may not necessarily be accurate renderings of the space they represent [10]. For example, cognitive maps of a geographical area often simplify by "mentally straightening" corners that are not at right angles [290]. People also have a preferred or "canonical" orientation by which they typically represent an environment [291]. This may often represent the direction in which you most frequently view the environment. For example, your cognitive map of a classroom may have the orientation of the direction you face when you sit in it. Reorienting one's perspective of a cognitive map through "mental rotation" requires mental effort because you must maintain the information in the visual spatial scratchpad of working memory [292]. As we discuss Chapter 8, this has implications for maps design.

6.5.4   Prospective Memory for Future Actions

Whereas failures of episodic memory are inaccurate recollection of things that happened in the past, failures of *prospective memory* are forgetting to do something in the future [293, 294]. Laura, in the story at the beginning of the chapter, forgot to activate the voice mode while the traffic was still light. In 1991, an air traffic controller positioned a commuter aircraft at the end of a runway and later forgot to move the aircraft to a different location. The unfortunate aircraft was still positioned there as a large transport aircraft was cleared to land on the same runway. Several lives were lost in the resulting collision. Failures of prospective memory are sometimes called absentmindedness, but actually reflect poor design.

Several system and task design strategies can support prospective memory, such as *reminders* [295]. Reminders include tying a string around your finger, setting a clock or programming a smartphone to sound an alarm, taping a note to the steering wheel of your car, or putting a package you need to mail in front of the door so that you will be sure to notice it (if not trip on it!) on your way out. In systems with multiple operators, sharing the knowledge of what one or the other is to do decreases the likelihood that both will forget that it is to be done. Also, verbally stating or physically taking some action (e.g., writing down or typing in) improves prospective memory. Checklists are also particularly powerful aids for prospective memory [296], showing particular benefits for healthcare safety in such contexts as surgical procedures and patient handovers [297, 298, 299].

6.5.5   Implications of Long-Term Memory for Design

Designers frequently fail to realize or predict the difficulty people will experience in using their system. One reason is that they are extremely familiar with the system and have a very detailed and complete mental model [16]. They know how the system works, when it will do various things, and how to control the system. They fail to realize that the average user does not have this mental model and may never interact with the system enough to develop one. When people have to do even simple tasks on an infrequent basis, they forget things. Manufacturers often write owners' manuals as if they will be read thoroughly and all of the information will be remembered for the life of the equipment. Neither is likely. Even with very clear instructions for using features of our new car, the typical driver is unlikely to read the owners' manual thoroughly.

The following are some ways that we can design the environment and systems within it so that people do not suffer inconveniences, errors, accidents due to poor retrieval from long-term memory.

1. **Encourage regular use** of information to increase frequency and recency.

> Like failures of working memory, problems with retrieval from long-term memory can undermine safety, performance, and satisfaction.

2. **Encourage active reproduction** or verbalization of information that is to be recalled. For example, taking notes in class or reading back verbal instructions increases the likelihood that the information will be remembered.

3. **Standardize**. One way that we can decrease the load on long-term memory is to standardize environments and equipment, including controls, displays, symbols, and operating procedures. The automotive industry has standardized the shift pattern, but not the location and operation of electronic windows and lighting. Standardization enables people to develop strong schemas and mental models that are applicable to a wide variety of circumstances. Of course, the standardizing across industries faces considerable challenges, from preserving uniqueness of product style to leaving room for innovation. In some cases, like aviation, the federal government can impose standardization through regulations. But in others, like the automobile dashboard, or mobile phone interfaces, it cannot.

4. **Use memory aids.** When a task will be performed infrequently or when correct task performance is critical, designers should provide computer-based or hardcopy memory aids or job aids, as discussed in Chapter 18. These consist of information critical for task performance and can be as simple as a list of procedures. Norman [16] characterizes memory aids as putting "knowledge in the world" (i.e., perception) so that the operator does not have to rely on "knowledge in the head" (i.e., long-term memory). In the context of command languages and menus, such aids often replace recall requirements with recognition opportunities. This important human factors topic is reconsidered in the discussion of human-computer interaction (Chapter 10).

5. **Design information to be remembered**. Information that must be remembered and later retrieved unaided should have the following characteristics:

   - Meaningful to the individual and semantically associated with other information.
   - Concrete rather than abstract words.
   - Distinctive concepts and information (to reduce interference).
   - Well-organized sets of information (grouped or otherwise associated).
   - Able to be guessed based on other information (top-down processing).
   - Little technical jargon.
   - Presented as a story that evokes vivid imagery.

6. **Design helpful habits.** Consider context, repetition, and reward to reinforce desired behaviors, and discourage bad habits.

7. **Support correct mental models.** One way to develop correct mental models is to apply the concept of visibility [300], and transparency [301]. A device has visibility if the user can immediately and easily determine the state of the device and the alternatives for action. For example, switches that have different positions when activated have visibility, whereas push/toggle switches do not. The concept of visibility also relates to the ability of a system to show variables intervening between an operator's action and the ultimate system response. An example is an oven display showing that an input has been read, the heat system is warming up, and when temperature will reach the target temperature. Visibility and transparency are topics of great relevance to human-automation interaction, and so will be discussed in more detail in Chapter 11.

6.6 Divided Attention and Time-Sharing

At the start of this chapter, we drew the distinction between attention as a filter and as a fuel. What these two metaphors have in common is that they both describe the limits of processing multiple entities. At selection, the filter forces us to selectively choose between multiple channels or external sources of information, as Laura choosing to focus attention on the navigational device, rather than the roadway. At the higher, or later stages of information processing, the fuel limits force us to selectively allocate resources between tasks, hence describing the limits of multi-tasking. Such tasks vary across a wide range of difficulty. For example some task pairs can be easily multi-tasked or time shared, such as driving while listening to music or walking while talking. Here we say there is no dual task decrement. Other pairs are harder and interfere more with each other, such as concentrating on a math problem while listening to a speaker, or driving while texting. Here the dual task decrement is large. In the following, we describe six factors that affect human multi-tasking capabilities. The first five of these are characteristic of the tasks, or task pairs themselves, and the sixth refers to characteristics of the individual.

6.6.1 Task difficulty and mental workload

Easy tasks require few mental resources. Hence, even if the brain has a limit on its resources, there are plenty of residual resources or spare capacity for other tasks, and hence these can be timeshared with little or no decrement. The lack of spare capacity and task complexity account for the mental workload imposed by a task, and this can be physically measured by electrophysilogocal, or other techniques, described in Chapter 13 (Stress). Such techniques often provide reliable measures of how hard the brain is working, as defined by cerebral blood flow [302].

High mental workload indicates reduced abilty to compensate for additional demands even if task performance is unaffected.

Habits and automatized tasks demand few resources and impose minimal workload [303, 304]. Such automaticity can be achieved through extensive training and practice, or in picking up a habit. For example, think of the difference in attention demand of speaking in your native language versus speaking in a second language you are just learning.

Cognitive complexity also affects attention demands, and like mental workload, can be objectively measured via task analysis. For example consider the difference in cognitive complexity between long division and multiplication; or between subtraction and addition [305]. Can you determine why the first arithmetic operation, in each pair, is more complex than the second? The same approach applies to a wide range of work environments and products, from laproscopic surgery to cooking [306, 307].

Finally, attention demands or mental workload can be determined by the working memory demands of the task, as we described earlier in the chapter. A computer task requiring you to retain five chunks of information concurrently while you access a new screen, has a higher mental workload than one that requires retaining only three chunks or, better yet, allows you to view the two screens side-by-side.

It is important to realize that for all three factors affecting attentional demand, the more automatized task may not be performed better as a single task. For example we can remember five chunks as well as three chunks (perfectly) if we are doing nothing else at the same time, but it will take more mental workload. So the mental workload of a task is an attribute besides performance that is of great importance to measure or predict. Tasks with high mental workload will undermine satisfaction. In all of these instances, lower mental workload of a component task, avails more spare capacity for other tasks, and hence a reduced dual-task decrement of one or both, thereby increasing multi-task efficiency.

6.6.2 Task resource structure

We can listen to a mobile phone while driving much better than texting while driving. Why? Driving is primarily a visual task and so is texting. But listening is auditory. Visual and auditory channels use separate resources, both in the senses (eyes versus ears) and in the brain itself (auditory versus visual cortex). The *theory of multiple resources* [308, 10] establishes that tasks using different resources interfere less with each other than tasks using the same resources. This is much like the economy in which financial resources used for social security cannot be shared with or re-allocated to other government agencies. Of course there are limits on this clear separation of resources. Both auditory and visual material involve higher levels of perception that are not distinct between the two modalities. Hence listening and reading compete for language-related resources, and there will be some interference between the two tasks, even if it is less than between simultaneous reading two texts or listening to two voices.

Thus far we have defined resources by the dimension of modality: auditory and visual (and tactile is sometimes included as a third modality [10]. There are three other dimensions along which resources can be defined, as shown in Figure 6.5.

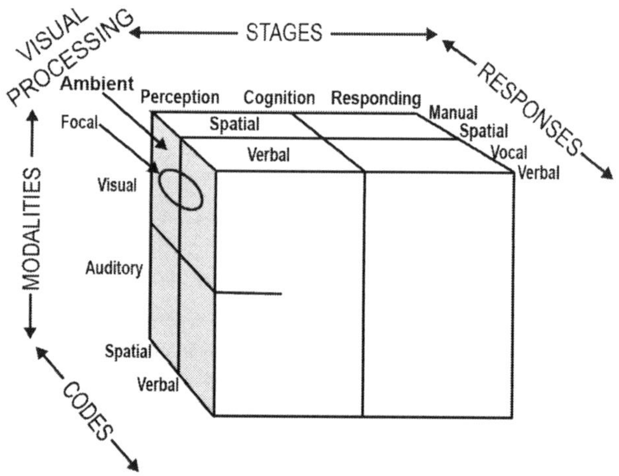

Figure 6.5 The multiple resource theory model of time-sharing and workload. Performing two tasks at the same time that share the same modes, codes, or stages is difficult and error prone (Adapted from Wickens Hollands et al, [10].)

The second dimension is spatial versus verbal (or linguistic) codes of processing. Language processing, in both reading and listening, as well as talking and keyboarding, and, in working memory, the phonetic loop, uses somewhat different resources from spatial processing (e.g., perceiving motion, using the visual-spatial scratchpad in working memory, or manually controlling a mouse). Because of the code separation, driving, a visual-spatial manual task, can be time-shared with conversation, a verbal task, with little dual task decrement. Like modalities, codes are also represented somewhat in different brain structures, here the right (spatial) and left (verbal) cerebral hemispheres.

The third dimension is stages of processing, as seen in Figure 6.5. Perception and cognition use somewhat different resources from action and responding. Here again this distinction partially corresponds to a brain distinction between posterior (back) cortex (perception) and anterior (forward) cortex (action). Thus perceiving while thinking is more difficult than perceiving while responding. Two simultaneous (and independent) responses are harder to execute than responding while perceiving. It is harder to doodle while taking notes in class, than it is to doodle while listening to a lecture.

The fourth dimension of these multiple resources lies within the visual system: Object recognition, such as reading print or interpreting symbols uses different resources (focal vision) from motion and spatial orientation perception, as in standing upright, or in driving straight (ambient vision), a distinction that was made in Chapter 4. Hence, without much difficulty or dual task decrement, a postal worker can read the address on the envelope while walking the street. The distinction of levels of the visual system dimension can be associated with different neurophysiological structures

Dimension	Levels	Examples
Modalities	Auditory	Synthesized voice display, spatially localized tones
	Visual	Print, electronic map
	Tactile	Tactile: vibration to the body signaling an alert
Codes	Spatial	Tracking, hand pointing, mental rotation, imaging
	Verbal	(visuospatial scratchpad)
		Listening to speech, rehearsing (phonetic loop)
Stages	Perceptual	Searching, imaging, reading
	Working Memory	Rehearsing, listening
	Response	Pushing, speaking, pointing, manipulating
Visual Channels	Focal	Reading, interpreting symbols
	Ambient	Processing flow fields, visual perception to maintain balance

Table 6.2 Multiple resource theory dimensions,levels and examples.

within the brain [309]. Figure 6.5 shows four dimensions of the multiple resource model as a cube.

Table 6.2 also presents examples of different levels along the four dimensions. The design implications of the model are straightforward in predicting task interference. To the extent that two tasks share more common levels on more dimensions of the model, they will be more likely to interfere with each other; their dual task decrement will be larger; and, of course this interference will also be greater with the amount of demand for common resources, the mental workload of each task. But if they interfere, will they both share the same decrement? This is determined by the third factor of task interference: resource allocation policy.

Given that two tasks compete for common resources, which task suffers the greater decrement is determined by the *resource allocation strategy* or policy that guides a person's decision about what tasks to invest in, and which to sacrifice. This is analogous to the limited federal budget, policy will determine which Department may get spared, and which suffers the brunt of the budget cuts [310]. In safety critical systems, this resource allocation policy is of tremendous importance. We know that texting and driving compete for a lot of shared visual resources. But texting would not be a hazard if the driver always chose to allocate nearly all visual resources to driving. However we know that this is not always done, and tragic accidents result.

6.6.3 Confusion

We noted that the similarity between items in working memory leads to confusion. We also presented a corresponding argument regarding similarity-based confusion in our discussion of visual sensation in Chapter 4. In the context of multi-tasking, we find that concurrent performance of two tasks that both have similar material increases task interference [311, 10, 254]. For example, monitoring basketball scores while doing mental arithmetic will probably lead to disruption as digits from one task become confused with digits relevant to the other. Correspondingly, listening to a voice navigational display of turn-by-turn directions instructing a left turn, while the automobile passenger says, "right... that's what I thought," could lead to the unfortunate wrong turn. Auditory background information, because of its intrusiveness, may be particularly likely to cause confusion even if it is not part of an ongoing task [174, 312]. Note the relevance of confusion to the task resource structure. Greater similarity of material within a shared resource amplifies dual-task interference.

6.6.4   Task switching

The factor of task switching emerges from the first two described above. Sometimes the joint competition for resources between the two tasks is simply so great as to make concurrent processing impossible. In such cases, we say that workload "exceeds the red line". Under such circumstances, the multi-tasker must choose to continue one task and abandon the other altogether, thus entering a sequential mode of multi-tasking, where the multitasker must now switch between tasks. Discrete task switching then leads to two sequential multi-tasking phenomena: *voluntary switching* [313] and *interruption management* [314, 293].

Voluntary Task Switching. The allocation policy of concurrent processing was manifest as a graded degree of emphasis. But when concurrence is impossible and we must choose, task switching is the discrete all-or-none analog of an allocation policy. Like the eyeball "deciding" which of several areas of interest to look at (Chapter 4), here the "mindball" needs to decide which of several tasks to do, abandon, or resume. Being able to predict task switching is important because it can predict or account for the phenomenon of task neglect, or *cognitive tunneling*. Cognitive tunneling occurs when one task grabs a user's attention for far longer than others, often with serious implications for safety for the neglected task. In aviation, pilots once overflew a Minneapolis destination, as they became engrossed in their laptop applications; hence neglecting the task of navigation. More tragically, in 1979 a commercial airline crashed into the Everglades, with 99 lives lost, when pilots became preoccupied in fixing a burnout landing gear indicator light, ignoring their control of altitude and monitoring of automation [222].

Task switching is the extreme dual-task attention allocation policy.

Voluntary task switching depends on five factors that determine the "attractiveness" of a task, to be switched to (if it is not currently being performed), or "stayed on" if it is being performed. Attractiveness depends on: salience, priority, interest, difficulty, and time-on-task [313, 315]:

   1. **Salience:** A salient task is one that calls attention to its pres-

ence. For example the auditory ring of the phone is more salient than the visual appearance of a message, and the latter is more salient if it flashes, than if it slowly illuminates. But a visual event, in turn is more salient than a task that needs to be remembered to be performed, using prospective memory as discussed earlier. We can say that such a task has "zero salience".

2. **Priority:** A high priority task is of course one that, if not done, or not done on time imposes considerable costs. In aviation, the pilot must keep the plane aloft ("aviating task" with highest priority) before addressing the "communications task" with lower priority [316], and if the two tasks compete for attention, she should always choose the former over the latter.

3. **Interest:** Interest or "engagement" is self evident [317]. But it is interest in a lower priority mobile phone conversation or a text message that is sometimes allowed to dominate the higher priority task of attending to the roadway when driving. In class, the student may allow the interest in a text message to dominate the higher priority task of listening to the lecture; particularly if the latter is boring.

4. **Difficulty or mental workload of a task:** Data suggest that, in times of high workload overload above the redline, when people choose, they tend to choose easier rather than more difficult tasks [313].

5. **Time on task:** The changing attractiveness of staying with a task the longer it has been performed without a break, seems to depend on a number of factors [318, 319]. Clearly for boring tasks (lacking interest) or highly fatiguing ones, the longer it has been performed, the greater tendency to switch away. But for tasks that demand working memory, and accumulate information as task time goes on, there will be an increasing switch resistance with time on task, so that the information that has been compiled, or the mental computations made will not be lost by a switch away [320].

Together, these five factors combine to influence which a task will be chosen, when many compete for attention, just as the factors that drive scanning, discussed in Chapter 4, establish where the eyes look when. However the five factors seem to have different weights. In particular research suggests that task priority does NOT exert a heavy influence on task switching relative to the other attributes, even though it should. Its weak pull on the brain is often dominated by the attributes of interest and engagement.

Several design implications of departures from optimal task management can be identified. Salient task reminders can be provided for tasks that may be likely to be neglected, or are of high importance; just as reminders can support prospective memory [293]. To the extent that there are higher priority visual tasks that may be neglected, the displays for such tasks can be positioned closer to the normal line of sight in the workplace, keeping them more salient, and involving less effort to include them in an easy scan pattern. Also training can be effective for multi-task management [321], an approach sometimes adopted by the airlines, as a component of crew resource management (CRM) to be discussed further in chapter 19.

Interruption Management: The more general task management issues involved in multi-task switching can be distilled to a simpler environment of just two tasks, an *ongoing task* and an *interrupting task*. The concern is often on how people manage interruptions of an ongoing task by an interrupting task [314]. The consequence of interruptions can range from simple annoyance [322], to major disruptions in the work environment [323]. When driving, it is easy to imagine the consequences of an incoming mobile phone call interrupting your train of thought as you approach a busy intersection.

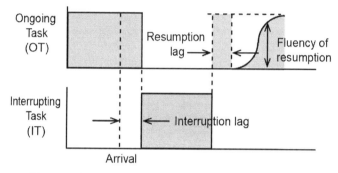

Figure 6.6 Representation of interruption management.

Figure 6.6 shows the interruption of an Ongoing Task (OT), which is the initial focus of attention, when an interrupting task (IT) arrives. The person eventually shifts attention to the IT, but the delay in doing so, known as the interruption lag, is an important measure. After dealing with the interruption for some period of time, the worker then resumes the OT, with a measure that can be described as the *fluency of return*. This fluency can be measured both by how long it takes to resume the OT, the resumption lag, and more importantly, the quality of performance after the return. Does one pick up the OT where it was left off? Or perhaps, does one have to "start from scratch"? Indeed, a worst-case scenario may be when a procedural OT is resumed at a point after where it was interrupted. Consider a safety critical checklist in the cockpit, or in medicine where interruptions occur frequently [323]. If the return point skips a safety critical step, disaster can result. This was the case when a pilot failed to extend the wing flaps, leading the plane to crash on a take off from Detroit Airport in 1987, killing 148 of the 149 on board [324].

Several factors affect the interruption lag:

- The interruption lag depends on the salience of the interruption. This is why warning and alerting systems are designed to capture attention. In aviation, the more salient auditory alert is used for more critical events than the less salient visual one.

- A more engaging OT will produce a longer interruption lag (if, indeed the IT is not ignored altogether).

- The interruption lag is likely to be shorter if the IT arrives during a "lull" or stopping point in the OT, such as after a paragraph has just been completed in an OT of reading [325].

The fluency of return depends on several factors including:

- The length and attention resource demands of the IT cause memory of the goals and material used in the OT to decay, and hence require a greater time to reinstate upon return [326, 325, 314].

- The timing of when OT was abandoned. If the OT was abandoned at a lull or stopping place (as we saw would shorten the Interruption lag), then the return is more rapid and fluent. One can often delay the attention switch until such a stopping point has been reached ("let me just finish the paragraph").

- The working memory demands of the OT. If the OT is using several items in working memory when the IT occurs (and is switched to), as when an interruption arrives just after 8 digits of a 10 digit spoken phone number have been encoded, a few of the first digits will be forgotten on resumption, and the number must be retrieved in full; or the number will be dialed incorrectly. Either way the fluency of return is degraded.

Designers can do several things to mitigate the unpleasant consequences of interruptions.

- Systems can be designed to postpone an interruption until a stopping place is inferred by an algorithm that identifies good times for notification delivery [327] or until workload driven by the OT is inferred to be low [328]. This possibility will be addressed further in chapters on human-computer interaction (Chapter 10) and human-automation (Chapter 11).

- Signal the importance of the interruption in a way that can be rapidly and pre-attentively processed by people, so that attention is not switched (or the interruption lag is delayed) for those interruptions of lesser importance [329]. Certainly smart phones can be coded to do this, by creating an increased salience from the e-mail, to the text, to the phone call, assuming that convention usually dictates that this sequence increases in importance.

- Teach strategies for responding to the arrival of the IT are effective, such as delay until a lull in OT demands, rehearsal, or place keeping. All of these promote a more fluent response to interruptions. [293, 321].

- Several strategies at the IT point can preserve the fluency of return. These may be mental, as when the OT is rehearsed, to boost the OT strength when it is returned to; they may also be physical, by setting a placeholder on the OT: a mark on the page of text or the paper checklist, or a cursor on the last line read; or they may be procedural, such as delaying the interruption lag until a stopping point has been reached [330].

As human-computer interaction moves away from traditional menu-driving systems to conversational agents interruptions become a critical design consideration. The value of conversational agents may hinge on their ability to delay interactions until the person is at a stopping place and to recognize the relative priority of the ongoing and interrupting task.

6.6.5 Training and Individual Differences

We have already suggested ways in which individual differences in training and skill can improve multi-tasking. For example extensive practice can produce automaticity, and hence produce zero resource demands of a component task (although some subtasks benefit more than others; See Chapter 18). Strategies for resource allocation and for task management can also be taught [331]. Closely related, returning to the filter of attention, there is good evidence that experts scan differently from novices in environments such as the cockpit [231], operating room [332], or roadway [333]. Such expert scan strategies can, to some extent, be trained [128].

Less firmly established, are the emerging findings that there are stable individual differences in multi-tasking abilities [334]. For example, some people appear to be more natural concurrent processors, while others adopt a more sequential switching process [335, 336]. Those with greater working memory capacity may be more effective at the executive control, desirable for adapting more effective task switching strategies [313]. If this is the case, then testing and selection for high-tempo multi-tasking jobs, such as aircraft piloting, may well be feasible.

6.7 Summary

In this chapter we discussed mental processes that lie at the core of much information processing in complex environments. The characteristics of human cognition presented here have many detailed implications for design, which we discuss in the following chapters. A few broad design considerations include the severe limits of attention and working memory: people can hold only three or four things in memory, and if they try to do two activities at once, one or both will suffer. Chunking, unitization, and automaticity mitigate these limits. The physical environment and technological aids (e.g., checklists, reminders, and designed habits) also make us smarter than our limited attention and working memory would suggest. Knowledge-in-the-world acts as a support structure to our short and long-term memory and appropriately salient alerts help us manage interruptions.

Many other chapters in this book relates this description to design. Our discussion of perception links to Chapters 4 and 5 on visual and auditory sensation, as well as to Chapter 8 on displays, where we consider how design can support perception. Our discussions of attention relate to topics in both Chapters 4 and 8 as well as to those of workload overload in Chapter 13. Cognition of all sorts is involved in computer usage (Chapter 10) and in dealing with automation and complex systems (Chapter 11). Cognition relies on long-term memory and knowledge, and knowledge is acquired through learning and training (Chapter 18). Finally, many aspects of cognition of perception and working memory are involved in the all-important task of decision making, the topic to which we turn in the next chapter.

Additional resources

Several useful resources that expand on the content touched on in this chapter include books that present design-relevant aspects of from cognitive psychology in an accessible manner:

1. **The science of learning:** This book integrates the most useful findings on learning and teaching from the last 100 years of research.

   Brown, P. C., Roediger, H. L., & McDaniel, M. A. (2014). *Make it Stick*. Harvard University Press.

2. **Tools to improve your memory:** A description of a how an average person became a world-class memory athlete over the course of a year.

   Foer, J. (2011). *Moonwalking with Einstein: Art and science of remembering everything*. New York: Penguin Books.

3. **The cognitive science behind design guidelines:** These books provide a more complete discussion of cognitive psychology applied to design.

   Johnson, J. (2013). *Designing with the Mind in Mind: Simple guide to understanding user interface design guidelines*. Elsevier.

   Weinschenk, S. (2011). *100 Things Every Designer Needs to Know about People*. Pearson Education.

Questions

Questions for 6.1 Cognitive Environment

P6.1   Describe the cognitive environment in terms of the three dimensions outlined at the start of the chapter.

Questions for 6.2 Information Processing Model of Cognition

P6.2   Based on the factors guiding attention, give two reasons why alarms are helpful in guiding people to notice critical events.

P6.3   Based on the concept of unitization, should you use all capitals or mixed case lettering in your resume?

P6.4   What is a requirement to develop unitization?

P6.5   Which of the following formats would most likely support the best recall: 1) 2F55T 2) 255 FT 3) 255FT 4) 2F 55T

P6.6   What is the role of selective and divided attention in guiding how people perceive and act on information in a multi-tasking environment, such as an emergency room?

Questions for 6.3 Selective Attention and Perception

P6.7   What are two bottom-up factors that guide selective attention?

P6.8   What are two top-down factors that guide selective attention?

P6.9   How would you change the design of a stop sign to make it more likely that drivers attend to it?

P6.10   From the perspective of selective attention, why would using a head up display be an ineffective way of reducing distraction associated with reading text messages while driving?

P6.11   Describe how the four factors that guide attention (salience, effort, expectancy, and value) might affect a driver's likelihood of attending to a pedestrian crossing.

Questions for 6.4 Working Memory

P6.12   What is the limit of working memory in terms of the number of chunks?

P6.13   Why can you remember a friend's birthday that consists of 8 digits despite the working memory limit of four chunks?

P6.14   In what way are telephone numbers designed to be consistent with the properties of working memory?

P6.15   How should the extremely volatile nature of short-term memory be considered in the design of a travel planning system, such as Orbitz? Give three specific design suggestions based on the implications of working memory limits.

Questions for 6.5 Long-Term Memory

P6.16   How does top-down processing affect what you perceive and recall?

P6.17   Top-down processing can lead to predictable errors. Describe the potential errors that might occur as a result of the following protocol for radio communication in a noisy stadium environment.

P6.18   How should the effects of top-down processing be considered in presenting negation in sentences?

P6.19   A friend asks you for advice on studying for a mid-term exam. Provide suggestions using the concepts of recall learning, spaced practice, and concept maps.

P6.20   How should characteristics of long-term memory guide the way you study for this exam?

P6.21   What are the characteristics of habitual behavior?

P6.22   What conditions lead to habit formation?

P6.23   If you have been charged to get more people to wear seat belts, describe how you would use the factors underlying habits to promote a lasting change in drivers' use of seatbelts.

Questions for 6.6 Divided Attention and Time-Sharing

P6.24   Based on the concepts of Multiple Resource Theory how would you design an entertainment and information system for a car?

Chapter 7

Decision Making and Macrocognition

At the end of this chapter you will be able to...

1. understand the elements of the cognitive environment that makes considering macrocognition important.

2. understand the difference between skill-, rule-, and knowledge-based behavior and the implications for design.

3. describe the reasons for decision-making heuristics and the associated biases.

4. demonstrate how task re-design, choice architecture, decision support systems, and training can influence behavior.

5. discuss the ethical considerations associated with designing for decisions.

6. guide design to support the elements of macrocognition: Situation awareness, decision making, planning, troubleshooting, and metacognition.

Amy, a relatively new internal medicine specialist treated a patient who exhibited a set of symptoms typical of a fairly common condition: rash, reported localized mild pain, headache, 102°F temperature, and chills. A localized skin discoloration near the rash was not considered exceptional or unusual ("just a bruise from a bump"), and a quick glance at the chart of the patient's history revealed nothing exceptional. Amy, already behind on her appointments, quickly and confidently decided, "that's flambitis" (a condition that was the subject of a recent invited medical seminar at the hospital), prescribed the standard antibiotics and dismissed the patient.

A day later the patient phoned the nurse to complain that the symptoms had not disappeared, but Amy, reading the message, instructed the nurse to call back and say that it would take some time for the medicine to take effect, and not to worry. Yet another 24 hours later, the patient appeared at the ER, with a temperature now of 104°F, and more intense pain. Amy was called in and a careful inspection revealed that the slight discoloration had darkened, and a prior condition in the medical chart had been overlooked in Amy's quick scan. These two newly appreciated symptoms or cues suggested that flambitis was not the cause, and led Amy to do a rapid, but intense and thoughtful, search of the medical literature to obtain reasonable evidence that the condition was a much less prevalent one called stabulitus. This was consistent with an earlier report in the patient's medical record that Amy, in her quick glance, had overlooked. Further research suggested a very different medication. After making that prescription, Amy now started monitoring the patient very closely and frequently, until she observed that, indeed, the symptoms were now diminished. Following this close call of a misdiagnosis and the resulting poor decision on treatment, the first serious decision error since her licensing, Amy vowed to double check her immediate instincts, no matter how much the symptoms looked like a common condition, to more thoroughly check the medical history, and to follow up on the patient's condition after the initial treatment.

Although this scenario happened to occur in the medical domain, each day people make many decisions in situations that range from piloting an aircraft and voting for a candidate to financial planning and shopping. Some of these decisions have life and death implications and other times a poor choice is just a minor annoyance. Generally, these decisions depend on understanding the situation by integrating multiple sources of information, determining what the information represents, and selecting the best course of action. This course of action might be simply dropping an item into your shopping cart or it might require a plan that coordinates other activities and people.

This chapter builds on the previous chapter's description of cognition. The elemental information processing stages of selective attention, perception, working memory, long-term memory, and mental workload all contribute to decision making. These concepts form the building blocks of cognition and can be thought of as elements of *microcognition*. In contrast, this chapter describes decision making in the context of *macrocognition,* or the high-level mental processes that build on the stages of information processing, which include situation awareness, decision making, problem solving, and metacognition. Macrocognition is defined by high-level processes that help people negotiate complex situations that are characterized by ambiguous goals, interactions over time, coordination with multiple people, and imperfect feedback [337]. Figure 7.1 highlights five elements of macrocognition, with the elements arrayed in a circle roughly in the order they might occur, but in reality, the process is more complex with all processes being linked to all other processes and occuring in a repeated cycle. At the center is meta-cognition—thinking about one's own thinking—which guides the individual macrocognitive processes. Microcognition and macrocognition offer complementary perspectives that suggest different ways to enhance safety, performance, and satisfaction.

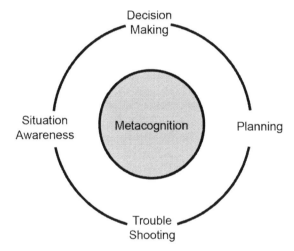

Figure 7.1 Five basic elements of macrocognition.

7.1 Cognitive Environment of Macrocognition

The cognitive environment governs how characteristics of microcognition, such as the limits of working memory, influence human performance. In a similar way, the elements of the cognitive environment govern how the characteristics of macrocognition influence performance, and when considering macrocognition is particularly important.

Many decisions take place in dynamic, changing environments, like those confronting the internal medicine specialist, Amy, described at the outset of the chapter [338, 339, 340]. Amy faced incomplete, complex, and dynamically changing information; time stress; interactions with others; high risk; uncertain outcomes, each with different costs and benefits. Not every situation is so

Characteristic	Example
Ill-structured problems with ambiguous goals	There is no single "best" way of responding to a set of a patient's symptoms.
Uncertain, dynamic environments	The situation at Amy's hospital is continually changing, presenting new decisions and considerations.
Information-rich environments	There is information on status boards, electronic patient records, and through talking with others.
Iterative perception-action-feedback loops	Any decision to regarding treatment, particularly after an initial misdiagnosis, is monitored and used decide what to do next.
Time pressure	Decisions often need to be made quickly because delays can jeopardize the outcome of a procedure.
High-risk situations	Loss of life can result from a poor decision.
Multiple shifting and competing individual and organizational goals	As the day evolves, the goals may shift from minimizing delays for routine procedures to responding to a major emergency. Also, what might be the top priority physician might not be the same for a nurse or patient.
Interactions with multiple people	Many people contribute information and perspectives to decisions: patients and nurses negotiate with Amy.

Table 7.1 Features of situations where macrocognition matters.

complicated, but those that include these elements indicate a need to consider the processes of macrocognition discussed in this chapter.

Table 7.1 summarizes features of the cognitive environment that makes it important to consider macrocognition. These features cause us to adopt different decision processes. Sometimes, particularly in high-risk situations, we carefully calculate and evaluate alternatives, but in many cases, we just interpret it to the best of our ability and make educated guesses about what to do. Some decisions are so routine that we might not even consider them to be decisions. Unlike the situations that influence microcognition, critical features associated with macrocognition include poorly defined goals that might not be shared by all involved. As in Amy's situation, concepts of macrocognition are particulalry important in situations that have multiple people interacting in an evolving situation where decisions and plans are made and then revised over time. In many cases, these features make decision making and problem solving difficult and error-prone. This makes macrocognition a central concern to human factors specialists working in complex systems, such as military operations, hospitals, aircraft cockpits, and process control plants.

We begin this chapter by describing the overall nature of skill and expertise in macrocognition, and how they change with practice and experience. We present three types of behavior that have implications for all elements of macrocognition, and then consider these behaviors with respect to decision making. Decision making highlights the challenges of engaging analytic thinking, the power of heuristics and the pitfalls of the associated biases. Principles to improve decision making are described in terms of task design, decision support systems, displays, and training. The final sections of the chapter addresses four closely related areas of macrocognition: situation awareness, troubleshooting, planning, and metacognition.

7.2 Levels of Behavior: Skill and Expertise

In understanding decision making over the last 50 years, there have been a variety of approaches to analyzing the skill or proficiency in reasoning that develops as the decision maker gains expertise. These are shown in Figure 7.2. To some degree, all of these approaches are related, but represent different ways that researchers have examined the processes underlying decision making and macrocognition. These approaches provide a framework for many of the sections to follow.

In the first row of Figure 7.2, Rasmussen [341] has proposed a three-level categorization of behavior. These levels evolve as the person develops progressively more skill or as the problems become progressively less complex. The progression from knowledge-based to rule-based to the more automatic skill-based reasoning and behavior parallels the development of automaticity described in the previous chapter. Closely paralleling this, in row 2 is the distinction that Hammond [342] has drawn between the careful analytic processing (describing all the options and factors that should enter into a choice), and the more "gut level" intuitive processing, often less accessible to conscious awareness. Here, as with Rasmussen's levels of behavior, more intuitive decisions are more likely to emerge with greater skill and simpler problems.

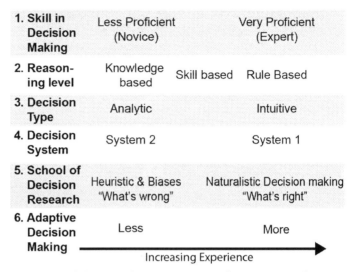

1. Skill in Decision Making	Less Proficient (Novice)		Very Proficient (Expert)
2. Reasoning level	Knowledge based	Skill based	Rule Based
3. Decision Type	Analytic		Intuitive
4. Decision System	System 2		System 1
5. School of Decision Research	Heuristic & Biases "What's wrong"		Naturalistic Decision making "What's right"
6. Adaptive Decision Making	Less		More

Increasing Experience

Figure 7.2 Parallel approaches to expertise and experience in decision making.

The third row shows different cognitive systems identified by Evans [343, 344] and highlighted as underlying differences in judgments by Kahneman [345]. Here System 2, like analytical judgments and knowledge-based reasoning, is considered to serve a deliberative function that involves resource-intensive effortful processes. In contrast, System 1 like intuitive judgments and the skill-based reasoning, engages relatively automatic "gut-feel" snap judgments. System 1 is highly guided by what is easy, effort-free and feels good or bad; that is, the emotional component of decision making. In partial contrast with skill-based behavior and intuitive judgments however, engaging System 1 does not necessarily represent greater expertise than engaging System 2. Instead, the two systems operate in parallel in any given decision, with System 1 offering a snap decisions of what to do, but then System 2, if time and cognitive resources or effort are available, overseeing and checking the result of System 1 to assure its correctness. System 1 also aids System 2 by focusing attention and filtering options—without it we would struggle to make a decision [346].

In the fourth row, we show two different "schools" of decision research that will be the focus of much of our discussion below. The "heuristics and biases" approach, developed by Kahneman and Tversky [347, 348] has focused on the kinds of decision shortcuts made because of the limits of reasoning, and hence the kinds of biases that often lead to decision errors. These biases identify "what's wrong" with decision making and what requires human factors interventions. In contrast, the naturalistic decision making school, proposed by Klein [349, 350] examines decision making of the expert, many of whose choices share features of skill-based behavior, intuitive decision making that are strongly influenced by System 1. That is, such decisions are often quick, relatively effort-free, and typically correct. While these two approaches are often set in contrast, it is certainly plausible to see both as being correct, yet applicable in different circumstances, and hence more complementary than competitive [351]. Heuristics and intuitive decision making work well for experienced people in familiar circumstances, but biases undermine performance of novices or experts in unfamiliar circumstances.

In the final row, we describe a characteristic of meta-cognition that appears, generally to emerge with greater skill. That is, it becomes increasingly adaptive, with the human better able to select the appropriate tools, styles, types, and systems, given the circumstances. That is, with expertise, people develop a larger cognitive toolkit, as does the wisdom regarding which tools to apply when.

To elaborate the first row of Figure 7.3, the distinctions of skill-, rule-, and knowledge-based (SRK) reasoning and behavior describe different cognitive processes that people use depending on their level of expertise and the situation [352, 341, 353]. This framework also also describes human error [354], which we discuss in Chapter 14). These distinctions are particularly important because the ways to improve decision making depend on supporting effective skill-, rule-, and knowledge-based behavior.

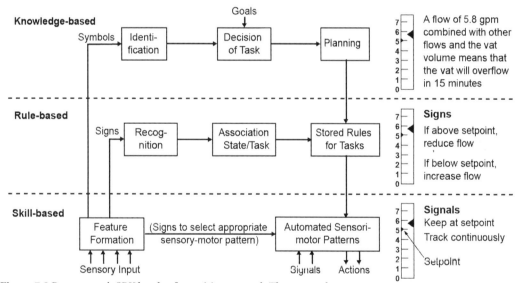

Figure 7.3 Rasmussen's SRK levels of cognitive control. The same physical cues (e.g., the meter in this figure) can be interpreted as signals, signs, or symbols. (Adapted from Rasmussen (1983). *Skills, rules, and knowledge: Signals, signs, and symbols, and other distinctions in human performance models. SMC-13(3), 257-266.*)

Figure 7.3 shows these three levels of behavior. Sensory input enters at the lower left, as a function of attentional processes. This input results in cognitive processing at either one of the three levels, depending on the operator's degree of experience with the particular situation and how information is represented [352, 342]. High levels of experience with analog representations promote relatively effortless skill-based behavior (e.g., riding a bicycle), whereas little experience with numeric and textual information will lead to knowledge-based behavior (e.g., selecting an apartment using a

△ Designs that enable skill-based behavior are "intuitive".

spreadsheet). In between, like the decision to bring a raincoat on a bike ride, are fairly straightforward rules: "if the forecast chance of rain is greater than 30%, then bring it."

People who are extremely experienced with a task tend to process the input at the skill-based level, reacting to the perceptual elements at an automatic, subconscious level. They do not have to interpret and integrate the cues or think of possible actions, but only respond to cues as signals that guide responses. Because the behavior is automatic, the demand on attentional resources described in Chapter 6 is minimal. For example, an operator might turn a valve in a continuous manner to counteract changes in flow shown on a meter (see bottom left of Figure 7.3).

When people are familiar with the task but do not have extensive experience, they process input and perform at the rule-based level. The input is recognized in relation to typical system states, termed signs, which trigger rules for accumulated knowledge. This accumulated knowledge can be in the person's head or written down in formal procedures. Following a recipe to bake bread is an example of rule-based behavior. The rules are "if-then" associations between cue sets and the appropriate actions. For example, Figure 7.3 shows how the operator might interpret the meter reading as a sign. Given that the procedure is to reduce the flow if the meter is above a set point, the operator then reduces the flow.

When the situation is new, people do not have any rules stored from previous experience to call upon, and do not have a written procedure to follow. They have to operate at the knowledge-based level, which is essentially analytical processing using conceptual information. After the person assigns meaning to the cues and integrates them to identify what is happening, he or she processes the cues as symbols that relate to the goals and decides on an action plan. Figure 7.3 shows how the operator might reason about the low meter reading and think about what might be the reason for the low flow, such as a leak. It is important to note that the same sensory input, the meter in Figure 7.3, for example, can be interpreted as a signal, sign, or symbol.

The relative role of skill-, rule-, and knowledge-based behavior, like the other elements in Figure 7.2, depends on characteristics of the person, the technology, and the situation [348, 355]. Characteristics of the person include experience and training. As we will see people can be trained to perform better in all elements of macrocognition; however, as with most human factors interventions, changing the task and tools is more effective.

In the following sections, we first discuss the cognitive processes in decision making: how it too can be described by stages, the normative approach to decision making (how it "should" be done to produce the best outcomes), and the reasons why people often do not follow the normative decision making processes. Two important departures from normative decision making, shown in the fourth row of Figure 7.2, receive detailed treatment: naturalistic decision making and heuristics and biases. Because decision errors produced by the heuristics and biases can be considered to represent human factors challenges, we complete our treatment of decision making by describing several human factors solutions to mitigate decision errors. Finally, our chapter concludes by describing four "close cousins" of decision making within the family of macrocognitive processes: situation awareness, troubleshooting, planning and meta-cognition.

7.3 Decision Making

What is a decision-making? Generally, it is a task in which (a) a person must select one option from several alternatives, (b) a person must interpret information for the alternatives, (c) the timeframe is relatively long (longer than a second), (d) the choice includes uncertainty; that is, it is not necessarily clear which is the best alternative. By definition, decision making involves risk—there is a consequence to picking the wrong alternative—and so a good decision maker effectively assesses risks associated with each alternative. The decisions discussed in this chapter range from those involving a slow deliberative process, involving how to allocate resources to those which are quite rapid, with few alternatives, like the decision to speed up, or apply the brakes, when seeing a yellow traffic light, or whether to open a suspicious e-mail [356].

Decision making can generally be represented by four phases as depicted in Figure 7.4: (1) acquiring and integrating information relevant for the decision, (2) interpreting and assessing the meaning of this information, (3) planning and choosing the best course of action after considering the costs and values of different outcomes, and (4) monitoring and correcting the chosen course of action. People typically cycle through the four stages in a single decision.

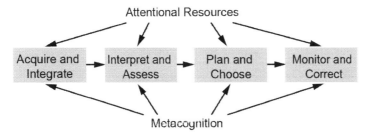

Figure 7.4 The four basic elements of decision making that draw upon limited attention resources and metacognition.

1. **Acquire and integrate** a number of cues, or pieces of information, which are received from the environment and go into working memory. For example, an engineer trying to identify the problem in a manufacturing process might receive a number of cues, including unusual vibrations, particularly rapid tool wear, and strange noises. The cues must be selectively attended, interpreted and somehow integrated with respect to one another. The cues may also be incomplete, fuzzy, or erroneous; that is, they may be associated with some amount of uncertainty.

2. **Interpret and assess** cues and then use this interpretation to generate one or more situation assessments, diagnoses, or inferences as to what the cues mean. This is accomplished by retrieving information from long-term memory. For example, an engineer might hypothesize that the set of cues described above is caused by a worn bearing. Situation assessment is supported by maintaining good situation awareness, a topic we discuss later in the chapter. The difference is that while maintaining SA refers to a continuous process, making a situation assessment involves a one time discrete action with the goal of supporting a particular decision.

3. **Plan and choose** one of alternative actions generated by retrieving possibilities from long-term memory. Depending on the time available, one or more of the alternatives are generated and considered. To choose an action, the decision maker might evaluate information such as possible outcomes of each action (where there may be multiple possible outcomes for each action), the likelihood of each outcome, and the negative and positive factors associated with each outcome. This can be formally done in the context of a decision matrix in which actions are crossed against the diagnosed possible states of the world that could occur, and which could have different consequences depending on the action selected.

4. **Monitor and correct** the effects of decisions. The monitoring process is a particularly critical part of decision making and can serve two general purposes. First, one can revise the current decision as needed. For example, if the outcomes of a decision to prescribe a particular treatment are not as expected, as was the case with Amy's patient is getting worse, not better, then the treatment can be adjusted, halted or changed. Second, one can revise the general decision process if that process is found wanting and ineffective, as Amy also did. For example, if heuristics are producing errors, one can learn to abandon them in a particular situation and instead adopt the more analytical approach shown to the left of Figure 7.2. In this way, monitoring serves as an input for the troubleshooting element of macrocognition.

Monitoring, of course, provides feedback on the decision process. Unfortunately, in decision making that feedback is often poor, degraded, delayed or non-existent, all features that undermine effective learning [10]. It is for this reason that consistent experience in decision making does not necessarily lead to improved performance [357, 351].

Figure 7.4 also depicts the two influences of attentional resources and meta-cognition. Many of the processes used to make ideal or "optimal" decisions impose intensive demands on perception and selective attention (for stage 1), particularly on the working memory used to entertain hypotheses in stage 2, and to evaluate outcomes in stage 4. If these resources are scarce, as in a multi-tasking environment, decision making can suffer. Furthermore, because humans are *effort conserving*, we often tend to adopt mental shortcuts or heuristics that can make decision making easier and faster, but may sacrifice its accuracy.

Meta-cognition describes our monitoring of all of the processes by which we make decisions, and hence is closely related to stage 4. We use such processes for example to assess whether we are confident enough in a diagnosis (stage 2) to launch an action (stage 3) without seeking more information. We describe meta-cognition in more detail near the end of the chapter.

7.3.1 Two Perspectives on Decision Making: Normative and Descriptive

Decision making has, for centuries, been studied in terms of how people should make optimal decisions: those likely to produce the best outcomes in the long run [358, 359]. This is called normative decision making. Within the last half century however, decision scientists have highlighted that humans often do not, in practice, adhere to such optimal norms for a variety of reasons, and so their decisions can be described in ways classified as descriptive decision making. We now discuss both the normative and descriptive frameworks.

Normative Decision Making considers the four phases of decision making in terms of an idealized situation in which the correct decision can be made by calculating the mathematical optimal choice. This mathematical approach is often termed *normative decision making*. Normative decision making specifies what people *should* do; they do not necessarily describe how people *actually* perform decision-making tasks. Importantly, these normative models make many assumptions that of incorrectly simplifies and limits their application to the decisions people actually face [360]. Normative models are important because they form the basis for many computer-based decision aids, and justify (often wrongly) that humans' fallible judgment should be removed from the decision process [361]. Although such normative models often outperform people in situations where their assumptions hold, many real-life decision cannot be reduced to a simple formula [362].

Normative decision making revolves around the central concept of *utility*, the overall value of a choice, or how much each outcome is "worth" to the decision maker. This model has application in engineering decisions as well as decisions in personal life. Choosing between different corporate investments, materials for product, jobs, or even cars are all examples of choices that can be modeled using *multiattribute utility theory*. The decision matrix described in Chapter 2 is an example of how multiattribute utility theory can be used to guide engineering design decisions. Similarly, it has been used to resolve conflicting objectives, to guide environmental cleanup of contaminated sites [363], to support operators of flexible manufacturing systems [364], and even to select a marriage partner [365].

The number of potential options, the number of attributes or features that describe each option, and the challenge in comparing alternatives on very different dimensions make decisions complicated. Multiattribute utility theory addresses this complexity, using a utility function to translate the multidimensional space of attributes into a single dimension that reflects the overall utility or value of each option. In theory, this makes it possible to compare apples and oranges and pick the best one.

Multiattribute utility theory assumes that the overall value of a decision option is the sum of the magnitude of each attribute multiplied by the utility of each attribute (Equation 7.1), where $U(v)$ is the overall utility of an option, $a(i)$ is the magnitude of the option on the ith attribute, $u(i)$ is the utility (goodness or importance) of the ith attribute, and n is the number of attributes.

Expected value of a choice based on utilities of attributes.

$$U(v) = \sum_{i=1}^{n} a(i)u(i) \qquad (7.1)$$

Figure 7.5 shows the analysis of four different options, where the options are different cars that a student might purchase. Each car is described by five attributes. These attributes might include the initial purchase price, the fuel economy, insurance costs, sound quality of the stereo, and maintenance costs. The utility of each attribute reflects its importance to the student. For example, the student cannot afford frequent and expensive repairs, so the utility or importance of the fifth attribute (maintenance costs) is quite high (8), whereas the student does not care as much about the sound quality of the stereo (4) or the fourth attribute (color), which is quite low (1). The cells in the decision table show the magnitude of each attribute for each option. For this example, higher values reflect a more desirable situation. For example, the third car has a poor stereo, but low maintenance costs. In contrast, the first car has a slightly better stereo, but high maintenance costs. Combining the magnitude of all the attributes shows that third car (option 3) is most appealing or "optimal" choice and that the first car (option 1) is least appealing.

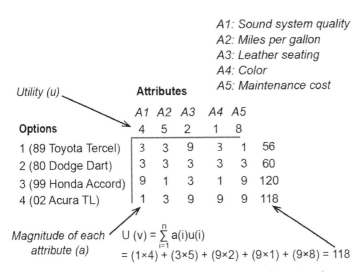

A1: Sound system quality
A2: Miles per gallon
A3: Leather seating
A4: Color
A5: Maintenance cost

Figure 7.5 Multi-attribute utility analysis combines information from multiple attributes of each of several options to identify the optimal decision.

Multi-attribute utility theory, shown in Figure 7.5, assumes that all outcomes are certain. However, life is uncertain, and probabilities often define the likelihood of various outcomes (e.g., you cannot predict maintenance costs precisely). Another example of a normative model is *expected value theory*, which addresses uncertainty. This theory replaces the concept of utility in the previous context with that of expected value. The theory applies to any decision that involves a "gamble" type of decision, where each choice has one or more outcomes and each outcome has a worth and a probability. For example, a person might be offered a choice between:

1. Winning $50 with a probability of 1.0 (a guaranteed win), or
2. Winning $200 with a probability of 0.30.

Expected value theory assumes that the overall value of a choice (Equation 7.1) is the sum of the worth of each outcome multiplied by its probability where $E(v)$ is the expected value of the choice, $p(i)$ is the probability of the ith outcome, and $v(i)$ is the value of the ith outcome.

$$E(v) = \sum_{i=1}^{n} p(i)v(i) \qquad (7.2)$$

(7.2) Expected value of a choice based on probabilities and values

The expected value of the first choice for the example is $50×1.0, or $50, meaning a certain win of $50. The expected value of the second choice is $200 × .30, or $60, meaning that if the choice were selected many times, one would expect an average gain of $60, which is a higher expected value than $50. Therefore, the normative decision maker should always choose the second gamble. In reality, people tend to avoid risk and go with the sure thing [366].

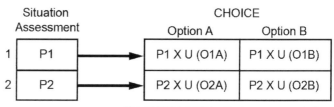

Figure 7.6 Expected value calculation of optimal choice.

Figure 7.6 shows two states of the world, 1 and 2, which are generated from situation assessment. Each has a probability, P1 and P2, respectively. The two choice options, A and B, may have four different outcomes as shown in the 4 cells to the right. Each option may also have different Utilities (U), which could be positive or negative, contingent upon the existing state of the world. The normative view of decision making dictates that the chosen option should be the one with the highest (most positive) sum of the products within the two different states.

Descriptive Decision Making accounts for how people actually make decisions. People can depart from the optimum, normative, expected utility model. First, people do not always try to maximize, *EV* nor should they because other decision criteria beyond expected value can be more important. Second, people often shortcut the time and effort-consuming steps of the normative approach. They do this because time and resources are not adequate to "do things right" according to the normative model, or because they have expertise that directly points them to the right decision. Third, these shortcuts sometimes result in errors and poor decisions. Each of these represents an increasingly large departure from normative decision making.

As an example of using a decision criterion different from maximizing expected utility, people may choose instead to minimize the possibility of suffering the maximum loss. This certainly could be considered as rational, particularly if one's resources to deal with the loss were limited. This explains why people purchase insurance; even though such a purchase decision does not maximize their expected gain. If it did, the insurance companies would soon be out of business! The importance of using different decision criteria reflects the mismatch between the simplifying assumptions of expected utility and the reality of actual situations. Not many people have the ability to absorb a $100,000 medical bill that might accompany a severe health problem.

Most decisions involve shortcuts relative to the normative approach. Simon [367] argued that people do not usually follow a goal of making the absolutely best or optimal decision. Instead, they opt for a choice that is "good enough" for their purposes, something satisfactory. This shortcut method of decision making is termed *satisficing*. In satisficing, the decision maker generates and evaluates choices only until one is found that is acceptable rather than one that is optimal. Going beyond this choice to identify something that is better is not worth the effort.

Real-life decision making is complex in ways that normative decision calculations cannot address: Imagine using a computer algorithm to pick your spouse based on expected value.

Satisficing is a very reasonable approach given that people have limited cognitive capacities and limited time. Indeed, if minimizing the time (or effort) to make a decision is itself considered to be an attribute of the decision process, then satisficing or other shortcutting heuristics can sometimes be said to be optimal—for example, when a decision must be made before a deadline, or all is lost. In the case of our car choice example, a satisficing would be to take the first car that gets the job done rather than doing the laborious comparisons to find the best. Satisficing and other shortcuts are often quite effective [360], but they can also lead to biases and poor decisions as we will discuss below.

Our third characteristic of descriptive decision making focuses on human limits that often cause decision errors. A general source of errors concerns the failure of people to recognize when shortcuts are inappropriate for the situation and adopt the more laborious decision processed. Because this area is so important, and its analysis generates a number of design solutions, we dedicate the next section to this topic.

7.4   Balancing Intuitive, Heuristic, and Analytic Decision Making

Consistent with our previous discussion of skill-, rule-, and knowledge-based performance, how people make decisions depends on the situation. People tend to make decisions at one of three ways: intuitive skill-based processing, heuristic rule-based processing, and analytical knowledge-based processing. Making decisions as described by the normative models is an example of analytic decision making and using satisficing heuristics is an example of rule-based decision making. Intuitive decision-making occurs when people recognize the required response without thinking.

As we learned in the context of Figure 7.2, people with a high degree of expertise often approach decision making in a fairly automatic pattern matching style, just as Amy did with her first diagnosis. *Recognition primed decision making* (RPD) describes this process in detail [368]. In most instances, experts simply recognize a pattern of cues and recall a single course of action, which is then implemented. In spite of the prevalence of rapid pattern-recognition decisions, there are cases where decision makers will use analytical methods, such as when the decision maker is unsure of the appropriate course of action. The decision maker resolves the uncertainty by imagining the consequences of what might happen if a course of action is adopted: a mental simulation, where the decision maker thinks: "if I do this, what is likely to happen" [369]. Mental simulation can help assess the alternatives, action, or plan under consideration [370]. In this process, the mental simulation can play out possible solutions based on information from the environment and their mental model. Mental simulation shows which options are the most promising, and also generates expectations for other cues not previously considered [371].

Induces intuitive skill and rule-based decisions	Induces analytical knowledge-based decisions
Familiar situations	Unusual situations
Time pressure	Abstract problems
Unstable conditions	Numbers and text rather than graphics
Ill-defined goals	Requirement to justify decision
Large number of cues	Integrated views of multiple stakeholders
Cues displayed simultaneously	Few relationships among cues
Need to conserve cognitive effort	Requires precise solution

Table 7.2 Features of situations that induce intuitive and analytical decision making.

Also, if uncertainty exists and time is adequate, decision makers will spend time to evaluate the current situation assessment, modify the retrieved action plan, or generate alternative actions [350]. Experts adapt their decision-making strategy to the situation. Table 7.2 summarizes some of the factors that lead to intuitive rule-based decision making and those that lead to analytical knowledge-based decision making. These characteristics of the person, task, and technology influence the use of heuristics as well as the prevalence of biases that sometimes accompany those heuristics, which we discuss in detail in the next section.

7.4.1 Vulnerabilties of Heuristics: Biases

Cognitive heuristics are rules-of-thumb that are easy ways of making decisions. Heuristics are usually very powerful and efficient [372], but they do not always guarantee the best solution [348, 373]. Unfortunately, because they represent simplifications, heuristics occasionally lead to systematic flaws and errors. The systematic flaws represent deviations from the normative model and are sometimes referred to as biases. Experts tend to avoid these biases because they draw from a large set of experiences and they are vigilant to small changes in the pattern of cues that might suggest the heuristic is inappropriate. To the extent a situation departs from these experiences, even experts will fall prey to the biases associated with various heuristics. Although the list of heuristics is large (as many as 37 [374]), the following presents some of the most notorious ones.

Acquire and Integrate Cues: Heuristics and Biases The first stage of the decision process begins with attending to information and integrating it to understand the situation or form a situation assessment (e.g., to support stage 2).

1. **Attention to a limited number of cues.** Due to working memory limitations, people can use only a relatively small number of cues to develop a picture of the world or system. This is one reason why configural displays that visually integrate several variables or factors into one display are useful (see Chapter 8 for a description).

2. **Anchoring and cue primacy**. When people receive cues over a period of time, there are certain trends or biases in the use of that information. The first few cues receive greater weight than subsequent information–cue primacy [375]. It often leads people to "anchor" on initial evidence and is therefore sometimes called the anchoring heuristic [348], characterizing the familiar phenomenon that first impressions are lasting. Amy anchored on the cues supporting her initial diagnosis, and gave little processing to additional information available in the phone call by the patient 24 hours later. Importantly, when assessing a dynamic changing situation, the anchoring bias can be truly detrimental because older information becomes progressively less reliable, even as the older information was, by definition, the first encountered and hence served as the anchor. The order of information has an effect because people use the information to construct plausible stories or mental models of the world or system. These models differ depending on which information is used first [376]. The key point is that, information processed early is often most influential.

3. **Cue salience.** Perceptually salient cues are more likely to capture attention and be given more weight [377, 10]; see also Chapter 6. As you would expect, salient cues in displays are things such as information at the top of a display, the loudest alarm, the largest display, the loudest most confident sounding voice in the room, and so forth. Unfortunately, the most salient cue is not necessarily the most diagnostic, and sometimes very subtle ones, such as the faint discoloration observed by Amy are not given much weight.

4. **Overweighting of unreliable cues.** Not all cues are equally reliable. In a trial, some witnesses, for example, will always tell the truth. Others might have faulty memories, and still others might intentionally lie. However, when integrating cues, people often simplify the process by treating all cues as if they are all equally valid and reliable. The result is that people tend to give too much weight to unreliable information [378, 379].

Interpret and Assess: Heuristics and Biases After a limited set of cues is processed in working memory, the decision maker generates and interprets the information, often by retrieving similar situations from long-term memory. These similar situations represent hypotheses about how the current situation relates to past situations. There are a number of heuristics and biases that affect this process:

1. **Availability.** The availability heuristic reflects people's tendency to make certain types of judgments or assessments, for example, estimates of frequency, by assessing how easily the state or event is brought to mind [380, 381, 382]. People more easily retrieve hypotheses that have been considered

recently and hence more available to memory. The implication is that although people try to generate the most likely hypotheses, the reality is that if something comes to mind relatively easily, they assume it is common and therefore a good hypothesis. As an example, if a physician readily thinks of a hypothesis, such as acute appendicitis, he or she will assume it is relatively common, leading to the judgment that it is a likely cause of the current set of symptoms. Unusual illnesses tend not to be the first things that come to mind to a physician. Amy did not think of the less likely condition. In actuality, availability to memory may not be a reliable basis for estimating frequency.

2. **Representativeness.** Sometimes people diagnose a situation because the pattern of cues "looks like" or is representative of the prototypical example of this situation. This is the representativeness heuristic [347, 383], and usually works well; however, the heuristic can bias decisions when a perceived situation is slightly different from the prototypical example even though the pattern of cues is similar or representative.

3. **Overconfidence.** People are often biased in their confidence with respect to the hypotheses they have brought into working memory [384, 345], believing that they are correct more often than they actually are and reflecting the more general tendency for overconfidence in metacognitive processes, as described in Chapter 6 [385]. Such overconfidence appears to grow when judgments are more predictive about the future (than of the current state) and when predictions become more difficult [10]. As a consequence, people are less likely to seek out evidence for alternative hypotheses or to prepare for the circumstances that they may be wrong. Less skilled people are more likely to overestimate their ability, even when they know about their relative ability [386].

4. **Planning bias.** Closely related, and directly expressing overconfidence is the planning bias [387, 388]. In planning on both a large scale, like the time required to complete a major construction project such as the Denver International Airport, or on a small scale, like the time required to write a quality paper before the deadline, people seem to assume that the best case scenario will unfold, and be cognitively blind to predicting unexpected delaying effects that can befall them; or at least to underestimate both the probability of those events, or their time costs.

5. **Cognitive tunneling.** As we have noted earlier in the context of anchoring, once a hypothesis has been generated or chosen, people tend to underutilize subsequent cues. We remain stuck on our initial hypothesis, a process introduced in the previous chapter as cognitive tunneling [389]. Examples of cognitive tunneling abound in the complex systems [390]. Consider the example of the Three Mile Island disaster in which a relief valve failed and caused some of the

displays to indicate a rise in the level of coolant [391]. Operators mistakenly thought that that emergency coolant flow should be reduced and persisted to hold this hypothesis for over two hours. Only when a supervisor arrived with a fresh perspective did the course of action get reversed. Notice that cognitive tunneling is different than the primacy, which occurs when the decision maker is first generating hypotheses.

Cognitive tunneling can sometimes be avoided by looking at the functionality of objects in terms beyond their normal use. The nearly catastrophic situation in a moon mission, well captured by the movie *Apollo 13*, demonstrated the ability of people to move beyond this type of functional fixedness. Recall that the astronauts were stranded without an adequate air purifier system. To solve this problem, the ground control crew assembled all of the "usable" objects known to be on board the spacecraft (tubes, articles of clothing, etc.). Then they did free brainstorming with the objects in various configurations until they had assembled a system that worked.

6. **Simplicity seeking and choice aversion.** Presenting people with more alternatives can make the decision harder and the result less satisfying [392]. More choice also resulted in fewer people participating in a retirement plan, and for those participating more choices led to less diversification because people distributed investments across the range of investment funds rather than distributing investments across the underlying asset classes (e.g., stocks and bonds)[393].

7. **Confirmation bias.** Closely related to cognitive tunneling are the biases when people consider additional cues to evaluate working hypotheses. People tend to seek out only confirming information and not disconfirming information, even when the disconfirming evidence can be more diagnostic [394, 395]. Amy did not carefully look for what might have been the disconfirming evidence in the patient's medical record. In a similar vein, people tend to underweight, or fail to remember, disconfirming evidence [396, 10] and fail to use the absence of important cues as diagnostic information. The confirmation bias is exaggerated under conditions of high stress and high mental workload [389, 397, 398]. Figure 7 shows confirmation bias at work.

Plan and Choose: Heuristics and Biases Choice of action is also subject to a variety of heuristics or biases. Some are based on basic memory processes that we have already discussed.

1. **Retrieve a small number of actions.** Long-term memory may provide many possible action plans, but people are limited in the number they can retrieve and keep in working memory. People tend to adopt a single course of action and fail to consider the full range of alternatives, even when time is available. Working-memory limits make it difficult to consider many alternatives simultaneously, and people to neglect cues after identifying a promising alternative.

Figure 7.7 Confirmation bias guides information seeking and interpretation. DILBERT ©2011 Scott Adams. Used By permission of ANDREWS MCMEEL SYNDICATION. All rights reserved.

2. **Availability of actions.** In retrieving possible courses of action from long-term memory, people retrieve the most "available" actions, just as they tend to do with hypotheses. In general, the availability of items from memory are a function of recency, frequency, and how strongly they are associated with the hypothesis or situational assessment that has been selected through the use of *if-then rules*. In high-risk professions like aviation, emergency checklists are often used to ensure that actions are considered, even if they may not be frequently performed [399].

3. **Availability of possible outcomes.** Other types of availability effects will occur, including the generation/retrieval of associated outcomes. As discussed, when more than one possible action is retrieved, the decision maker must select one based on how well the action will yield desirable outcomes. Each action often has more than one associated consequence, which are probabilistic. As an example, will a worker adhere to a safety procedure and wear a hardhat versus ignoring the procedure and going without one? Wearing the hardhat has some probability of saving the worker from death due to a falling object. A worker's estimate of this probability will influence the decision to wear the hardhat. The worker's estimate of these likelihoods will not be objective based on statistics, but are more likely to be based on the availability of instances in memory. It is likely that the worker has seen many workers not wearing a hardhat who have not suffered any negative effects, and so he or she is likely to think the probability of being injured by falling objects is less than it actually is. Thus, the availability heuristic will bias retrieval of some outcomes and not others. Chapter 14 describes how warnings can be created to counteract this bias by showing the potential consequences of not complying, thus making the consequences more available.

The decision maker is extremely unlikely to retrieve all of the possible outcomes for an action, particularly under stress [400]. Thus, selection of action suffers from the same cognitive limitations as other decision activities we have discussed (retrieval biases and working-memory limitations). Because of these cognitive limitations, selection of action tends to follow a satisficing model: If an alternative action passes certain criteria, it is selected. If the action does not work, another is considered. Again, this bias is much more likely to affect the performance of novices than experts [340].

4. **Hindsight bias.** After someone is injured because he or she did not wear a hardhat, people are quick to criticize because it was such an obvious mistake. The tendency for people to think "they knew it along" is called the hindsight bias [401, 402]. This process is evident in the "Monday morning quarterback phenomena" where people believe they would not have made the obvious mistakes of the losing quarterback. More importantly, hindsight bias often plagues accident investigators who, with the benefit of hindsight and the very available (to their memory) example of a bad outcome, inappropriately blame operators for committing errors that are obvious only in hindsight [403].

5. **Framing bias.** The framing bias is the influence of the framing or presentation of a decision on a person's judgment [404]. According to the normative utility theory model, the way the problem is presented should have no effect on the judgment. For example, when people are asked the price they would pay for a pound of ground meat that is 10% fat or 90 percent lean, they will tend to pay 8.2 cents per pound more for the option presented as 90% lean even though they are equivalent [405]. Likewise, students feel that they are performing better if they are told that they answered 80 percent of the questions on the exam correctly compared to being told that they answered 20% of the questions incorrectly. People also tend to view a certain treatment as more lethal if its risks are expressed as a 20% mortality rate than if expressed as 80% life saving and are thereby less likely to choose the treatment when expressed in terms of mortality [406]. Thus the direction of a choice can be influenced by the extent to which it is framed as a gain or a loss.

The framing bias is also clearly expressed when the choice is between a risky option and a sure thing. To provide examples of contrast between negative and positive frames, suppose the risky-sure thing choice is between positive outcomes: accept a sure gift of $100 or take a risky gamble with 50-50 odds to win $200 or nothing at all. In these circumstances, people tend to be risk-averse, more often choosing to take the $100 rather than the risky chance to get $200.

In contrast suppose you are late for a job interview across town. You can speed, with a high chance of getting to the appointment on time, but also incurring the risk of getting caught by the police, fined, and be very late for the appointment. Alternatively you can choose to drive the speed limit, and certainly be slightly late. Here the choice is between two negatives, a risky one and a sure thing. You are "caught

between a rock and a hard place", and under such circumstances people tend to be risk-seeking. [407, 404].

The second of these contexts, the negative frame of choice,
is often characteristic of real life decisions. For example, in
addition to the speeding choice above, consider a company
with major safety violations in its plant. Management can
choose to invest heavy funding into addressing them through
new equipment, hiring safety consultants, and pulling workers off the line for safety training, thus incurring the sure loss
of time and money. Alternatively they can chose to take the
risk that there will be neither a serious injury nor a surprise
inspection from federal safety inspectors. All too-often, the
framing bias will lead to an inclination toward the second
option, at the expense of worker safety.

A direct expression of this form of the framing bias is known
as the *sunk cost bias* [408, 409]. This bias affects individual
investors who hesitate to sell losing stocks (a certain loss),
but tend to sell winning stocks to lock in a gain. Likewise,
when you have invested a lot of money in a project that has
"gone sour", there is a tendency to keep it in the hopes that
it will turn around. Similarly, managers and engineers tend
to avoid admitting a certain cost when replacing obsolete
equipment. The sunk cost bias describes the tendency to
choose the risky loss over the sure one, even when the rational, expected value choice should be to abandon the project.
Because people tend to incur greater risk in situations involving losses, decisions should be framed in terms of gains to
counteract this tendency.

> Sunk cost bias makes it difficult
> for you to make money in the
> stock market.

6. **Default heuristic**. Faced with uncertainty regarding what
   choice to make people often adopt the default alternative
   [410]. Most countries use their drivers' liscences to allow
   people to specify whether to donate their organs or not in the
   event of a fatal crash. Countries differ according to whether
   people need to opt in and decide to donate, or opt out and
   decide not to donate. Over 70% people follow the default and
   let the designers of the form decide for them. A similarly large
   effect is seen for people choosing to enroll in a retirement
   savings plan or having to opt out. Defaulting people into a
   retirement plan increased participation from about 50% to
   about 90% [411, 412]

7.4.2   Benefits of Heuristics and the Cost of Biases

The long list of decision-making biases and heuristics above may
suggest that people are not very effective decision makers in everyday situations, and might suggest that human contributions
to decision making are a problem that should be fixed. However,
this perspective neglect the fact that most people do make good
decisions most of the time, and have the flexibility to deal with
situations that can't be reduced to an equation. The list of biases

accounts for the infrequent circumstances, like the decision makers in the Three Mile Island nuclear plant, when decisions produce bad outcomes.

One reason that most decisions are good, is that heuristics are accurate most of the time. A second reason is that people have a profile of resources: information-processing capabilities, experiences, and decision aids (e.g., a decision matrix) that they can adapt to the situations they face. Experts are proficient in adjusting their decision strategies. To the extent that people have sufficient resources and can adapt them, they make good decisions. When people are not able to adapt, such as where people have little experience with the situations, poor decisions can result [351].

While the focus can be either on the general high quality of most decisions, or on the errors due to biases associated with heuristics. Both of these approaches are equally valid, but focusing on the errors supports the search for human factors solutions to eliminate, or at least mitigate those biases that do show. It is to this that we now turn.

7.4.3 Principles for Improving Decision Making

Decision making is often an iterative cycle in which decision makers are often adaptive, adjusting their response according to their experience, the task situation, cognitive ability, and the available decision-making aids. It is important to understand this adaptive decision process because system design, training, and decision aids need to support it. Attempts to improve decision making without understanding this process tend to fail. In this section, we briefly discuss some possibilities for improving human decision making: task redesign, including choice architecture and procedures; training; displays; and automated decision support systems.

Task Redesign. We often jump to the conclusion that poor performance in decision making means we must do something "to the person" to make him or her a better decision maker. However, sometimes a change in the system can support better decision making, eliminating the need for the person to change. As described in Chapter 1, decision making may be improved by task design. Changing the system should be considered before changing the person through training or even providing a computer-based decision aid. For example, consider the situation in which the removal of a few control rods led to a runaway nuclear reaction, which resulted in three deaths and 23 cases exposure to high levels of radioactivity. Learning from this experience, reactor designers now create reactors that remain stable even when several control rods are removed [220]. Creating systems with greater stability leaves a greater margin for error in decisions and can also make it easier to develop accurate mental models.

Choice architecture describes how choices can be influenced by design in much the same way architecture of a building influences the movement of people through buildings [413]. Choice architects influence decisions by recognizing the natural cognitive tendencies we have discussed and presenting people with information and options that will take advantage of these tendencies to generate good decisions. The following principles represent some of the more effective ways of nudging people towards a decision [414].

1. **Limit the number of options.** Because too many options place a high burden on the decision maker, the number of options should be limited to the fewest number that will encourage exploration of options. Although the appropriate number depends on the specific elements of the decision maker and situation, four to five options where none is better on all dimensions. Fewer options should be offered if decision makers are less capable, such as older people, those in a time pressured situation, or less numerate decision makers faced with numerical options [415, 414].

2. **Select useful defaults.** The effect of defaults on organ donation rates demonstrates the power of defaults: People often choose default options. Options for designing defaults include random, uniform choice for all users, forced choice, persistent default where the system remembers previous settings, and predictive default where the system picks based on user characteristics. If there is no time pressure and the choice is important then active choice should be used. If there is an obvious benefit to a particular choice then a uniform default for all users should be used, such when organizations select double-sided printing as the default [416]. As laptops, tablet and desktop computers, as well as phones, TVs and cars become more integrated predictive defaults become more feasible and valuable.

3. **Make choices concrete.** People focus on concrete immediate outcomes and tend to be overly optimistic about future regarding available time and money. To counteract people's tendency to neglect the abstract future situation a limited window on opportunity can focus their attention like:"offer ends midnight tonight". Another approach is to translate the abstract future value choices into immediate, salient consequence. One example of this strategy is to show people their future self so they can invest for that future self [417]. People who saw realistic computer renderings of older version of themselves invested more.

4. **Create linear, comparable relationships.** People tend struggle to consider complex transformations and non-linear relationships. Transforming variables to their concrete linear equivalent promotes better decisions. For example, describing interest rates in terms of the number of payments to eliminate debt in three years is more effective than expecting people to calculate the non-linear, compounding effect of interest. Likewise, presenting fuel economy data in terms of gallons per 100 miles rather than miles per gallon, eliminates the mental transformation that is needed to compare

Listen carefully to infomercials to see how they guide decisions.

vehicles [418]. The units presented should be those directly relevant to the decision.

5. **Sequence and partition choices.** The sequence and grouping of choices includes decisions: People are more likely to choose defaults if they first have to select an option from many choices. Creating categories influences distribution of selections: People are biased towards an even distribution across categories. This general tendency guides the selection of investment options in retirement funds, and so options should be groups to avoid biasing investors towards creating a risky portfolio simply because there are more categories of risky funds. When presenting food choices aggregating unhealthy options into one category and disaggregating healthy options will lead people to pick more of the healthy options.

Proceduralization outlines processes to improve the quality of decision-making [419]. This may include for example prescriptions of following the decision decomposition steps of multiattribute utility theory. Such a technique has been employed successfully in certain real world decisions which are easily decomposable into attributes and values, such as selecting the location of the Mexico City airport [420], or coordinating environmental and energy policy [421]. The formal representation of fault tree and failure modes analysis [46], is a procedure that can assist the decision-maker in diagnosing the possibility of different kinds of system failures. A study of auditors has recommended a procedure by which evidence, accumulated by a junior auditor, is compiled and presented to a senior auditor who makes decisions, in such a way as to avoid the sequential biases often encountered in processing information [422].

One widely used procedural approach has been designed to support the traditional "decision-analysis" cognitive process of weighing alternative actions. This method is popular with engineers and business managers and uses a decision table or decision matrix. It supports the normative multiattribute utility theory described at the start of this chapter and in Chapter 2. Decision tables are used to list the possible outcomes, probabilities, and values of the action alternatives. The decision maker enters estimated probabilities and values into the table. Computers are programmed to calculate and display the utilities for each possible choice. A decision table is helpful because it reduces the working-memory load. By deflecting this load to a computer, it encourages people to consider the decision space more broadly. More generally, tools for multiattribute utility theory succeed not by making the decision for people, but by helping people think through the decision[423]

Decision trees are useful for guiding decisions that involve evaluating information in a sequence to make a decision. With this method, a branching point is used to represent the decision alternatives; this is followed by branching points for possible consequences and their associated probabilities. This sequence is repeated as far as necessary to make the decision, so the user can see the overall probability for each entire action-consequence sequence. An important challenge in implementing these techniques is user acceptance [424]. The step approach is not how people typically make decisions, and so the approach can seem foreign. However, for those tasks where choices involve high risk and widely varying probabilities, such as types of treatment for cancer, it can be worth training users to be more comfortable with this type of aid.

Figure 7.8 shows a *fast and frugal decision tree* [425], which provide a fast and understandable way to guide decisions. The defining feature of these trees is that each branch leads to a decision. In the figure, the first node of the tree indicates that a case should not be diagnosed as cancer if "cellsize.u" is greater than 3. Fast and frugal decision trees helped caregivers direct people suffering from chest pain and other symptoms of a heart attack towards the coronary care unit or a regular bed. The tree reduced the number of people who were mistakenly sent to the coronary care unit that were not suffering from a heart attack, while also reducing the number of heart attack victims wrongly turned away from the coronary care unit [372]. Decision trees were also easy to use and more understandable than a logistic regression model that performed similarly well from a statistical perspective.

> Fast and frugal decision trees are not as mathematically precise as other statistical models, but people understand them and so are more likely to use them.

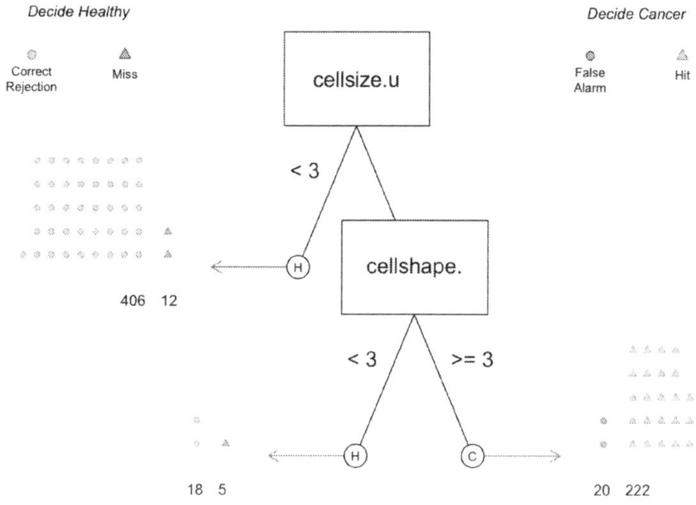

Figure 7.8 Fast and frugal decision tree for hypothetical cancer diagnosis. The tree provides a diagnosis after considering each indicator (e.g., cellsize.u). (Created with R package FFTrees [426].)

Training decision making. Pure practice does not necessarily lead to better decision making, because of the poor feedback offered in many naturalistic decision environments, such a health care, or legal decisions, where the feedback is often delayed, missing, or incorrect. However in a relatively predictable environment, with reliable cues, extensive practice and expertise can lead to relatively rapid and accurate diagnosis, in the process described earlier as recognition primed decision making [368]. The conditions and job environments that possess such cues are well articulated by

Kahneman [345], as well as those which are not. And which the findings seem to indicate that expertise does not produce better decision making.

Training has a mixed record of success in improving decision making. For example, training decision makers on how to use some of the procedures and tools, such as decision trees discussed in the previous section, can improve decisions. Training and instructions to remove or reduce many of the biases discussed above, a technique known as debiasing [427], has mixed results. Instructions or exhortations to avoid biases are ineffective[355]. Similarly, simply teaching people about biases (e.g., reading this chapter) in only moderately effective. Education may produce *inert knowledge* which can be understood, but not transferred to practice. Instead, effective techniques focus not only on instructing the nature of a particular bias in a particular context [340], but also providing feedback to show how better outcomes are produced when the trained strategy is followed, and worse outcomes when it is not. The following are some specific examples of success.

Hunt and Rouse [428] trained operators to extract diagnostic information from the absence of cues. Some success in reducing the confirmation bias has also been observed by the training strategy of "consider the opposite" [429], such as forcing forecasters to entertain reasons why their forecasts might not be correct reduced their biases toward overconfidence in the accuracy of the forecast [430] .

Also successful is a training aid that provides more comprehensive and immediate feedback, so that operators are forced to attend to the degree of success or failure of their rules. We noted that the feedback given to weather forecasters is successful in reducing the tendency for overconfidence in forecasting [431]). Similarly people think of events in terms of probability rather than frequency because probabilities account for events that did not occur (negative evidence) as well as those that did [348].

Perhaps the most effective training approach for decision making is termed the *pre-mortem*. Rather than examining the reasons why a decision was poor after the damage has been done—post-mortem analysis— the pre-mortem analysis encourages decision makers to consider everything that might go wrong before if the candidate decision was made [432].

Displays. There is good evidence that displays can influence the front end of decision processes (cue integration and diagnosis), by guiding selective attention. Items at the start and end of a menu receive more attention and are ordered more frequently [433]. Pictorial representations of risk data led people to decisions that reflects a more calibrated sense of risk than did numerical or verbal statements[434]. Similarly, loan application information structured as a list was perceived as more demanding information structured as a matrix. This perceived demand influenced judgments, suggesting that people minimized the amount of attentional effort required for information integration [435]. Cook and Smallman [436] found that an integrated graphical display of intelligence cues shown to professional intelligence analysis reduced the confirmation bias, relative to a text-based presentation which implicitly suggested a sequential ordering and so invited sequential biases.

As we will discuss in the following chapter, sources of information that need to be integrated in diagnosis, should be made available simultaneously (not sequentially; to mitigate anchoring), and in close display proximity so that all can be accessed with minimal effort. Emergent features of object displays can sometimes facilitate the integration process in diagnosis [437, 438, 439].

Automation and decision support tools. Finally, automation and expert systems have offered promise in supporting human decision making. This is described in much more detail in chapter 12, but to provide a link here, such support can be roughly categorized into front end (diagnosis and situation assessment) and back end (treatment, choice and course-of-action recommendations) support. This dichotomy is well illustrated in the two major classes of medical decision aids [440, 441], because automation is so closely bound to decision support tools and expert systems decision advisors, we postpone further discussion of this topic until Chapter 11, where the entire chapter is devoted to human-automation interaction.

7.5   Situation Awareness

The diagnosis error made by the medical specialist, Amy in our vignette can be examined more thoroughly using the concept of situation awareness (SA). Situation awareness, or SA, characterizes people's awareness and understanding of dynamic changes in their environment [442, 443, 444]. A pilot loses SA whenever he or she suffers a catastrophic controlled-flight into terrain [445, 222], and as we shall see later in Chapter 16, control room operators at the Three Mile Island nuclear power plant lost SA when they believed the water level in the plant to be too high rather than too low, a misdiagnosis that led to a catastrophic release of radioactive material [391].

SA is "the perception of the elements in the environment within a volume of time and space, the comprehension of their meaning, and the projection of their status in the near future" [446] (p 36). These three stages, perception (and selective attention), understanding, and prediction, must be applied to a specific situation. Thus, a person cannot be said to have SA without specifying what that awareness is (or should be) about. A car driver might have good awareness of navigational information and time (where I am and how much time it will take me to drive to my destination), but poor awareness of the vehicle ahead that is merging onto the highway. Improving situation awareness for navigation and for the merging vehicle would require very different designs. Note that SA does not define nor incorporate action. That concerns the decisions made from one's awareness or assessment of the situation.

Many elements of microcognition support SA and were covered in the previous chapter. Selective attention is necessary for the first stage, while the second stage of understanding depends very much upon both working memory and long-term memory. The third stage, projection and prediction, has not yet been discussed but will be considered in more detail when we discuss planning and scheduling. In addition, mental models guide SA development by defining what information people pursue and the interpretation of that information. For example, Amy's mental model of the operating room procedures might guide her to ask a nurse for estimated completion time for the perforated viscus procedure. She only asks about this procedure because her mental model of the other procedures gives her a good sense of when they would be done and so she only needs information about the procedure with an uncertain completion time.

As noted above, situation awareness is not the same as performance. One can have good performance (a lucky decision outcome that was correct) without good awareness. Correspondingly, the pilot of an out-of-control aircraft may have very good situation awareness of the loss of stability; but be unable to perform the necessary actions to recover.

7.5.1 Measuring Situation Awareness

The importance of SA can often be realized after an accident by inferring that the loss of SA was partially responsible. In controlled-flight-into-terrain accidents it is almost always assumed that the pilot lost awareness of the aircraft's altitude over the terrain [445]. However, "measuring" SA after the fact by assuming its absence is not the same as measuring how well a particular system or operator maintains SA in the absence of an unexpected event [447]. A popular technique for SA measurement is the SA global assessment technique (SAGAT); [448] in which the operator is briefly interrupted in the performance of a dynamic task and asked questions about it; for example, asking a driver to identify the location of other road traffic [449] or asking an anesthesiologist about the patient's state [450] or asking the pilot to identify the direction to the nearest hazardous terrain [451]. Sometimes the display is blanked after the question, to assure that the information is stored in memory. One can then assess the accuracy of answering such questions. Alternatively, one can assess the time required to retrieve the correct answer off of a display that remains visible, in a technique called SPAM (Situation Present Assessment Method) [452].

SA can sometimes be measured by a subjective evaluation ("rate your SA on a scale of 1 to 10" [453]), which has been embodied in a well used measurement tool called SART (situation awareness rating technique; Vidulich Tsang, Vu et al., 2015). However a concern about the validity of such self-rating techniques is that people are not always aware of what they are not aware. This issue of metacognition is addressed at the end of this chapter.

SA can be an important tool for accident analysis, understanding when its loss was a contributing factor [445]. To the extent that accidents may be caused by SA loss, an added implication is that systems should be designed and, when appropriate, certified to support SA. This becomes important when federal regulators are responsible for certification, such as the case with new aircraft or nuclear power plants.

7.5.2 Principles for Improving SA

Specific principles that follow from these considerations and from a recent review include [444]:

1. **Create displays that help people notice changes (Stage 1 SA).** Particularly in multitasking situations with dynamic systems, displays should highlight changes to make them easy for people to notice. Chapter 8 addresses issues of display layout to support SA.

2. **Make the situation easy to understand (Stage 2 SA).** Present information about the state of the system relative to the person's goals rather than require that they interpret and mentally combine and transform information. This might also mean bringing together there are several display elements that might otherwise be placed in different locations.

3. **Keep the operator somewhat "in the loop".** This issue will be addressed in more detail in Chapter 11 (Automation). The critical concept introduced here is related to the generation effect. People are more likely to remember actions, and the consequence of actions, if they themselves have generated the action, than if they were watching another agent generate the same action. Automobile manufacturers of self driving cars are struggling to find ways of keeping the driver somewhat in the loop (e.g., hands on the wheel), even as automation is imposing steering actions, in order to preserve SA, should automation fail.

4. **Help people project the state of the system into the future (Stage 3 SA),** This is particularly important when the system responds slowly, like a supertanker, industrial oven, or air traffic system. Here create a display that shows the future state, such as the predictive displays we discuss in Chapter 8. This relieves the person of mentally simulating and projecting future states.

5. **Organize information around goals.** Rather than arbitrary or technology oriented placement of information, displays should cluster information according to the goals the person is trying to achieve.

6. **Display to broaden attention.** Recognizing that SA may be most critical for dealing with unexpected situations, displays should avoid narrowing people's attention to a limited array of information that is specific to a particular task or limited to routine situations. Supporting SA when unexpected things happen typically means adding information to the display.

This information must be carefully integrated to avoid issues of clutter.

7. **Train for SA.** When training for SA, it is important to realize that training for routine performance may conflict with training to maintain SA [454]. The former will focus on the information needed for the task as it was intended to be performed. The latter should focus on what is often a broader scope of selective attention, to be aware of the state of the world should the system fail.) Many of the biases relevant to diagnosis, discussed above, are paralleled by biases in situation awareness: for example the confirmation bias or anchoring. Hence debiasing training, can be effective here.

7.6 Problem Solving and Troubleshooting

Many of the decision tasks studied in human factors require diagnosis, which is the process of inferring the underlying or "true" state of a system. Examples of inferential diagnosis include medical diagnosis, fault diagnosis of a mechanical or electrical system, inference of weather conditions based on measurement values or displays, and so on. Sometimes this diagnosis is of the current state, and sometimes it is of the predicted or forecast state, such as in weather forecasting or economic projections.

The cognitive processes of problem solving and troubleshooting are often closely linked because they have so many overlapping elements. Both start with a difference between an initial "state" and a final "goal state" and typically require a number of cognitive operations to reach the latter. The identity of those operations is often not immediately apparent to the human engaged in problem-solving behavior. Troubleshooting is often embedded within problem solving in that it is sometimes necessary to understand the identity of a problem before solving it. Thus, we may need to understand why our car engine does not start (troubleshoot) before trying to implement a solution (problem solving). Although troubleshooting may often be a step within a problem-solving sequence, problem solving may occur without troubleshooting if the problem is solved through *trial and error* or if a solution is accidentally encountered through serendipity.

While both problem solving and troubleshooting involve attaining a state of knowledge, both also typically involve performance of specific actions. Thus, troubleshooting usually requires a series of tests whose outcomes are used to diagnose the problem, whereas problem solving usually involves actions to implement the solution. Both are considered to be iterative processes of perceptual, cognitive, and response-related activities.

Both problem solving and troubleshooting impose heavy cognitive demands, and human performance is therefore often limited [455, 456]. Many of these limits are manifest in the heuristics and biases discussed earlier in the chapter, in the context of decision making. In troubleshooting, for example, people usually maintain no more than two or three active hypotheses in working memory as to the possible source of a problem [457]. More than this number overloads the limited capacity of working memory, since each hypothesis is complex enough to form more than a single chunk. Furthermore, when testing hypotheses, there is a tendency to focus on only one hypothesis at a time in order to confirm it or reject it. Thus, the engine troubleshooter will probably assume one form of the problem and perform tests specifically defined to confirm that it is the problem.

Naturally, troubleshooting success depends on attending to the appropriate cues and test outcomes. This dependency makes troubleshooting susceptible to attention and perceptual biases. The operator may attend selectively to very salient outcomes (bottom-up processing) or to outcomes that are anticipated (top-down processing). As we consider the first of these potential biases, it is important to realize that the least salient stimulus or event is the *nonevent*. People do not easily notice the absence of something [428]. Yet the absence of a symptom can often be a very valuable and diagnostic tool in troubleshooting to eliminate faulty hypotheses of what might be wrong. For example, the fact that a particular warning light might not be on could eliminate from consideration a number of competing hypotheses.

> Because it is hard to perceive the absence of something, people respond the presence of cues, but neglect their absence. People neglect the absence of pain when their hands freeze due to sever frostbite.

7.6.1 Principles for Improving Problem Solving and Troubleshooting

The systematic errors associated with troubleshooting suggest several design principles.

1. **Present alternate hypotheses**. An important bias in troubleshooting, resulting from top-down or expectancy-driven processing, is often referred to as cognitive tunneling, or confirmation bias [458, 401]. In troubleshooting, this is the tendency to stay fixated on a particular hypothesis (that chosen for testing), look for cues to confirm it (top-down expectancy guiding attention allocation), and interpret ambiguous evidence as supportive (top-down expectancy guiding perception). In problem solving, the corresponding phenomenon is to become fixated on a particular solution and stay with it even when it appears not to be working. Decision aids can challenge the persons' hypothesis and highlight disconfirming evidence.

2. **Create displays that can act as an external mental model.** These cognitive biases are more likely to manifest when two features characterize the system under investigation. First, high system complexity (the number of system components and their degree of coupling or links) makes troubleshooting more difficult [459]. Complex systems are more likely to produce incorrect or "buggy" mental models [460], which can hinder the selection of appropriate tests or correct interpretation of test outcomes. Second, intermittent failures of a

given system component turn out to be particularly difficult to troubleshoot [456]. A display that shows the underlying system structure, such as flow through the network of pipes in a refinery, can remove the burden of remembering that information.

3. **Create systems that allow people to post alternate hypotheses**. People generate a limited number of hypotheses because of working memory limitations [384]. Thus, people will bring in somewhere between one and four hypotheses for evaluation. Because of this people often fail to consider all relevant hypotheses [345]. Under time stress, decision makers often consider only a single hypothesis [461]. This process degrades the quality of novice decision makers far more than expert decision makers. The first option considered by experts is likely to be reasonable, but not for novices. Systems that make it easy for people to suggest many alternate hypothesis make it more likely a complete set of hypotheses will be considered.

7.7   Planning and Scheduling

The cognitive processes of planning and scheduling are closely related to those discussed in the previous section, because informed problem solving and troubleshooting often involve careful planning of future tests and activities. However, troubleshooting and diagnosis generally suggest that something is "wrong" and needs to be fixed. Planning and scheduling do not have this implication. That is, planning may be invoked in the absence of problem solving, as when a routine schedule of activities is generated. Planning often accompanies decision making to implement the course of action decided upon.

In many dynamic systems, the future may be broken down into two separate components: the predicted state of the system that is being controlled and the ideal or command state that should be obtained. Thus, a factory manager may have predicted output that can be obtained over the next few hours (given workers and equipment available) and a target output that is requested by external demands (i.e., the factory's client). When systems cannot change their state or productive output easily, we say they are sluggish, or have "high inertia." In these circumstances of sluggish systems, longer range planning becomes extremely important to guarantee that future production matches future demands. This is because sudden changes in demand cannot be met by rapid changes in system output. Examples of such sluggish systems–in need of planning–are the factory whose equipment takes time to be brought online, the airspace in which aircraft cannot be instantly moved to new locations, or any physical system with high inertia, like a supertanker or a train.

In time critical operations effective planning depends vitally upon anticipating events in the world that might derail the plan implementation. Unfortunately people are not very good at envisioning such events [345], nor the time required to address them. Hence the planning bias, discussed earlier in the chapter, is often manifest.

You will recognize the importance to planning of two concepts discussed earlier in this chapter. First, stage 3 situation awareness is another way of expressing an accurate estimate of future state and future demands. Second, skilled operators often employ a mental model of the dynamic system to be run through a mental simulation in order to infer the future state from the current state [369]. Mental simulation imposes heavy demands on cognitive resources. If these resources have been depleted or are diverted to other tasks, then prediction and planning may be poor, or not done at all, leaving the operator unprepared for the future.

7.7.1   Principles for Improving Planning and Scheduling

Human limits in the area of planning and scheduling are often addressed with automation. Operations research offers many approaches to design the best plan given certain assumptions. Unfortunately, reality often violates these assumptions and people must intervene.

1. **Create contingency plans and plan to re-plan.** In general, people tend to avoid complex planning schedules over long time horizons [462], a decision driven both by a desire to conserve the resources imposed by high working memory load and by the fact that in an uncertain world accurate planning is impossible, and plans may need to be revised or abandoned altogether as the world evolves in a way that is different from what was predicted. Re-planning is essential. Here, unfortunately, people sometimes fail to do so, creating what is known as a plan continuation error [463, 464], a form of behavior that has much in common with cognitive tunneling, the confirmation bias and the sunk cost bias. Contingency plans and planning to re-plan can avoid these tendencies.

2. **Create predictive displays.** As with problem solving and troubleshooting, a variety of automation tools are proposed to reduce these cognitive demands in planning [465]. Most effective are predictive displays that offer visual representations of the likely future, reducing the need for working memory [466]. We discuss these in the next chapter. Also potentially useful are computer-based planning aids that can either recommend plans [467] or allow fast-time simulation of the consequence of such plans to allow the operator to try them out and choose the successful one [468]. Air traffic controllers can benefit from such a planning aid known as the User Request Evaluation Tool (URET) to try out different routes to avoid aircraft conflicts [469].

7.8 Metacognition

Throughout this chapter we have cited the importance of meta-cognition: thinking about ones' own thinking and cognitive processes. Metacognition influences the decision-making process by guiding how people adapt to the particular decision situation. Here we highlight five of the most critical elements of meta-cognition for macrocognition.

1. **Knowing what you don't know.** That is, being aware that your decision processes or those necessary to maintain adequate situation awareness are inadequate because of important cues that are missing, and, if obtained, could substantially improve situation awareness and assessment.

2. **The decision to "purchase" further information.** This can be seen as a decision within the decision. Purchasing may involve a financial cost, such as the cost of an additional medical test required to reduce uncertainty on a diagnosis. It also may involve a time cost, such as the added time required before declaring a hurricane evacuation, to obtain more reliable information regarding the forecast hurricane track. In these cases, meta-cognition is revealed in the ability to balance the costs of purchase against the value of the added information [470]. The meta-cognitive skills here also clearly involve keeping track of the passage of time in dynamic environments, to know when a decision may need to be executed even without full information.

3. **Calibrating confidence in what you know.** As we have described above, the phenomenon of overconfidence is frequently manifest in human cognition [345], and when one is overconfident in ones' knowledge, there will be both a failure to seek additional information to reduce uncertainty, and also a failure to plan for contingencies if the decision maker is wrong in his/her situation assessment.

4. **Choosing the decision strategy adaptively.** As we have seen above, there are a variety of different decision strategies that can be chosen; using heuristics, holistic processing, System 1, recognition primed decisions, or deploying the more elaborate effort-demanding algorithms, analytic decision strategies using System 2. The expert has many of these in her toolkit, but meta-cognitive skills are necessary to decide which to employ when, as Amy did in our earlier example, by deciding to switch from an RPD pattern match, to a more time analytical strategy when the former failed.

5. **Processing feedback to improve the toolkit.** Item 4 relates to a single decision requirement—in Amy's case, the diagnosis and choice of treatment for one patient. However meta-cognition can and should also be employed to process the outcome of a series of decisions, realize from their negative

outcomes that they may be wanting, and learning to change the rules by which different strategies are deployed, just as the student, performing poorly in a series of tests, may decide to alter his/her study habits. To deploy such meta-cognitive skills here obviously requires some effort to obtain and process the feedback of decision outcomes, something we saw was relatively challenging to do with decision making.

7.8.1 Principles for Improving Metacognition

As with other elements of macrocognition, metacognition can be improved by some combination of changing the person (through training or experience) or changing the task (through task and technology).

1. **Ease information retrieval** Requiring people to engage in manual activity to retrieve information is more effortful than simply requiring them to scan to a different part of the visual field [131, 471], a characteristic that penalizes the concepts of multilevel menus and decluttering tools. Solutions to this problem are offered by pop-up messages and other automation features that can infer a user's information needs and provide them without imposing the effort cost of access [472].

2. **Highlight benefits and minimize effort of engaging decision aids** Designers must understand the effort costs generated by potentially powerful features in interfaces. Such costs may be expressed in terms of the cognitive effort required to learn the feature or the mental and physical effort and time cost required to load or program the feature. Many people are disinclined to invest such effort even if the anticipated gains in productivity are high. The feature will go unused as a result.

3. **Manage cognitive depletion.** An extended series of demanding decisions can incline people towards an intuitive approach to decisions, even when an analytic one would be more effective. Coaching people on this tendency might help them take rest breaks, plan complicated decisions early rather than late in the day, and avoid systems that introduce unnecessary decisions. People tend to make the easy or default decision as they become fatigued. As an example, Figure 7.9 shows how cognitive depletion changes the ruling of Israeli judges making parole decisions [473]. The timeline starts at the beginning of the day and each open circle represents the first decision after a break. The pattern cannot be explained by obvious confounding factors such as the gravity of the offense or time served. Similar effects are seen in other domains such as physicians choosing to prescribe more antibiotics as they become cognitively depleted over the day [474].

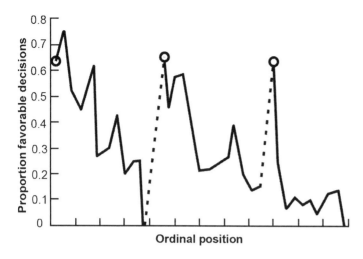

Figure 7.9 Effect of cognitive depletion on rulings in favor of prisoners (Source: Reprinted with permission from *Proceedings of National Academy of Sciences*, Dantziger, Levav, and Pesso (2011), Extraneous factors in judicial decisions. *PNAS*, 108, 17, Figure 1, p. 6890. [473].)

4. **Training metacognition.** For example, Cohen, Freeman, and Thompson [430] suggest training people to do a better job at metacognition, teaching people how to (1) consider appropriate and adequate cues to develop situation awareness, (2) check situation assessments or explanations for completeness and consistency with cues, (3) analyze data that conflict with the situation assessment, and (4) recognize when too much conflict exists between the explanation or assessment and the cues. Training in metacognition also needs to consider when it is appropriate to rely on the automation and when it is not.

7.9 Summary

We discussed decision making and the factors that make it more and less effective. Normative mathematical models of utility theory describe how people should compare alternatives and make the "best" decision. However, limited cognitive resources, time pressure, and unpredictable changes often make this approach unworkable, and people use simplifying heuristics, which make decisions easier but also lead to systematic biases. In many situations people often have years of experience that enables them to refine their decision heuristics and avoid many biases. Decision makers also adapt their decision making by moving from skill- and rule-based decisions to knowledge-based decisions according to the degree of risk, time pressure, and experience. This adaptive process must be considered when improving decision making through task redesign, choice architecture, decision-support systems, or training.

Techniques to shape decision making discussed in this chapter offer surprisingly powerful ways to affect decisions and so the ethical dimensions of these choices should be carefully considered. As an example, should the default setting be designed to provide people with the option that aligns with their preference, what is best for them, what is likely to maximize profits, or what might be best for society [413]? The concepts in this chapter have important implications for safety and human error, discussed in Chapter 14. In many ways the decision-support systems described in this chapter can be considered as displays or automation—Chapter 11 addresses automation, and we turn to displays in the next chapter.

Additional resources

Several useful resources that expand on the content touched on in this chapter include books that address decision making and its implications for daily life and for design:

1. **Decision making in daily life:** Ariely, D. (2008). *Predictably Irrational.* New York: HarperCollins.

   Duhigg, C. (2013). *The Power of Habit: Why we do what we do and how to change.* Random House.

   Gilbert, D. (2009). *Stumbling on Happiness.* Vintage Canada.

   Kahneman, D. (2011). *Thinking, Fast and Slow.* New York: Macmillan.

   Webb, A. (2013). *Data, a Love Story: How I cracked the online dating code to meet my match.* New York: Penguin.

2. **Cognitive engineering resources:** These books provide a more complete discussion of cognitive engineering and its applications to design.

   Hollnagle, E., & Woods, D. D. (2005). *Joint Cognitive Systems: Foundations of cognitive systems engineering.* Boca Raton: CRC Press.

   Lee, J. D., Kirlik, A. (2013). *The Oxford Handbook of Cognitive Engineering.* New York: Oxford University Press.

   Thaler, R. H., Sunstein, C. R. (2008). *Nudge: Improving Decisions about Health, Wealth, and Happiness.*

Questions

Questions for 7.1 Cognitive Environment of Macrocognition

P7.1   Describe the cognitive environment in terms of the three dimensions outlined at the start of the chapter.

P7.2   What is the relationship between microcognition and macrocognition?

P7.3   Describe five elements of macrocognition and how they relate to each other.

P7.4   What features of the cognitive environment make macrocognition important to consider, particularly compared to situations where the concepts of microcognition might dominate?

Questions for 7.2 Levels of Behavior: Skill and Expertise

P7.5   Use the task of driving to give examples of skill-, rule-, and knowledge-based behavior and the associated intuitive, heuristic, and analytical decision processes.

P7.6   What role do skill-, rule-, and knowledge-based behavior play in experts making decisions? What are the associated implications for interface design?

P7.7   Why might a normative approach, such as multi-attribute utility theory, indicate an optimal decision that is different from that of a person? Consider the decision to buy a lottery ticket.

Questions for 7.3 Decision Making

P7.8   Following a similar process outlined in the book, use multi-attribute utility theory to select between five potential places to live. Would you be happy to rely on the result?

P7.9   How does the concept of satisficing relate to heuristic decision making?

P7.10   What factors lead to intuitive, heuristic, and analytical decision making?

P7.11   Identify a situation where intuitive, heuristic, and analytical decision making would dominate.

P7.12   How does metacognition influence the type of decision making a person might adopt?

P7.13   What are the four main elements of the decision making process?

Questions for 7.4 Balancing Intuitive, Heuristic, and Analytic Decision Making

P7.14   Describe how primacy can undermine decisions, particularly in dynamic environments?

P7.15   What are two practical implications of the choice paradox?

P7.16   How do the availability and representative heuristics benefit experienced decision makers?

P7.17   Describe two elements of confirmation bias that might distort a decision maker’s understanding of the situation.

P7.18   Describe a situation in which the availability heuristic might work well and one where it might fail?

P7.19   How would a student frame his performance on the exam to most impress his parents?

P7.20   Describe a situation where the default heuristic affects millions of lives.

P7.21   How might you use task redesign to improve decisions?

P7.22   From the perspective of choice architecture how would you design a web application to enhance the retirement savings of employees?

P7.23   How is the gas consumption of cars best represented to present people with an easy to understand number?

P7.24   What is the decoy effect and how might you use it to guide people to an option that produces more revenue for your company?

P7.25   Describe how you might frame the decision to replace an expensive injection molding machine to help managers avoid the sunk cost bias.

P7.26   Why are heuristics important in making decisions and why do biases sometimes undermine their effectiveness?

P7.27   Given the diagnostic value of the following features, describe how you might design a display to minimize and maximize decision bias.

P7.28   Describe the benefits of a fast-and-frugal decision tree in supporting decisions.

Questions for 7.5 Situation Awareness

P7.29   Define situation awareness in terms of its three stages.

P7.30   Give examples of why situation awareness cannot be defined by performance on a given task.

P7.31   What method would you recommend to measure situation awareness? Defend your answer in terms of practicality and validity of the resulting data.

P7.32   One of the principles for supporting SA suggests that it should be supported in a broad fashion rather than narrowly. Describe a situation where this would be particularly important.

P7.33   Describe a principle for supporting each of the three levels of SA.

Questions for 7.7 Planning and Scheduling

P7.34   What is a danger of what-if analyses that can be performed with a spreadsheet?

P7.35   What type of design approach might you take to help people develop a good mental model to support trouble shooting?

P7.36   What kind of display might be particularly helpful in supporting planning and scheduling?

Questions for 7.8 Metacognition

P7.37   Give an example of task redesign that can help people become more effective decision makers.

P7.38   Why is an expert system often a poor way to support decision making?

P7.39   Describe how you might design a spreadsheet to support better decisions.

P7.40   What role does metacognition play in decision making?

P7.41   How might knowing about cognitive depletion change how you might manage a long series of important decisions?

Chapter 8

Displays

At the end of this chapter you will be able to...

1. explain how the power of representation can affect human performance

2. define information requirements and identify displays that support them best

3. apply 15 display design principles based on attention, perceptual, memory, and mental models

4. support tasks with appropriate displays from labels and graphs to complex systems, such as nuclear power plants

The operator of an energy-generating plant is peacefully monitoring its operation when suddenly an alarm sounds to indicate that a failure has occurred. Looking up at the top panel of the display warning indicators, he sees several warning tiles flashing, some in red, some in amber. Making little sense out of this "Christmas tree" pattern, he looks at the jumbled array of steam gauges and strip charts that show the continuously changing status of the plant. Some of the indicators appear to be out of range, but do not show any coherent pattern, and it is not easy to see which ones are associated with the warning tiles, arrayed in the separate display region above. He turns to the operating manual, which contains a well-laid-out flow diagram of the plant on the early pages. However, he must search for a page at the back to find information on the emergency warning indicators and locate still a different page describing the procedures to follow. Scanning rapidly between these five disconnected sources of information in an effort to understand what is happening within the plant, he ultimately fails and the plant fails catastrophically.

Our unfortunate operator could easily sense the changes in display indicators and read the text and diagrams in the manual. He could perceive individual elements, but the poorly integrated displays made it difficult to interpret the overall meaning. In Chapters 4 and 5 we described how the various sensory systems (primarily the eyes and ears) process the raw sensory information (light and sound). In Chapters 6 and 7 we described how this perceived information is processed further and stored temporarily in working memory, or more permanently in long-term memory, and used for diagnosis and decision making. This chapter focuses on *displays*, which are artifacts designed to guide attention to relevant system information, and then support its perception and interpretation (Figure 8.1). A speedometer in a car; a warning tone in an aircraft, a text message on a mobile phone, an instruction panel on an automatic teller, a gauge in an industrial plant, a PowerPoint slide, a web page, a graph, a map, and a nutrition label on a food package are all examples of displays.

The concept of the display is closely linked with that of the user interface. A user interface differs from a display in that it both displays information and accepts input to manipulate the display and control the system. Chapter 9 focuses on input devices and controlling a system, as in using a steering wheel to control the path of a car. Chapter 10 focuses on human-computer interaction, where the where displays and controls are integrated into a user interface, as in the gestural interface of a tablet computer.

Figure 8.1 shows the role of displays. Displays show information that describes the state of a system or action requested of the person. People perceive this information through top-down processing guided by their mental model, and through bottom-up processing driven by the displayed information. Through this perceptual process, people become aware of what the system is doing and what needs to be done. Displays connect people to the system by bridging the *gulf of evaluation*—the difference between actual state of the system and the people's understanding of the system relative to their goals [16]. Consequently, effective displays must include the information from the system relevant to the intended tasks of people and represent this information in a manner that is compatible with the perceptual and cognitive properties of people. To ensure this compatibility, we present 15 human factors principles to guide the design of displays. This compatibility also depends on presenting the information needed for people's tasks, and hence should follow a careful task analysis. We describe six categories of tasks that displays can support, demonstrating the application of the 15 principles. Before we present and apply these principles, we describe types of displays.

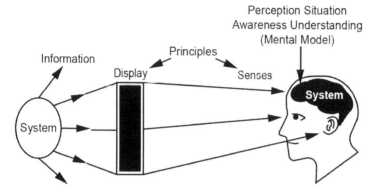

Figure 8.1 Displays convey information about the system state to the person. A system generates information, some of which must be processed by the operator to perform a task. That information is presented on a display in a way that supports perception, situation awareness, and understanding. Often, an accurate mental model of the system facilitates this understanding.

8.1   Types of Displays and Tasks

We can classify displays along at least three dimensions: their physical features, the tasks they support, and the properties of people that dictate the best match of display and task. The *physical features* of displays are what the designer has to work with in building a display. For example, a designer might code information with color, shape, position, motion, or angle; or she might use some combination of visual, auditory, or haptic modalities; or even place display elements in different locations.

These features of displays are mentioned at various points in the chapter. However, the choice of these features depends on understanding the task the display is intended to support: navigation, control, decision making, training and so forth. However, defining the task is only a first step. Once the task is defined (e.g., navigate from point A to point B) we must complete an *information analysis*, as described in Chapter 2, to identify what the person needs to know for the task.

Successful displays consider the task *and* cognitive capabilities.

Task	Static or Dynamic	Data elements	Important principles
Alert	Dynamic	Few	Attention
Label	Static	Few	Perception
Monitor	Dynamic	Many	Attention, Mental model
Integrate	Dynamic	Very many	Memory, Mental model
Navigate, Guide	Dynamic	Many	Perception, Mental model
Visualize	Static	Many	Attention, Perception

Table 8.1 Tasks that define display types, along with their features and related principles.

Finally, and most importantly, no single display type is best suited for all tasks. For example, a digital display that is best for reading the exact value of an indicator is not good for assessing, at a quick glance, the approximate value or rate of change of the indicator. As Figure 8.1 shows, the best mapping between the physical form of the display and the task requirements depends on principles of human perception and information processing. These *principles* are grounded in the strengths and weaknesses of human perception and cognition, as described in Chapters 4 through 7), and it is through the careful application of these principles to the information analysis that the best displays emerge. Table 8.1 shows types of displays defined by the tasks they support and the most relevant categories of principles that can guide their design.

Figure 8.2 shows the Chernobyl nuclear power plant control room. Similar to the opening vignette, these displays failed to provide the operators with a complete picture of the plant's state and contributed to the catastrophic failure of the plant [475, 476]. On the wall are alerts in the form of annunciator panels that indicate when variables exceed permissible levels and gauges that show pressures and flow rates. Representing this information in an integrated display can help operators understand the state of the plant so that they can respond effectively. Without an effective display even simple systems can quickly overwhelm people—think about playing tic-tac-toe without a pencil and paper.

Figure 8.2 Reactor operator's console and core display in the Unit 1 main control room, Chernobyl nuclear power plant. (Source: (WT-shared) Carlwillis at wts wikivoyage/CC BY-SA 3.0.)

8.2 Fifteen Principles of Display Design

One of the basic tenets of human factors is that lists of longer than four or five items are not easily retained unless they are given with some organizational structure. To help retention of the otherwise daunting list of 15 principles of display design, we put them into four categories: (1) those that relate to *attention*, (2) those that directly reflect *perceptual* operations, (3) those that relate to *memory*, and (4) those that can be traced to the concept of the *mental model*. Some of these principles have been introduced in previous chapters (4, 5, and 6) and others will be discussed more fully later in this chapter. Each can be applied to a variety of displays.

8.2.1 Principles Based on Attention

Complex multi-element displays require two components of attention to process [477]. As discussed in Chapter 6, *selective attention* may be necessary to choose the displayed information sources necessary for a given task, allowing those sources to be perceived without distraction from neighboring sources. *Divided attention* may allow parallel processing of two (or more) sources of information if a task requires it such as when processing displays of time and speed to understand the distance traveled. The four attentional principles described next characterize ways of capitalizing on attentional strengths or minimizing their weaknesses in designing displays.

1. **Salience compatibility**. Important and urgent information should attract attention. Critical information that fails to attract people's attention will not be processed. As described in Chapter 6, features of the display such as contrast, color, and flashing increase salience. Auditory alerts are also highly salient and can attract attention no matter where attention

might be directed [213]. As noted in Chapter 4, highly urgent sounds are a particularly powerful way to attract attention [478, 479]. To avoid alarm fatigue and annoyance, salience should be compatible with the importance of the information: highly salient indicators should be used for highly important information.

2. **Minimize information access cost.** It costs time and effort to "move" selective attention from one display location to another to access information [224]. The operator in the opening story wasted valuable time going from one page to the next in the book and visually scanning from there to the instrument panel. The information access cost may also include the time required to proceed through a computer menu to find the correct "page." Thus, good designs are those that minimize this cost by keeping frequently accessed sources in locations where the cost of traveling between them is small. We discuss this principle again in the context of workplace layout in Chapter 10. One direct implication of minimizing access cost is to keep displays small so that little scanning is required to access all information. Such a guideline should be employed carefully because, as we learned in Chapter 4, very small size can degrade legibility [480].

3. **Proximity compatibility.** When two or more sources of information are related to the same task and must be mentally integrated to complete the task (e.g., a graph line must be related to its legend, or the plant layout must be related to the warning indicator meanings in our opening story) these information sources are defined to have close mental proximity. Good displays should arrange information so their mental proximity is reflected in their display proximity, producing high proximity compatibility [481]. Placing two sources that need to be integrated close to each other is one way to increase display proximity; however, there are other ways of obtaining close display proximity than nearness in space including: displaying in a common color, using a common format, linking them with lines, or by configuring them in a pattern. These five techniques are shown in Figure 8.3a.

However, as Figure 8.3b shows, display proximity is not always good, particularly if one element is the focus of selective attention. In this case, the elements have low mental proximity and so they should have low display proximity. Overlapping images can make their individual perception hard. "Low mental proximity" of tasks is best served by "low display proximity" where the elements are separated, not overlaid or tightly packed together. In summary, if mental proximity is high (information must be integrated), then display proximity should also be high (close in space, color, format, linkage, or configuration). If mental proximity is low (elements require focused attention), the display proximity can, and sometimes should, be lower.

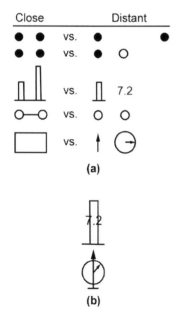

Figure 8.3 Proximity compatibility. a) Five examples of close display proximity on the left. Proximity defined by (1) space, (2) color (or intensity), (3) format, (4) links, and (5) object configuration. (b) Two examples of close spatial proximity (overlay) that make it hard to focus on one indicator and ignore the other.

4. **Avoid resource competition.** Multiple resource theory describes information processing demands. As we discussed in Chapter 6, sometimes processing a lot of information can be facilitated by dividing that information across resources—presenting visual and auditory information concurrently—rather than presenting all information as visually or auditorily. For example, present the speed the car on the speedometer and indicate an unbelted seat belt with an auditory alert.

8.2.2 Perceptual Principles

5. **Make displays legible (or audible).** This guideline is not new. It integrates nearly all of the information discussed in Chapters 4 and 5, relating to issues such as contrast, visual angle, illumination, noise, masking, and so forth. Legibility is so critical to the design of good displays that it is essential to restate it here. Legible displays are necessary, although not sufficient, for creating usable displays. The same is true for audible displays. Once displays are legible, additional perceptual principles should be applied.

6. **Avoid absolute judgment limits.** As we noted in Chapters 4 and 5 when discussing alarm sounds, we should not require the operator to judge the level of a represented variable based on a single sensory variable, like color, size, or loudness, which contains more than five to seven possible levels (Figure 8.4). To require greater precision, as in a color-coded map with nine hues, is to invite errors of judgment. More generally, people are much more sensitive to some variables, such as position along an axis than they are size, which is discussed in detail in the context of graphs at the end of this chapter.

Absolute Judgment:
"If the light is amber, proceed with caution."

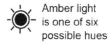
Amber light is one of six possible hues

Figure 8.4 Avoid absolute judgment, such as requiring people to associate labels to levels of color.

7. **Support top-down processing.** People perceive and interpret signals in accordance with what they expect to perceive based on their past experience. If a signal is presented that is contrary to expectations, like the warning or alarm for an unlikely event, then more physical evidence of that signal must be presented to guarantee that it is interpreted correctly. Sometimes expectancies are based on long-term memory. However, in the example shown in Figure 8.5, these expectations are based on the immediate context of encountering a series of "on" messages, inviting the final line to also be perceived as on. In such circumstances the word **OFF** should be made more salient.

Top-Down Processing:
A Checklist

| A should be on |
| B should be on |
| C should be on |
| D should be off |

Figure 8.5 Consider how top-down processing guides perception (a tendency to perceive as "D should be on")

8. **Exploit redundancy gain.** When the viewing or listening conditions are degraded, a message is more likely to be interpreted correctly when the same message is expressed more than once (Lu et al., 2013). This is particularly true if the same message is presented in multiple physical forms (e.g., tone and voice, voice and print, print and pictures, color and shape); that is, redundancy is not simply the same as

Redundancy Gain:
The Traffic Light

Position and hue are redundant

Figure 8.6 Use redundancy gain to avoid confusion by presenting the same information through different channels.

Similarity:
Confusion

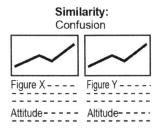

Figure 8.7 Avoid perceptual confusion associated with making important distinctions with subtle differences.

repetition. When alternative physical forms are used, there is a greater chance that the factors that might degrade one form (e.g., noise degrading an auditory message) will not degrade the other (e.g., printed text). The traffic light (Figure 8.6) is a good example of redundancy gain. Unlike the principle of avoiding resource competition, exploiting redundancy gain reinforces a single message by presenting through multiple modalities, whereas the principle of avoiding resource competition uses multiple modalities to communicate two different messages (e.g., speed and set belt status).

9. **Make discriminable.** Similarity causes confusion. Similar appearing signals are likely to be confused either at the time they are perceived or after some delay, if the signals must be retained in working memory before action is taken. Similarity is the ratio of similar features to different features. Thus, AJB648 is more similar to AJB658 than is 48 similar to 58, even though in both cases only a single digit is different. Where confusion could be serious, the designer should delete unnecessary similar features and highlight dissimilar ones in order to make them distinctive. Note, for example, the high degree of confusability of the two captions in Figure 8.7. You may need to look very closely to see its discriminating feature ("l" versus "t") in the figure caption. Poor legibility also amplifies the negative effects of poor discriminability. The "tallman" method, discussed in Chapter 4, that highlights difference in drug names by capitalizing discriminating letters and making them bold is an example of making labels more discriminable.

8.2.3 Memory Principles

Human memory is vulnerable, particularly working memory because of its limited capacity, as discussed in Chapter 6: We can keep only a small number of "mental balls" in the air at one time, and so, for example, we may easily forget a phone number before we have had a chance to dial it or write it down. Our operator in the opening vignette had a hard time remembering information on one page of the manual while reading the other. Our long-term memory is vulnerable because we forget certain things or sometimes because we remember other things too well and persist in doing them when we should not. The final three principles address different aspects of these memory processes.

10. **Knowledge in the world.** Replace memory with visual information. The importance of presenting knowledge in the world is the most general memory principle, echoing guidelines presented in Chapter 6[16]. People ought not be required to retain important information solely in working memory or retrieve it from long-term memory. There are several ways that this is manifest: the visual echo of a phone number (rather than reliance on the fallible phonetic loop), the checklist (rather than reliance on prospective memory),

and the simultaneous rather than sequential display of information to be compared as dictated by the proximity compatibility principle (Principle 3). Of course, sometimes too much knowledge in the world can lead to clutter, and systems designed to rely on knowledge in the head are not necessarily bad.

11. **Support visual momentum.** Displays that include multiple separated elements, such as windows or pages, require people to remember information from one display so they can orient to another. This is particularly true for sequentially viewed displays. Supporting visual momentum helps reduce this memory load and makes it easier to integrate inform across display elements. Methods to enhance visual momentum include: providing context for the detail (e.g., map of the US highlighting the location of the state map that is shown in detail) and including perceptual landmarks and a fixed structure that is used for multiple display elements (e.g., the consistent layout of each page of most websites). Figure 8.8 shows examples perceptual landmarks and fixed structure that makes it easier to see how the pattern across months differs across years.

12. **Provide predictive aiding.** Humans are not very good at predicting the future. This limitation results mostly because prediction relies heavily on working memory. We need to think about current conditions, possible future conditions, and then "run" the mental model by which the former may generate the latter. When our mental resources are consumed with other tasks, prediction falls apart and we become reactive, responding to what has already happened, rather than proactive, responding in anticipation of the future. Because proactive behavior is usually more effective than reactive, displays that can explicitly predict what will (or is likely to) happen are generally quite effective in supporting human performance. A predictive display replaces a resource-demanding cognitive task and with a simpler perceptual one. Figure 8.4 shows some examples of effective predictor displays.

13. **Be consistent.** When our long term-memory works too well, it may continue to trigger actions that are no longer appropriate, and this an automatic human tendency. Old habits die hard. Because there is no way to avoid this, good designs should try to accept it and design displays in a manner that is consistent with other displays that the user might use at the same time (e.g., a user alternating between two computer systems) or might be familiar with from other experiences. Hence, the habits from those other displays will transfer positively to support processing of the new displays. Thus, for example, color coding should be consistent across a set of displays so that red always means the same thing. To be consistent with most other displays, red should be used to indicate danger or states to be avoided. As another exam-

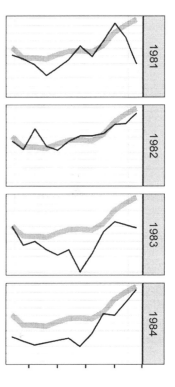

Figure 8.8 Features that support visual momentum: the common structure (consistent axis range) and visual landmark (grey line that shows the mean values across the years).

ple, a set of display panels should be consistently organized, thus reducing information access cost each time a new set is encountered.

8.2.4   Mental Model Principles

When operators perceive a display, they often interpret what the display looks like and how it moves in terms of their expectations or mental model of the system being displayed, a concept discussed in Chapter 6 [16]. The information presented to our reactor operator in the opening story was not consistent with the mental model of the operator. Hence, it is good to format the display to capture aspects of a correct mental model in a way that reflects user's experience with the system. Principles 14 and 15 illustrate how this can be achieved.

14. **Principle of pictorial realism.** A display should look like (i.e., be a picture of) the variable that it represents [482]. Thus, if we think of temperature as having a high and low value, a thermometer should be oriented vertically. If the display contains multiple elements, these elements can sometimes be configured in a manner that looks like how they are configured in the environment that is represented (or how the operator conceptualizes that environment).

15. **Principle of the moving part.** The moving element(s) of any display of dynamic information should move in a spatial pattern and direction that is compatible with the user's mental model of how the represented element actually moves in the physical system [482]. Thus, if a pilot thinks that the aircraft moves upward when altitude is gained, the moving element on an altimeter should also move upward with increasing altitude.

8.2.5   Summary of principles

In concluding our discussion of principles, it should be immediately apparent that principles sometimes conflict or "collide." Making all displays consistent, for example, may sometimes cause certain displays to be less compatible than others, just as making all displays compatible may make them inconsistent. Putting too much knowledge in the world or incorporating too much redundancy can create cluttered displays, which can undermine attention to some display elements. Minimizing information access effort by creating very small displays can reduce legibility. Alas, there is no easy resolution when two or more principles collide. But creative design can sometimes avoid conflicts, and Table 8.1 can help identify the most critical principles for a given display. We now turn to a discussion of various categories of displays, illustrating how certain principles have been applied. As we encounter each principle in an application, we place a reminder of the principle number in parentheses, for example, (A4) refers to the principle

Display Design Principles

Attention principles
A1 Salience compatibility
A2 Minimize information access cost
A3 Proximity compatibility
A4 Avoid resource competition

Perception principles
P5 Make displays legible (or audible)
P6 Avoid absolute judgment limits
P7 Support top-down processing
P8 Exploit redundancy gain
P9 Make discriminable

Memory principles
M10 Knowledge in the world
M11 Support visual momentum
M12 Provide predictive aiding
M13 Be consistent

Mental model principles
MM14 Pictorial realism
MM15 Moving part

Table 8.2 Display design principles.

of multiple resources, the fourth principle discussed under attention. The letter refers to the category: attention (A), perception (P), memory (M), and mental model (MM). See Table 8.2 for a list.

In the following sections we apply these principles to specific applications. With these applications we use the term "guidelines" to distinguish them from the 15 principles; the guidelines are more specific design suggestions derived from the principles.

8.3 Alerts

We discussed alerting displays to some extent in Chapter 4 in the context of noticing and detection, and in Chapter 5 in the context of auditory warnings, and shall do so again when we discuss both automation (Chapter 11). If it is critical to alert the operator to a particular condition, then the omnidirectional auditory channel is best. However, there may well be several different levels of seriousness of the condition to be alerted, and not all of these need or should be announced auditorily, thus matching the salience of the alert to the importance of the information is critical (A1). For example, I do not need a time-critical and intrusive auditory alarm to tell me that my car has passed a mileage level at which a particular service is needed.

Conventionally, system designers have classified three levels of alerts—warnings, cautions, and advisories—which can be defined in terms of the severity of consequences of failing to heed their indication. Warnings, the most critical category, should be signaled by salient auditory alerts; cautions may be signaled by auditory alerts that are less salient (e.g., softer voice signals); advisories need not be auditory at all, but can be purely visual. Both warnings and cautions can clearly be augmented by redundant visual or tactile signals as well (P8). When using redundant vision for alerts, flashing lights are effective because the onsets that capture attention occur repeatedly. Each onset is itself a redundant signal. In order to avoid possible confusion of alerting severity, the aviation community has also established explicit guidelines for consistent color coding (M13), such that warning information is always red; caution information is yellow or amber; advisory information should be other colors (e.g., blue, white), which is clearly discriminable (P6) from red and amber.

Note that the concept of defining three levels of condition severity is consistent with the guidelines for "likelihood alarms" discussed in Chapter 5 [483], in which different degrees of danger or risk are signaled to the user, rather than simply as a binary safe versus unsafe indicator.

Salience compatibility—matching salience of an alert display with the urgency of the information—is particularly critical in alert design.

8.4 Labels and Icons

Labels may also be thought of as displays, although they are generally static and unchanging features for the user. Their purpose is to unambiguously signal the identity or function of an entity, such as a control, display, piece of equipment, entry on a form, or other system component; that is, they present knowledge in the world (M10) of what something is. Labels are usually presented as print but may sometimes take the form of icons. The four design criteria for labels, whether presented in words or pictures, are visibility, discriminability, meaningfulness, and location. Figure 8.9 shows icons from Google's design guide. Simple images, with little detail and consistent, symmetrical forms and makes these icons easy to see and discriminate.

Figure 8.9 Typical icons from Google's design specifications. https://www.google.com/design/spec/style/icons.html#icons-product-icons.

1. **Visibility and legibility (P5).** This criterion relates directly back to issues of contrast sensitivity, discussed in Chapter 4. Stroke width of lines (in text or icons) and contrast from background must be sufficient so that the shapes can be discerned under the poorest expected viewing conditions. This means that the icons should not include high spatial frequency components (e.g., small features and thin lines).

2. **Discriminability (P9).** This criterion dictates that any feature that is necessary to discriminate a given label from an alternative is clearly and prominently highlighted. We noted that confusability increases with the ratio of shared to distinct features between potential labels.

   As described in Chapter 6, a special "asymmetrical" case of confusion is the tendency to confuse negative labels ("no exit") with positive ones ("exit"). Unless the negative (e.g., "no," "do not," and "don't") is clearly and saliently displayed, it is very easy for people to miss it and assume the positive version, particularly when viewing the label (or hearing the instructions) under degraded sensory conditions. Even if understood correctly people will be slower to respond.

3. **Meaningfulness.** Even if a word or icon is legible and not confusable, this is no guarantee that it triggers the appropriate meaning in the mind of the viewer when it is perceived. Unfortunately, too often icons, words, or acronyms that are meaningful in the mind of the designer are meaningless in the mind of the actual users. This mismatch reflects the

mismatch of designers' expectations and those of the users. Icons should be designed with the expectations and associated top-down processing of actual users in mind (P7). Because this unfortunate situation is far more likely to occur abbreviations and icons than with words, we argue that labels based only on icons or abbreviations should be avoided where possible [484]. Icons may well be advantageous where the word labels may be read by those who are not fluent in the language (e.g., international highway symbols) and sometimes under degraded viewing conditions; thus, the *redundancy gain* (P8) that such icons provide is usually of value. But the use of icons alone appears to carry an unnecessary risk when comprehension of the label is important. The same can be said for abbreviations. When space is small—as in the label of a key that is to be pressed—effort should be made to perceptually "link" the key to a verbal label that may be presented next to the key..

Pair labels with icons to ensure people interpret the icon correctly.

4. **Location.** An obvious but sometimes overlooked feature of labels: They should be physically close to and unambiguously associated with the entity that they label, thereby adhering to the proximity compatibility principle (A3). Note how the placement of labels in Figure 8.10 violates this. While the temperature label is close to the display indicating temperature, the speed label is far from the speed display. A similar issue concerns the location of displays relative to controls, termed *stimulus-response compatibility*, which we discuss in Chapter 9.

Temp   Speed

Figure 8.10 Labeling displays. It is important to have clear associations between displays and labels.

As described in Chapter 5, many of the considerations associated with icons also apply to designing sounds in creating earcons, synthetic sounds that have a direct, meaningful association with the thing they represent. In choosing between icons and earcons, it is important to remember that earcons (sound) are most compatible for indicating actions or events that play out over time (e.g., informing that a computer command has been accomplished), whereas icons are better for labeling *states or variables*.

8.5 Monitoring Displays

Displays for monitoring are those that support the viewing of potentially changing quantities, usually represented on some analog or ordered value scale, such as speed, temperature, noise level, or changing machine status. A variety of tasks may need to be performed based with such displays. A monitored display may need to be set, as when an appropriate frequency is dialed in to a radio channel. It may simply need to be watched until it reaches a value at which some action is taken, or it may need to be tracked, in which case another variable must be manipulated to follow the changing value of the monitored variable. (Tracking is discussed in considerably more detail in Chapter 9.) Whatever the action to be taken based on the monitored variable, discrete or continuous, immediate or delayed, four important guidelines can guide the design of monitoring displays.

1. **Legibility.** Display legibility (P5) is of course the familiar criterion we revisited in the previous section, and it relates to the issues of contrast sensitivity discussed in Chapter 4. If monitoring displays are digital, the issues of print and character resolution must be addressed. If the displays are analog dials or pointers, then the visual angle and contrast of the pointer and the legibility of the scale against which the pointer moves become critical. A series of guidelines may be found in Helander [485] and MIL-STD-1472G [49] to assure such legibility. Designers must consider degraded viewing conditions (e.g., low illumination) under which such scales may need to be read, and they must design to accommodate such conditions.

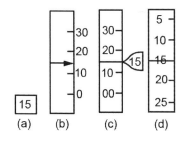

Figure 8.11 Analog and digital displays for monitoring. (a) digital display; (b) moving pointer analog display; (c) moving scale analog display with redundant digital presentation; (d) inverted moving scale display adheres to principle of the moving part. Both (b) and (c) adhere to the principle of pictorial realism.

2. **Analog versus digital.** Most variables to be monitored are continuously changing quantities. Furthermore, users often form a mental model of the changing quantity. Hence, adhering to the principle of pictorial realism (MM14, [482]) would suggest the advantage of an analog (rather than digital) representation of the continuously changing quantity [486]. In comparison to digital displays (Figure 8.11a), analog displays like the moving pointer in Figure 8.11b can be more easily read at a short glance; the value of an analog display can be more easily estimated when the display is changing, and it is also easier to estimate the rate and direction of that change. At the same time, digital displays do have an advantage if very precise "check reading" or setting of the exact value is required. But unless these are the only tasks required of a monitoring display, and the value changes slowly, then if a digital display is used, it should be redundantly paired with its analog counterpart (P8), like the altitude display shown in Figure 8.11c.

3. **Analog form and direction.** If an analog format is chosen for display, then the principle of pictorial realism (MM14; [482]) would state that the orientation of the display scale should be in a form and direction congruent with the operator's mental model of the displayed quantity. Cyclical or circular variables (like compass direction or a 24-hour clock) share an appropriate circular form of a round dial or "steam gauge" display, whereas linear quantities with clearly defined high and low points should be reflected by linear scales. These scales should be vertically arrayed so that high is up and low is down. This orientation feature is easy to achieve for a fixed-scale moving pointer display (Figure 8.11b) or a moving scale fixed-pointer display shown in Figure 8.11c.

Many displays are fairly dynamic, showing visible movement while the operator is watching or setting them. The principle of the moving part (MM15) suggests that displays should move in a direction consistent with the user's mental model:

An increase in speed or any other quantity should be signaled by a movement upward on the moving element of the display (rightward and clockwise are also acceptable, but less powerful movement stereotypes for increase). Although the moving pointer display in Figure 8.11b adheres to this stereotype, the moving scale display in Figure 8.11c does not. Upward display movement will signal a decrease in the quantity. The moving scale version in Figure 8.11d, with the scale inverted, can restore the principle of the moving part, but only by violating the principle of pictorial realism (MM14) because the scale is now inverted. Moving scale displays are also hard to read if the quantity is changing rapidly.

Despite its advantages of adhering to the principles of both pictorial realism and the moving part, there is one cost with a linear moving pointer display (Figure 8.11b). It cannot present a wide range of scale values in a small physical space. If the variable travels over a large range and the required reading precision is also high (a pilot's altimeter, for example), this can present a problem. One answer is the moving scale display, which can present a wide range of numbers with precision. If the variable does not change rapidly (i.e., there is little motion such as a fuel gauge), then the principle of the moving part has less relevance, and so its violation imposes less of a penalty. A second option is to use circular moving pointer displays that take less space, as with a speedometer. While circular displays are less consistent with the principle of pictorial realism (if displaying linear quantities), they are consistent with the stereotype of increase clockwise [487].

A third possibility is a hybrid scale in which high-frequency changes of the displayed variable drive a moving pointer against a stable scale, while sustained low-frequency changes can gradually shift the scale quantities to the new (and appropriate) range of values as needed (maintaining high numbers at the top) [482, 10]. Such a display adheres to the principles of the moving part and pictorial realism, but is not consistent. When the pointer is in the middle of the scale it indicates a different value in different situations.

Clearly, as in any design solution, there is no "magic layout" that is ideal for all circumstances. As always, task analysis is important, and should identify the rate of change of the variable, its needed level of precision, and its range of possible values. This information can help identify a display format that fits the task. The exercise in Table 8.3 highlights this challenge.

One final factor influencing the choice of display concerns the nature of control that may be required to set or to track the displayed variable. Fortunately for designers, many of the same laws of display expectations and mental models apply to control; that is, just as people expect (MM15) that an upward (or clockwise) movement of the display signals an increasing quantity, the user also expects that an upward

Design exercise: Design an monitoring display for a drone

Consider a display to indicate the altitude of a remotely operated drone. A task analysis identified that the altitude of the drone can vary from 0 to 10,000 ft above the ground and this information is needed with a precision of 100 to coordinate with other drones. At the same time the operator must tell at a glance whether it is high (greater than 5000ft) or low (less than 500ft). The drone climbs and descends relatively slowly, with a maximum change of 500 ft/minute.

Design an altimeter that considers the principles of the moving part, pictorial realism, and consistency.

Table 8.3 Design challenge: An altimeter for a drone.

(or clockwise) movement of the control will be required to increase the displayed quantity [487]. We revisit this issue in more detail in Chapter 9 when we address issues of display-control compatibility.

4. **Prediction and sluggishness.** Many monitored variables in high-inertia systems, like ships or chemical processes, are sluggish in that they change relatively slowly. But as a consequence of the dynamic properties of the system that they represent, the slow change means that their future state can be known with some degree of certainty. Such is the case of the supertanker, for example: Where the tanker is now in the channel and how it is moving (speed and turn rate) will quite accurately predict where it will be several minutes into the future. Another characteristic of such systems is that efforts to control them which are executed now will also not have an influence on their state until much later. Thus, the shift in the supertanker's rudder will not substantially change the ship's course until minutes later, and the adjustment of the heat delivered to a chemical process will not change the process temperature until much later (Chapter 16). Hence, control should be based on the operator's prediction of future state, not present conditions. But as we discussed in Chapter 7, prediction is not something we do very well, particularly under stress; hence, good *predictive displays* (M12) can be a great aid to human monitoring and control performance (Figure 8).

Predictive displays of physical systems are typically driven by a computer model of the dynamics of the system under control and by knowledge of the current and future inputs (forces) acting on the system. Because, like the crystal ball of the fortune-teller, these displays are driven by automation making inferences about the future, they may not always be correct and are less likely to be correct the further into the future the prediction [488]. People tend to treat predictions as actual future states. Hence, the designer should be wary of showing a prediction further forward than is reasonable and might consider depicting limits on the degree of certainty of the predicted variable. For example, a display could predict the most likely state and the 90 percent confidence interval around possible states that could occur a certain time into the future. This confidence interval will grow as that time—the *span of prediction*—increases. This is what is displayed in the "cone of uncertainty" surrounding forecast hurricane tracks (Figure 8.12).

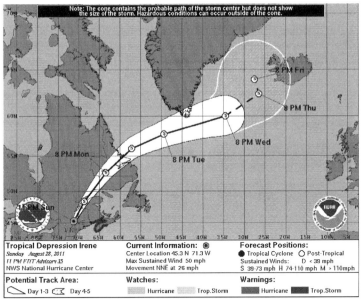

Figure 8.12 Predictive display of the track of tropical depression Irene with a cone of uncertainty. (Source: National Hurricane Center, www. nhc.noaa.gov)

8.6 Integrative Displays

Many real-world systems are complex. The typical nuclear reactor may more than 35 variables that are considered critical for its operation, and an aircraft has at least seven that must be monitored in even the most routine situation. Hence, an important issue in designing multiple displays is to decide where they go, that is, what should be the layout of the multiple displays [489, 490]. In the following section we discuss several guidelines for display layout, and while these are introduced in the context of monitoring displays, these guidelines apply to nearly any type of display, such as the layout of elements of a Web page [69]. Following the discussion of display layout we address similar issues related to head-up displays and configural displays.

8.6.1 Display Layout

In many work environments, the designer may be able to define a primary visual area (PVA) (see Chapter 12). For the seated user, this maybe the region of forward view as the head and eyes look straight forward. For the vehicle operator, it may be the direction of view of the highway (or runway in an aircraft approach). The PVA defines the reference point for many display layout guidelines.

1. **Frequency of use** dictates that frequently used displays should be closer to the PVA. This makes sense because their frequent access dictates a need to "minimize the travel time"

between them and the PVA (A2). Note that sometimes a very frequently used display can itself define the PVA. With the conventional aircraft display suite shown in Figure 8.13, this principle is satisfied by positioning the most frequently used instrument, the attitude indicator, at the top and center, closest to the view out the windshield on which the pilot must fixate to land the aircraft and check for other traffic.

2. **Importance of use**, is closely related to frequency of use, but dictates that important information, even if it may not be frequently used, be displayed so that attention will be captured when it is presented. While displaying such information within the PVA often accomplishes this, other techniques, such as auditory alerts coupled with guidance of where to look to access the information, can accomplish the same goal.

3. **Display relatedness or sequence of use** dictates that related displays and those pairs that are often used in sequence should be close together. (Indeed, these two features are often correlated. Displays are often consulted sequentially because they are related, like the commanded setting and actual setting of an indicator.) This principle captures the key feature of the proximity compatibility principle (A3) [481]. We saw the manner in which it was violated for the operator in our opening story. As a positive example, the aircraft cockpit display layout in Figure 8.13, the vertical velocity indicator and the altimeter, in close spatial proximity on the right side, are also related to each other, since both present information about the vertical behavior of the aircraft. The figure caption also describes other examples of related information in the instrument panel.

4. **Consistency** is related to both memory and attention. If displays are always consistently laid out with the same item positioned in the same spatial location (M13), then our memory of where things are serves us well, and memory can easily and automatically guide selective attention to find the items we need. Stated in other terms, top-down processing can guide the search for information in the display. Thus, for example, the Federal Aviation Administration provides strong guidelines that even as new technology can revolutionize the design of flight instruments, the basic form of the four most important instruments in the panel in Figure 8.13 those forming a T—should always be preserved. Thus visual scanning skills will transfer from one cockpit to another.

5. **Phase-related displays** are needed because the guideline of consistency conflicts with those of frequency of use and relatedness. Phase-related operations are situations where the variables that are frequently used (or related and used in sequence) during one phase of operation very different from those during another phase. In nuclear power-plant

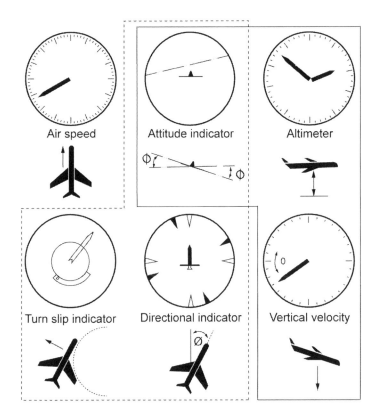

Figure 8.13 Display layout of a conventional aircraft instrument panel. The attitude directional indicator is in the top center. The outlines surround displays that are related in the control of the vertical (solid outline) and lateral (dashed box) position of the aircraft. Note that each outline surrounds physically proximate displays. The three instruments across the top row and that in the lower center form a T shape, which the FAA mandates as a consistent layout for the presentation of this information across all cockpit designs.

monitoring, the information that is important in startup and shutdown is different from what is important during routine operations. Under such circumstances, a totally consistent layout for all phases may be unsatisfactory, and current, "soft" computer-driven displays allow flexible formats to be created in a phase-dependent layout. Phase-dependent layouts should adhere to three design guidelines: (1) Salient visible signals should clearly indicate the current configuration; (2) Where possible, enforce consistency (M13) across all configurations; (3) Resist the temptation to create excessive number of configurations [491]. Remember that as long as a display design is consistent, the user's memory will help guide attention to find the needed information rapidly, even if that information may not be in the best location for a particular phase.

6. **Organizational grouping** is a guideline that can be used to contrast the display array in Figure 8.14a with that in Figure

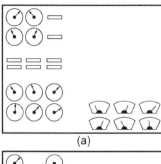

(a)

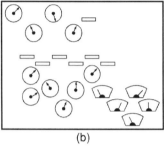

(b)

Figure 8.14 Display grouping. (a) high; (b) low. All displays within each physical grouping and thus have higher display proximity must be somehow related to each other for the display layout on the left to be effective (P9).

8.14b. An organized, "clustered" display, such as that seen in Figure 8.14a, provides an aid that guides visual attention to particular groups as needed (A2), as long as all displays within a group are functionally related and their relatedness is clearly indicated to the user. If these guidelines are not followed and unrelated items are placed in a common spatial cluster, then such organization can be counterproductive because it violates the principle of proximity compatibility (A3).

7. **Control-Display compatibility** dictates that displays should be close to their associated controls (A3).

8. **Clutter avoidance** dictates that there should ideally be a minimum visual angle between all pairs of related displays, and much greater separation between unrelated displays. We discuss stimulus-response compatibility in Chapter 9 and clutter avoidance in the following sections.

8.6.2   Head-Up Displays and Display Overlays

We have seen that one important display layout guideline involves moving important information sources close to the PVA. The ultimate example of this approach is to actually superimpose the displayed information on top of the PVA creating what is known as the head-up display, or HUD [492, 493, 494]. HUDs typically display near domain information—instrument readings or text—over the far domain—the driving or flying environment (see Figure8.15). HUDs are often used for vehicle control but may have other uses as well when the PVA can be clearly specified. For example, a HUD might be used to superimpose a computer graphics designer's palette information over the design workspace [495].

Figure 8.15 A head-up display (HUD) for the copilot of a C-130 aircraft (Source: By Telstar Logistics (flickr) [CC BY 2.0 (http://creativecommons.org/licenses/by/2.0)], via Wikimedia Commons).

HUDs promise three advantages. First, assuming that the driver or pilot should spend most of the time with the eyes directed outward at the far domain, then overlapping the HUD imagery should allow both the far-domain environment and the near-domain instrumentation to be monitored in parallel with little information access cost (A2). Second, particularly with aircraft HUDs, it is possible to present *conformal imagery*–imagery that has a direct spatial counterpart in the far domain. Such imagery, like the horizon line that overlays the actual horizon. Conformal imagery that overlays the far domain supports divided attention between the two domains because it adheres to the proximity compatibility principle (A3). Third, many HUDs are projected via collimated imagery, which essentially reorients the light rays from the imagery in a parallel fashion, thereby making the imagery to appear to the eyes to be at an accommodative distance of optical infinity. The advantage of this is that the lens of the eyeball accommodates to more distant viewing than the nearby windshield and so does not have to reaccommodate to shift between focus on instruments and on far domain viewing (see Chapter 4).

Against these advantages, is one important cost. Moving imagery too close together (i.e., superimposed) can create excessive clutter (A3) and the associated proximity-compatibility principle (A3). Hence, it is possible that the imagery may be difficult to read against the background of varied texture and that the imagery itself may obscure the view of critical visual events in the far domain. The issue of overlay-induced clutter is closely related to that of map overlays, discussed later in this chapter.

The three overall benefits tend to outweigh the clutter costs. In aircraft, flight control performance is generally better when critical flight instruments are presented head-up (and particularly so if they are conformal [496, 494]). In driving, the digital speedometer instrument is sampled for a shorter time in the head-up location [497], although in both driving and flying, speed control is not substantially better with a HUD than with a head-down display [497, 498, 496]. There is also evidence that relatively expected discrete events (like the change in a digital display to be monitored) are better detected when the display is in the head-up location [499, 498, 494].

Nevertheless, cost of clutter can be devastating in some situations. HUD imagery can undermine detection of unexpected events in the far domain, such an aircraft taxiing out onto the runway toward which the pilot is making an approach [500, 496, 494]. These lapses would likely be worse as the HUD information becomes more complex and is less related to the far domain activities, such as reading long text messages while driving.

> Head up displays have most benefits, and least costs, when used to project conformal imagery.

8.6.3   Head-Mounted Displays

A close cousin to the HUD is the head-mounted or helmet-mounted display in which a display is rigidly mounted to the head so that it can be viewed no matter which way the head and body are oriented as characteristic of Google glass or virtual reality systems Such a display has the advantage of allowing the user to view superimposed imagery across a much wider range of the far domain than is possible with the HUD. In an aircraft or helicopter, the head-mounted displays (HMDs) can allow the pilot to retain a view of HMD flight instruments while scanning the full range of the outside world for threatening traffic or other hazards. For other mobile operators, the HMD can reduce information access costs while keeping the hands free for other activities, such as the mountaineering rock climber [501]. For example, consider a maintenance worker, operating in an awkward environment in which the head and upper torso must be thrust into a tight space to perform a test on some equipment. Such a worker would greatly benefit by being able to consult information on how to carry out the test, displayed on an HMD, rather than needing to pull his head out of the space every time he must consult a manual. The close proximity thus created between the test space and the instructions makes it easier to integrate these two sources of information (A3). The use of a head-orientation sensor with conformal imagery can also present information on the HMD specifying the direction of particular locations in space relative to the momentary orientation of the head; for example, the location of targets, the direction to a particular landmark, or due north [502, 503].

HMDs can be either monocular (presented to a single eye), bi-ocular (presented as a single image to both eyes), or binocular (presented as a separate image to each eye); furthermore, monocular HMDs can be either opaque (allowing only the other eye to view the far domain) or transparent (superimposing the monocular image on the far domain). Opaque binocular HMDs are part of virtual reality systems. Each version has its benefits and costs. The clutter costs associated with transparent HUDs may be mitigated somewhat by using a monocular HMD, which gives one eye unrestricted view of the far domain. However, presenting different images to the two eyes can sometimes create problems of binocular rivalry or binocular suppression in which the two eyes compete to send their own image to the brain rather than fusing to send a single, integrated image [504].

To a greater extent than is the case with HUDs, efforts to place conformal imagery on HMDs can be problematic because of potential delays in image updating. Such conformal displays, have been termed augmented reality, are used to depict spatial positions in the outside world, they must be updated each time the display moves (i.e., head rotates) relative to that world. Hence, conformal image updating on the HMD must be fast enough to keep up with potentially rapid head rotation. If it is not, then the image can become disorienting and lead to motion sickness [505]; alternatively, it can lead users to adopt an unnatural strategy of reducing the speed and extent of their head movements [506, 502].

At present, the evidence is mixed regarding the relative advantage of presenting information head-up on an HMD versus head-down on a handheld display [502, 503]. Often, legibility issues (P5) may penalize the small-sized image of the handheld display, and if head tracking is available, then the conformal imagery that can be presented on the HMD can be very valuable for integrating near- and far-domain information (A3). Yet if such conformal imagery or augmented reality cannot be created, the HMD value diminishes, and diminishes still further if small targets or high detail visual information must be seen through a cluttered HMD in the world beyond [503]. In short, the imagery of transparent HMDs in safety critical environments should be kept to a minimum.

8.6.4 Configural Displays

Sometimes, multiple displays of single variables can be arrayed in both space and format so that certain properties relevant to the monitoring task will emerge from the combination of values on the individual variables. As an example, a patient-respiration monitoring display [507], creates a rectangle where the height indicates the volume or depth of patient breathing, and the width indicates the rate. Therefore, the total area of the rectangle indicates the total amount of oxygen respired by the patient (right rectangle) and imposed by the respirator (left rectangle). This relationship holds because the amount = depth rate and the rectangle area = height width. Thus, the display has been configured to produce an emergent feature [508, 509]; that is, a property of the configuration of individual variables (in this case depth and rate) emerges on the display to signal a significant, task-relevant, integrated variable (the rectangle area or amount of oxygen A3). Note also in the figure that a second emergent feature may be perceived as the shape of the rectangle—the ratio of height to width that signals either shallow rapid breathing or slow deep breathing (i.e., different "styles" of breathing, which may indicate different states of patient health).

The rectangle display can be widely used because of the number of other systems in which the product of two variables represent a third, important variable [437]. Examples are distance = speed time, amount = rate time, value (of information) = reliability diagnosticity, and expected value (in decision making) = probability value. Figure 8.16 shows such a display for distance traveled by moving fast for a short period as in the Hare from the fable of the Tortoise and the Hare. On the bottom is the distance covered by moving slowly for longer period.

Another example of a configural display shown in is the safety-parameter monitoring display developed by Woods, Wise, and Hanes [510] for a nuclear power control room. The eight critical safety parameters are configured in an octagon such that when all are within their safe range, the easily perceivable emergent feature of symmetry is observed. Furthermore, if a parameter departs from its normal value as the result of a failure, the distorted shape of the polygon can uniquely signal the nature of the underlying fault, a feature that was sadly lacking for our operator in the story at the beginning of the chapter. Such a feature would also be lacking in more conventional arrays of displays like those shown in Figure 8.14.

Configural displays generally consider space and spatial relations in arranging dynamic displayed elements. Spatial proximity

Hare

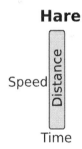

Tortoise

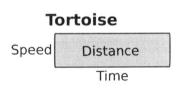

Figure 8.16 A configural display combines two variables to create a third. Here the emergent variable of distance is the area formed by plotting speed and time, Showing how the slow and steady Tortoise goes further than the fast, but inconsistent Hare.

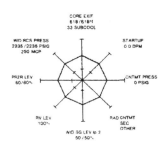 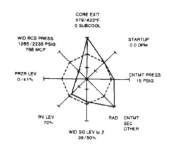

(a) Normal Condition

(b) During loss of coolant accident

Figure 8.17 Wide-range iconic configural displays. (Source: Woods, D. D., Wise, J., and Hanes, L. An Evaluation of Nuclear Power Plant Safety Parameter Display Systems. Reproduced with permission from *Proceedings of the Human Factors Society 25th Annual Meeting.*, pp 110-114. Copyright 1981 by the Human Factors Society. All rights reserved.)

may help monitoring performance, and object integration may also help, but neither is sufficient or necessary to support information integration from emergent features. The key to such support lies in emergent features that map to task-related variables [511, 512]. The direct perception of these emergent features can replace the more cognitively demanding computation of derived quantities, effectively placing knowledge in the world (M10). Will such integration hinder focused attention on the individual variables? In general it does not [511, 512].

8.6.5   Putting It All Together: Supervisory Displays

In many large systems, such as those found in the process-control industry (see Chapter 16), dynamic supervisory displays are essential to guarantee appropriate situation awareness and to support effective control. As such, several of the display principles and guidelines discussed in this chapter should be applied and harmonized. Figure 14 provides such an example. In the figure, we noted the alignment of the parallel monitoring displays to a common baseline to make their access easy (A2) and their comparison or integration (to assure normality) also easy by providing the emergent feature (A3). The display provides redundancy (P8) with the digital indicator at the bottom and a color change in the bar when it moves out of acceptable range. A predictor (M12), the white triangle, shows the trend. The fixed-scale moving pointer display conforms to mental model principles MM14 and MM15. Finally, the display replaced a separate, computer-accessible window display of alarm information with a design that positioned each alarm directly under its relevant parameter (A3).

One of the greatest challenges in designing such a display is to create one that can simultaneously support monitoring in routine or modestly nonroutine circumstances as well as in abnormal circumstances requiring diagnosis, problem solving, and troubleshooting, such as those confronting the operator at the beginning of the chapter. The idea of presenting totally different display suites to support the two forms of behavior is undesirable, because in complex systems, operators may need to transition back and forth between them; and because complex systems may fail in many ways, the design of a display to support management of one form of failure may harm the management of a different form.

In response to this challenge, human factors researchers have developed what are called *ecological interfaces* [513, 514, 515]. See the *Handbook of Cognitive Engineering* for guidelines [12]. The design of ecological interfaces is complex and beyond the scope of this textbook. However, their design capitalizes on graphical representation of the process, which can produce emergent features that perceptually signal the departure from normality, and in some cases help diagnose the nature of a failure. Considered in terms of the levels of behavior introduced in Chapter 7, the intent of ecological interface design is to support skill-, rule-, and knowledge-based behavior, allowing people to use the relatively effortless skill-based behavior most of the time, but engage knowledge-based behavior in abnormal situations.

A particular feature of ecological interfaces is their complete representation of system variables that allow the operator to reason at various levels of abstraction abnormal situations [352]. (1) Where is a fault located? (2) Is it creating a loss of energy or buildup of excessive pressure in the plant? (3) What are its implications for production and safety? These three questions represent different levels of abstraction, ranging from the physical (very concrete, like question 1) to the much more conceptual or abstract (question 3). An effective manager of a fault in a high-risk system must be able to rapidly switch attention between various levels. Despite their added complexity, ecological interfaces are more effective in supporting fault management than other displays, while not harming routine supervision [513, 439].

Different displays may be needed for different aspects of the task, such as phase-related displays. If possible, these should be visually available at the same time. Doing so adds visual clutter, but keeps knowledge in the world (M10) rather than forcing people to remember information, or page through many screens [516].

8.7 Navigation Displays and Maps

A navigational display (the most familiar of which is the map) should serve four fundamentally different classes of tasks: (1) provide guidance about how to get to a destination, (2) facilitate planning, (3) help recovery if the traveler becomes lost, and (4) maintain situation awareness regarding the location of a broad range of objects [517]. For example, a pilot map might depict other air traffic or weather in the surrounding region, or the process controller might view a "mimic diagram" or map of the layout of systems in a plant (Chapter 16). The display itself may be paper or electronic. Environments in which these tasks should be supported range from cities and countrysides to buildings and malls, to 3D brain

Creating useful map displays requires consideration of the full range of tasks they support, not just guidance.

maps to support neurosurgery. Recently, these environments have also included spatially defined "electronic environments" such as databases, hypertext, and large menu systems (see Chapter 15). Navigational support also may be needed in multitask conditions while the traveler is engaged in other tasks, like driving the vehicle.

8.7.1 Route Lists and Command Displays

The simplest form of navigational display is the route list or command display. This display typically provides the traveler with a series of commands (turn left, go straight, etc.) to reach a desired location. In its electronic version, it may provide markers or pointers of where to turn at particular intersections. Furthermore, most navigational commands can be expressed in words, and if commands are issued verbally through synthesized voice they can be easily processed while the navigator's visual/spatial attention is focused on the road or desired course [518], following the attention principle of multiple resources (A4) described in Chapter 6. Command displays are generally easy to use. Still, to be effective, timing of command displays is critical: each command should be given at the right place and time. Well-timed commands require understanding time to prepare for the maneuver, as well as the speed and position of the vehicle.

A printed route list is vulnerable if the traveler strays off the intended route, and any sort of electronically mediated command display will suffer if navigational choice points (i.e., intersections) appear in the environment that were not in the database (our unfortunate traveler turns left into the unmarked alley). Thus, command displays are not effective for depicting where one is (allowing recovery if lost), and they are not very useful for planning and maintaining situation awareness. In contrast, spatially configured maps do a better job of supporting planning and situation awareness. There are many different possible design features within such maps, and we now consider.

8.7.2 Maps

1. **Legibility.** To revisit a recurring theme (P5), maps must be legible to be useful. For paper maps, care must be taken to provide necessary contrast between labels and background and adequate visual angle of text size. If color-coded maps are used, then low-saturation coding of background areas enables text to be more visible [519, 520]. However, colored text may also lead to poor contrast (Chapter 4). In designing such features, attention should also be given to the conditions in which the maps may need to be read (e.g., poor illumination, as discussed in Chapter 4). Unfortunately, legibility may sometimes suffer because of the need for detail (a lot of information) or because limited display size forces the use of a very small map. With electronic maps, detail can be achieved without sacrificing legibility by incorporating zooming capabilities.

2. **Clutter and Overlay.** Another feature of detailed maps is their tendency to become cluttered. Clutter has two negative consequences: It slows down the time to access information (A2) (i.e., to search for and find an item) and it slows the time to read the items as a consequence of masking by nearby items (the focused attention disruption resulting from close proximity, A3). Clutter is a critical impediment to all displays including those that are not maps, such as the cluttered warning label, the medical chart [521], or the clutter of multiple highway signs. It has been the focus of recent efforts to develop computational models that can predict those costs [521, 522, 523].

Besides the obvious solution of creating maps with minimal information, three possible solutions avail themselves. First, effective color coding can present different classes of information in different colors. Hence, the human selective attention mechanism is more readily able to focus on features of one color (e.g., roads), while filtering out the temporarily unneeded items of different colors (e.g., text symbols, rivers, terrain [131]). Care should be taken to avoid too many colors (if absolute judgment is required, P6) and to avoid highly saturated colors [520]. Second, with electronic maps, it is possible for the user to highlight (intensify) needed classes of information selectively while leaving others in the background [131]. The enhanced intensity of target information can be a more effective filter for selective and focused attention than will be the different color. Third, carrying the concept of highlighting to its extreme, decluttering allows the user to simply turn off unwanted categories of information [524]. One problem with both highlighting and decluttering is that the more flexible the options are, the greater is the burden of choice imposed on the user, and this may impose unnecessary decision load [131] and interface requirements. Thus. in some environments, such as a vibrating vehicle, the control interface necessary to accomplish the choice is vulnerable and automatic decluttering provides a benefit [525].

3. **Position Representation.** People benefit in navigational tasks if they are presented with a direct depiction of where they are on the map. This feature can be helpful in normal travel, as it relieves the traveler of the mental demands of inferring the direction and rate of travel. In particular, however, this feature is extremely critical in aiding recovery from getting lost. This, of course, is the general goal of providing "you are here" maps in malls, buildings, and other medium-scale environments [526].

4. **Map Orientation.** A key feature of good maps is their ability to support the navigator's rapid and easy cross-checking between features of the environment (the forward view) and the map [527]. This can be done most easily if the map is

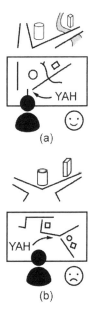

Figure 8.18 Map orientation. Good (a) and poor (b) mounting of "you are here" map. In (b) the observer must mentally rotate the view of the map by 90 °so that left and right in the world correspond to left and right in the map.

oriented in the direction of travel so that *up* on the map is forward and, in particular, left on the map corresponds to left in the forward view. Otherwise, time-consuming and error-prone mental rotation is required [528]. To address this problem, electronic maps can be designed to rotate so that up on the map is in the direction of travel and "you are here" maps can be mounted so that the top of the map corresponds to the direction of orientation as the viewer observes the map [526, 529], as shown in Figure 8.18a. When this correspondence is achieved, the principle of pictorial realism (MM14) is satisfied. When the map is not oriented in this manner, as in Figure 8.18b, the person needs to mentally rotate the map, which is effortful and error prone.

Despite the advantages of map rotation for navigation, however, there are some costs. For paper maps, the text will be upside down if the traveler is headed south. Furthermore, for some aspects of planning and communications with others, the stability and universal orientation of a fixed north-up map can be useful [528]. Thus, electronic maps should be designed with the task in mind and so should have a fixed-map option available.

5. **Scale.** In general, we can assume that the level of detail, scale, or availability with which traveler information needs to be presented becomes less of a concern in direct proportion to the distance away from the traveler and falls off more rapidly behind the traveler than in front. This is because the front is more likely to be in the future course of travel. Therefore, electronic maps often position the navigator near the bottom of the screen (see Figure 8.18a). The map scale should be user-adjustable if possible, not only because of clutter, but because the traveler's needs can vary from planning, in which the location of a route to very distant destinations may need to be seen (at a small scale), to guidance, in which only detailed information regarding the next choice point is required (large scale).

One possible solution to addressing the issue of scale are dual maps in which local information regarding one's momentary position and orientation is presented alongside a large-scale map. The former can be ego-referenced and correspond to the direction of travel, and the latter can be world-referenced. Figure 8.19 shows some examples. Dual maps are particularly valuable if the user's momentary position and/or orientation is highlighted on the wide-scale, world-referenced map [528], adhering to the principle of visual momentum (M11), which serves to visually and cognitively link two related views [530]. Both maps in Figure 8.19 indicate the position of the local view within the global one.

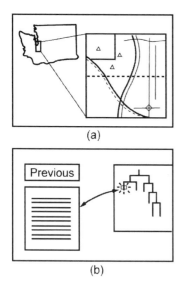

(a)

(b)

Figure 8.19 Visual momentum and multi-window displays.

6. **Three-Dimensional Maps.** Increasing graphics capabilities have enabled the creation of effective and accurate 3-D or perspective maps that depict terrain and landmarks [527]. If

it is a rotating map, then such a map will nicely adhere to the principle of pictorial realism (P6). But are 3-D maps helpful? The answer depends on the extent to which the vertical information, or the visual identity of 3-D landmark objects, is necessary for navigation. For the pilot flying high over flat terrain or for the driver navigating a gridlike road structure, vertical information is likely to play little role in navigation. But for the hiker or helicopter pilot in mountainous terrain, for the pilot flying low to the ground, or the vehicle driver trying to navigate by recognizing landmark objects in the forward field of view, the advantages of vertical (i.e., 3-D) depiction of the environment and landmarks become far more apparent [527]. This is particularly true given the difficulties that unskilled users have reading 2-D contour maps. Stated simply, the 3-D display usually looks more like a picture of the area that is represented (MM14), and this is useful for maintaining situation awareness.

The costs and benefits of 3-D maps tend to be task-specific. For maps to support 3-D visualization (like an architect's plan), 3-D map capabilities can be quite useful [531]. In tasks such as air traffic control, where very precise separation along lateral and vertical dimensions must be judged, however, 3-D displays may impose costs because of the ambiguity with which they present this information (see Chapter 4). Perhaps the most appropriate guidance that should be given is to stress the need for careful task and information analysis before choosing to implement 3-D maps: 1) How important is vertical information in making decisions? 2) Does that information need to be processed at a very precise level? If yes, then 3-D representations of the vertical dimensions are not good [532, 533]. If no and if all is needed is some global information regarding "above" or "below," then the 3-D displays can be effective.

3-D maps don't support precise judgments of distance, but 2-D maps do.

If a 3-D perspective map is chosen, then two important design guidelines can be offered [534]. First, as noted in Chapter 4, the greater number of natural depth cues that can be rendered in a synthetic display, the more compelling will be the sense of depth or three dimensionality. Stereo, interposition and motion parallax, which can be created by allowing the viewer to rotate the display, are particularly valuable cues. [534, 535]. Second, if display viewpoint rotation is an option, it is worthwhile to have a 2-D viewpoint (i.e., overhead lookdown) as a default option.

7. **Planning Maps and Geographic Data Visualization.** Our discussion of maps has assumed the importance of a traveler at a particular location and orientation in the map-depicted database. But there are several circumstances in which this is not the case; the user does not "reside" within the database. Here we consider examples such as air traffic control displays, vehicle dispatch displays, process-control mimic diagrams, construction plans, wiring diagrams, and the display of 3-D

scientific data spaces. The user is more typically a "planner" who is using the display to understand the spatial relations between its elements.

Many of the features we have described apply to these "maps for the non-traveler" as well (e.g., legibility and clutter issues, flexibility of scale). But since there typically is no direction of travel, map rotation is less of an issue. For geographic maps, north-up is typically the fixed orientation of choice. For other maps, the option of flexible, user-controlled orientation is often desirable.

8.8   Data Visualization and Graph Design

Some displays are designed to present a range of numbers and relationships between these numbers. These numbers may represent things as varied as nutrition and cost of different products, the range of desired values for different maintenance testing outcomes, and economic or scientific data. How such data are depicted has a strong influence on their interpretation [536, 10]. An initial choice is whether to represent the data via tables or graphs.

As with our discussion of dynamic displays, when the comparison was between digital and analog representation, one key consideration is the precision with which a value must be read. If high precision is required, the table may be a wise choice. Furthermore, unlike dynamic digital displays, tables do not suffer the problems of reading digital information while it is changing. However, as shown in the top of Figure 8.20, tables do not support perception of change; that is, trends of values across the table are hard to see compared to the same data presented as a graph, as in the bottom of Figure 8.20. Tables offer even less support for seeing the rate of change (acceleration or deceleration) less so still for trends over two dimensions (e.g., an interaction between variables), which can be easily seen by the divergence of the two lines on the right side of the graph in Figure 8.20, but not in the table above it.

Thus, if precision is not required and trend information is important, the graph represents the display of choice. If so, then the questions remain: What kind of graph? Bar or line? Pie? 2-D or 3-D? and so on. While Tufte [537, 538, 539], Kosslyn [540], Munzner [541], and Few [542] offer comprehensive treatments of graphic presentation, we provide several straightforward guidelines.

8.8.1   Matching types of graphs to questions

Graphs answer questions about data by showing relationships and making comparisons easier. Before creating a graph it is critical to specify the questions and comparisons of interest. Table 8.4 shows common graphs and general questions they might answer. For example, in the upper left is a graph that shows the association between variables. This type of graph answers questions such as how does X influence Y, as in "does increasing the prices of gas

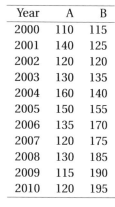

Year	A	B
2000	110	115
2001	140	125
2002	120	120
2003	130	135
2004	160	140
2005	150	155
2006	135	170
2007	120	175
2008	130	185
2009	115	190
2010	120	195

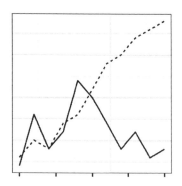

Figure 8.20 A tabular representation of data and a graphical representation of the same data. Note how much easier it is to see the trend in the graph, but how much easier it is to read off specific values from the table.

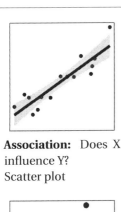

Association: Does X influence Y?
Scatter plot

Fluctuation: Does X change over time?
Timeline

Distribution: What is the spread of X?
Histogram

Proportion: What amount of X is Y?
Stacked bar chart

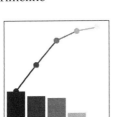

Comparison: Is X greater than Y?
Boxplot

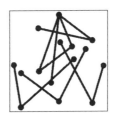

Contribution: How much of X is due to Y?
Pareto chart

Connection: Is X connected to Y?
Network diagram

Hierarchy: Is X part of group Y?
Dendrogram

Table 8.4 Types of graphs and the questions they answer. (Created using R.)

reduce the amount of driving?". A scatter plot shows the strength and nature of this association.

The examples for each type of graph in Table 8.4 represent one of many possible representations. For example, the stacked bar chart addresses questions of proportion, but so can pie charts and 3-D pie charts. How do you choose between these alternatives? One consideration is to select display dimensions that make it easy for people to make comparisons needed to answer the questions—identify effective mapping between data and display dimensions—which we turn to in the following section.

8.8.2 Mapping data to display dimensions

For the purposes of display design, three different data types guide the choice of display dimensions: interval, ordinal, and nominal [543]. Interval data include real or integer numbers (e.g., height and weight), ordinal data are categories that have a meaningful order (e.g., compact, mid-size, and full-size cars), and nominal data are categories that have no order (e.g., male, female). Each data type can be represented with one of several display dimensions, such as color or position, but certain mapping support more accurate judgments.

For all three types of data, position, such as the horizontal or vertical placement of a point in a graph, support the most precise judgments. The other ways of coding information depend on the type of data: hue is a poor choice for interval data, but a good choice for nominal data, as is shape. Because shape and color have no natural mapping to magnitude, they are a poor choice for interval and ordinal data. Magnitude is best represented by position on a common scale, followed by position on an unaligned scale, length and then angle, followed by size [541, 543]. Because size and angles are relatively hard to judge, pie charts are not a good way to represent proportions.

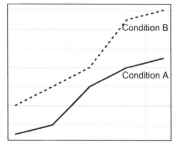

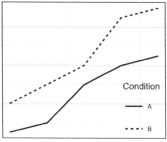

Figure 8.21 Legend proximity. Close proximity of label to line (top): Good. Low proximity of label to line (bottom): Poor.

Limits of absolute judgment underlie the effectiveness of coding data with various display dimensions. Coding nominal data with more than seven hues will exceed people's ability and so they would not be able to reliably link lines on a graph to categories. Data presented on aligned scales, such as the bottom category in a stacked bar chart, can be judged very precisely, but the limits of absolute judgment make interpreting the upper categories more difficult. This means that the bottom category of a stacked bar chart should be chosen carefully. Generally, avoid placing data on unaligned scales. Instead, support relative judgments based on a common scale. The circular format of pie charts means that there are no aligned scales and is another reason why they are not as effective as stacked bar charts.

Because visualization involves multiple conceptual dimensions, a natural choice is to use three-dimensional Euclidian space. However, three-dimensional figures make accurate comparisons difficult due the ambiguity of rendering three dimensions on a two dimensional plane. Of all the ways to represent a quantity, the volume of a three-dimensional object leads to the most inaccurate judgments [541].

Another important conceptual dimension is time. Time, like space, is compatibly mapped to display dimension of position, often advancing from left (past) to right (future). Time can also be directly mapped to display time via animation. Animated graphs can be compelling, but they require working memory to track objects across the display and so severely limit the number of data points that can be compared. Interactive visualization described in Chapter 10 can give control with a slider and avoids this limit to some degree.

8.8.3 Proximity

Visual attention must sometimes do a lot of work, traveling from place to place on the graph (A2), and this effort can hinder graph interpretation. Hence, it is important to construct graphs so things that need to be compared (or integrated) are either close together in space or can be easily linked perceptually by a common visual code. This, of course, is a feature for the proximity compatibility principle (A3) and can apply to keeping legends close to the lines that they identify, rather than in remote captions or boxes (Figure 8.21). Similarly, keeping two lines whose slopes and intercepts need to be compared on the same panel of a graph rather than on separate panels (Figure 8.22. The problems of low proximity will be magnified as the graphs contain more information—more lines.

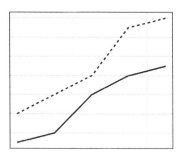

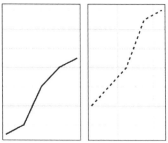

Figure 8.22 Data proximity. Close proximity of lines to be compared (top): Good. Low proximity of lines to be compared (bottom): Poor.

Similarly, in a box plot with many categories people will be able to compare categories that are close to each other more precisely than those that are separated. You should order categories so that those to be compared are closest.

Proximity goes beyond physical distance. A line linking points on a timeline can enhance proximity as can color and shape. Lines and color can be effective way of making groups of points in a network diagram "closer", and easier to interpret as a group. Objects with identical colors tend to be associated together, even when they are spatially separated. Furthermore a unique color tends to stand out. It is also the case that space is compatibly mapped to space, so that visualization of geographic areas is best accomplished when the dimensions of rendered space correspond to the dimensions of displayed space–a map.

As with its application to other display designs, the proximity compatibility principle means that the visual proximity of elements of the graph need to correspond to the mental proximity needed to interpret this information. For graphs, this means the questions and comparisons the graph is intended to address should specify what is "close" in the graph.

8.8.4   Legibility

As with other types of displays, issues of legibility are again relevant. However, in addition to making lines and labels of large enough to be readable, a second critical point relates to discriminability (P9). Too often, lines that have very different meanings are distinguished only by points that are highly confusable, as in the graph on the left of Figure 8.23. Here is where attention to incorporating redundant coding (P8) of differences can be quite helpful. In modern graphics packages, color is often used to discriminate lines, but it is essential to use color coding redundantly with another salient cue. Why? As we noted in Chapter 4, not all viewers have good color vision, and a non-redundant colored graph printed from a black and white printer or photocopied may be useless.

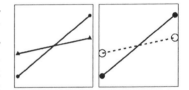

Figure 8.23 Confusable lines distinguished by small symbols (on the left) and lines made more discriminable (on the right) by redundantly coding the lines with larger symbols and line type.

8.8.5   Clutter

Graphs can easily become cluttered by presenting more lines and marks than the actual information they convey. As we know, excessive clutter can be counterproductive [544, 521], and this has led some to argue that the data-ink ratio should always be maximized [539]; that is, the greatest amount of data should be presented with the smallest amount of ink. While adhering to this guideline is a valuable safeguard against the excessive ink of "chart junk" graphs, such as those that unnecessarily put a 2-D graph into 3-D perspective, the guideline of minimizing ink can however be counterproductive if carried too far. Thus, for example, the "minimalist" graph in center of Figure 8.24, which maximizes data-ink ratio, gains little by its decluttering and loses a lot in its representation of the trend, compared to the line graph on the right Figure 8.24. The line graph contains an emergent feature—slope—which is not visible in the dot graph. The latter is also much more vulnerable to the conditions of poor viewing (or the misinterpretation caused by the dead bug on the page!).

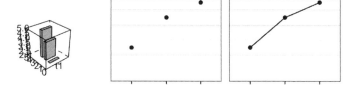

Figure 8.24 Chart clutter and data to ink ratio. (left) Example of a boutique graph with a low data-ink ratio. The 3-D graph contains the unnecessary and uninformative representation of depth; (middle) minimalist graph with very high data-ink ratio; (right) line graph with intermediate data-ink ratio. Note the trend information added by the line.

Figure 8.24 shows that you can increase the data-to-ink ratio by reducing the "ink" devoted to non-data elements. Another way to increase the data-to-ink ratio is to include more data. More data can take the form of reference lines and multiple small graphs, as in Figure 8.8. More data can also take the form of directly plotting the raw data rather than summary data.

Figure 8.25 shows an extreme version, in which each data point represents one of approximately 693,000 trips reported in the 2009 travel survey [545]. The horizontal axis indicates the duration and the vertical axis shows distance of each trip. The diagonal lines of constant speed place these data in context by showing very slow trips—those under the 3mph line—and very fast trips—those over the 90mph line. Histograms at the top and side show the distribution of trip duration and distance. The faint vertical and horizontal lines show the mean duration and distance. Like other visualizations that include the raw data, this visualization shows what is behind the summary statistics, such as mean trip distance and duration.

Showing the underlying data has the benefit of providing a more complete representation, but it can also overwhelm people. Data can create clutter. One way to minimize clutter is by grouping and layering the data. In the case of Figure 8.25 this means making the individual data points small and faint.

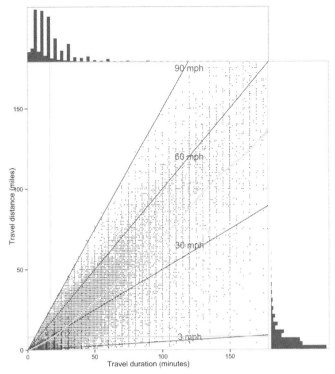

Figure 8.25 An example of extreme data-to-ink with over 693,000 data points. Reference lines and marginal histograms aid interpretation.

8.8.6 Interactive Data Visualization

Vastly increased computer graphics power has enabled the construction of complex data visualizations, sometimes containing thousands of data points, which people can manipulate, not just look at [546, 10, 547]. The design of such interactive data visualization should consider many of the principles discussed in the current chapter, such as the proximity compatibility principle, the map frame of reference, many are also embodied in an understanding of control—the manual interactions to be taken in manipulating the data representation. Hence, we defer discussion of interactive visualization until after we have discussed control in the next chapter, and present it in our chapter on Human-Computer Interaction, Chapter 10.

8.9 Summary

We presented a wide range of display principles designed to facilitate the transmission of information from the senses, discussed in Chapters 4 and 5, to cognition, understanding, and decision making, discussed in Chapters 6 and 7. Good displays minimize the gulf of evaluation—the gap between the state of the system and the person's interpretation of the state relative to goals and intentions. There is no single "best" way to do this, but consideration of the

The Dutch Target (Urinals in Schiphol Airport)

Displays do more than convey information, they also guide behavior. Urinals that include a fly painted in the enamel substantially improve mens' aim.

A Dutch maintenance worker, Jos Van Bedoff, suggested the idea for urinals in Amsterdam's Schiphol airport. "When flies were introduced at Schiphol Airport, spillage rates dropped 80 percent, says manager Aad Keiboom. A change like that, of course, translates into major savings in maintenance costs."

Creating displays that guide people to act effectively depends as much on the design of displays as it does the design of controls.
http://www.npr.org/ templates/story/ story.php? storyId=121310977

15 principles presented above can certainly help to rule out bad displays. Much of the displayed information eventually leads to action—to control some aspect of a system or the environment or otherwise to respond to a displayed event. Good controls minimize the gulf of execution—the gap between the intended state of the system and the actions needed to reach that state. In the next chapter, we discuss some of the ways in which the human factors engineer can assist with that control process.

Additional Resources

Several useful resources that expand on the content touched on in this chapter include books that address graph design and display design in more detail:

1. **Display design:** Burns, C. M., & Hajdukiewicz, J. R. (2004). *Ecological Interface Design.* New York: CRC Press.

2. **Graph design:** These books provide a more complete discussion of graph design and data visualization.

   Few, S. (2012). *Show Me the Numbers: Designing tables and graphs to enlighten.* Oakland, CA: Analytics Press.

   Tufte, E. R. (1983). *The Visual Display of Quantitative Information.* Cheshire, CT: Graphics Press.

Questions

Questions for 8.1 Types of Displays and Tasks

P8.1   How does a display differ from an interface?

P8.2   Describe three fundamental considerations of display design that can also be used to classify displays.

P8.3   Give one example of why it is critical to understand the system and tasks before you can develop a good display.

P8.4   Describe two elements of information analysis that are important for classifying displays.

P8.5   Give an example of how the power of representation can make a difficult task easier.

Questions for 8.2 Fifteen Principles of Display Design

P8.6   If you were able to redesign the speedometer for cars, describe how the attention-based principles might lead you to something very different than what is in today's vehicles.

P8.7   How would you minimize information access cost and what might be one of the negative consequences?

P8.8   Proximity compatibility describes the need for what two types of proximity to be compatible?

P8.9   According to the proximity compatibility principle, would you expect people to perform better or worse when a display places information needed for two separate tasks next to each other?

P8.10   How might reducing information access cost interfere with legibility?

P8.11   According to absolute judgment limits why would it be a bad idea to differentiate 15 different lines by hues in a graph?

P8.12   How might you make similarly spelled drug names more discriminable?

P8.13   What is an example of principle of using knowledge in the world applied to enhancing prospective memory?

P8.14   For what type of display might the principle of visual momentum be particularly important?

P8.15   What type of situation is predictive aiding most important? Give a specific example.

P8.16   Describe how the principle of pictorial realism might be used to orient the direction of movement of a temperature indicator.

P8.17   Describe how the principle of the moving part might be used to orient the direction of movement of a temperature indicator.

P8.18   With a moving scale indicator what would you need to do to ensure that it adheres to the principle of the moving part?

P8.19   Why is the navigation display of an oil tanker (supertanker) a particularly good application of a predictive display?

P8.20   For what situations would a digital display be superior to an analog display?

Questions for 8.3 Alerts

P8.21   Describe how you might use salience compatibility to adjust the display of an alert for a fire alarm compared to an incoming text message; for a low fuel warning compared to a forward collision warning.

P8.22   For what type of display is salience compatibility most important?

Questions for 8.4 Labels and Icons

P8.23   Describe how the Tall Man labeling approach works.

Questions for 8.5 Monitoring Displays

P8.24   Why does the FAA mandate a consistent arrangement of primary flight displays?

P8.25   Describe how phase-related displays might be needed in a nuclear power plant and why adopting phase-related displays would conflict with the principle of consistency.

P8.26   How would "principles of attention" guide the choice of location when configuring a display layout that contains many individual displays?

P8.27   In what situations would a digital or analog display be better?

P8.28   In what situations would a HUD not be appropriate?

P8.29   What safety issue with HUDs does the concept of proximity compatibility point to?

Questions for 8.6 Integrative Displays

P8.30   Describe an emergent feature display that helps caregivers integrate respiration rate and respiration volume in tracking patientsâĂŹ respiration.

P8.31   Give an example of a configural displays and describe why it is useful.

Questions for 8.7 Navigation Displays and Maps

P8.32   What application might be best supported by a north-up map?

P8.33   Identify when a 3-D map display would be useful and when it would not.

P8.34   Why is it impossible to define the one best projection that should be used in all maps?

Questions for 8.8 Data Visualization and Graph Design

P8.35   In the context of a graph, describe the tradeoff between analog and digital representation (e.g., a table), particularly for detecting trends.

P8.36   What is the best way to code any of the three main types of data: interval, ordinal, and nominal data?

P8.37   What is a particularly poor way of coding interval data?

P8.38   What is a particularly poor way of coding ordinal data?

P8.39   How would you use the principle of redundancy gain in creating a graph?

P8.40   How would you reconfigure the space shuttle O-ring data to make the critical information more apparent?

Chapter 9

Controls

At the end of this chapter you will be able to...

1. understand basic control tasks and the control devices that support them best

2. link information theory to the design of discrete control devices

3. apply 15 principles to the design of control devices based on attention, perceptual, memory, mental models, and response selection and execution

4. link control theory to design of continuous control devices

5. describe the role of stability in continuous control and how to improve it

Exiting the rental car lot, he pulled onto the freeway entrance ramp at dusk, he started to reach for what he thought was the headlight control. Suddenly, however, his vision was obscured by a gush of washer fluid across the windshield. As he reached to try to correct his mistake, his other hand twisted the very sensitive steering wheel and the car started to veer off the ramp. Quickly, he brought the wheel back but overcorrected, and then for a few terrifying moments the car seesawed back and forth along the ramp until he brought it to a stop in a ditch, his heart pounding. He cursed himself for failing to learn the location of controls before starting his trip. Reaching once more for the headlight switch, he now activated the flashing hazard light. Fortunately, this time, a very appropriate error.

Our hapless driver experienced several difficulties in control that can be placed in the context of the information-processing model discussed in Chapter 6. This model can be paraphrased by "knowing what's happening, deciding what to do, and then doing it." Control is the "doing it" part of this description. It is both a noun (a control) and a verb (to control). Referring to the model of information processing presented in Chapter 6, we see that control primarily involves the selection and execution of responses—that is, the last two stages of the model—along with the feedback loop that allows people to determine that the control action has been executed as intended.

In this chapter, we begin by describing a range of control devices and indicate which tasks they serve best. Then we describe principles to guide design of these controls. These principles build on the categories that guide display design—attention, perception, memory, and mental model—but also includes a new category that is based on response selection and execution. Information theory describes much of what makes response selection and execution difficult and so forms the basis for several of these principles. We then apply these principles to address discrete controls, such as buttons and switches; knobs and levers, keyboards and voice commands for verbal or symbolic input (e.g., typing).

In the second part of the chapter we describe continuous control. This includes the selection and execution of responses over time in response to the changing state of the system. Examples of continuous control include steering cars and piloting planes. For continuous control, we identify design principles and apply them to positioning devices and to teleoperation and remote manipulation, such as with drones and robots.

9.1 Types of Controls and Tasks

Table 9.1 shows an array of *control tasks* and potential controls. The symbols indicate the controls most suited to the various tasks. For example, for the task of turning on or off a system, such as a smartphone, a button or toggle switch would be best. The use of the push buttons, selector switch, and round knob reflects lessons from information theory—the number of the states of the con-

	Push button	Toggle switch	Lever	Selector switch	Keyboard	Voice	Round knob	Joystick or Mouse
Discrete control task								
Two states (On-Off)	●	◑	●L	○				
Three states		◑	●L	●				
Sequential states				●				
>24 discrete states	●	◑						
Continuous setting			◑				●	
Entering text					●	◑		
Continuous control task								
Point and select								●
Track values (1D)			●				○	◑
Track values (2D)								●

Table 9.1 Pairing controls with tasks. First choice (●), second choice (◑), and third choice (○). L-for large movement or force. (Adapted from MIL-STD 1472 [49]; Ely, Thomson, and Orlansky [548].)

troller should be as small the task allows. Consistent with information theory, choosing a knob that allows continuous adjustment when the person needs only to discriminate between three states would make the person slower and more error prone. The ranking of the options in Table 9.1 also reflect the physical properties of the tasks. Where large motion and force is needed, a lever would be superior to a button. Although Table 9.1 indicates the "best" control option for each task, the actual best option depends on many other considerations discussed in terms of 15 principles for control design.

9.2 Information Theory: Response Selection and Execution

In Chapters 4, 6, and 8 we learned that we perceive rapidly (and accurately) information that we expect. In a corresponding manner, we select more rapidly and accurately those actions we expect to carry out than those that are surprising to us. We do not, for example, expect the car in front of us to come to an abrupt halt on a freeway. Not only are we slow in perceiving its expansion in the visual field, but we are much slower in applying the brake (selecting the response) than we would be when the light unexpectedly turns yellow at an intersection that we are approaching. Uncertainty increases response time.

The effect of expectations and uncertainty on response time is a general phenomenon that can be described using the mathematics of information theory. Information theory quantifies the uncertainty and complexity that people face in selecting and executing a response. This uncertainty explains why it takes longer to respond to many equally likely options, compared to a situation with one or two likely options. The Hick-Hyman Law quantifies this relationship. Information theory also describes the complexity of movements. Complex movements, such as reaching across the table to pick up a needle, take longer than reaching for an apple that is right in front of you. Fitts's Law quantifies this relationship between precision and distance of movements. Both Fitts's Law and the Hick-Hyman law support important principles described in the following section. More generally, information theory describes the complexity of the response options and the control task. This general principle is what guides many of the pairings of controls and tasks in Table 9.1. Simple controllers, such as a push button, are paired with simple tasks, such as turning something on and off. Careful selection of controls can reduce complexity and produce faster and more accurate responses.

The complexity of controller should match the complexity of the control task: Don't use a knob when a button will do.

9.3   Fifteen Principles for Discrete Controls

Selecting the appropriate control device for a task is more complicated than the description in Table 9.1 might suggest. In addition, each control device has many different features that contribute to how well it supports a task. The following principles describe how to select features that will make selecting and executing a response simpler and so will reduce errors and response time. Hence, there is a positive correlation between response time and error rate or, in other words, a positive correlation between speed and accuracy. Good designs tend to increase both speed and accuracy.

If the consequences of errors are great, we will need to design so that responses are very accurate. Such designs might sacrifice speed. As we will see below, this *speed-accuracy tradeoff* is a general consideration in selecting and designing control devices because one device might induce faster, but less precise behavior and the other more precise but slower behavior. The value of greater precision or greater speed depend on the task situation. The speed-accuracy tradeoff is a specific way in which the details of the tasks must be considered in selecting an appropriate control device.

9.3.1 Attention Principles

1. **Proximity compatibility.** The concept of proximity compatibility for displays discussed in Chapter 8 also applies to controls. Similar to the benefit displaying information that must be mentally integrated in close physical proximity, performance benefits if controls are near other controls whose activation needs to be mentally integrated as part of a sequence or coordinated combination of control input, as in the buttons on a computer mouse. Similar to displays, close proximity can be achieved by using a common color, configuring them in a coherent pattern, linking them with lines, or by minimizing the physical distance between them. Controls should also be located close to relevant displays. Figure 9.1 shows the ribbon bar in Microsoft Word grouping the controls by general function, such as *Home* and *Insert*, and within each of these categories mentally proximate controls are grouped together, such as in the font section, which includes size, color, and highlighting. Each of these icons acts in a manner similar to a physical on-off button. When they are clicked on they invoke a function.

Figure 9.1 Proximity compatibility applied to control grouping in Microsoft Word.

2. **Avoid resource competition.** When people are performing more than one task at the same time, multiple resource theory (Chapter 6) predicts a benefit of dividing the tasks across different mental resources. For control, this can mean enabling voice control so that tasks that might otherwise compete for the hands can be completed by speaking. The same logic can be applied to allocating controls to the right and left hands and to hands and feet. For drivers this might be voice control for placing a telephone call rather than forcing the drivers' hands off the steering wheel and eyes off the road to dial the phone. Similarly, stalks and buttons on the steering wheel divide controls across the two hands and the steering wheel and pedals divides vehicle control between the hands and feet.

9.3.2 Perceptual Principles

3. **Make accessible.** A central perceptual principle for displays concerns making them legible or audible. With controls making them accessible is a similar and equally important. Accessible controls must be easily reached, which is a concern for locating buttons in the cockpit of a plane or the interior

of a car, and also in designing control panels. We address this aspect of accessibility in Chapter 12.

Accessibility also concerns the ease with which people can identify the control relevant to their current task. A general rule that benefits those with permanent vision disabilities and those that have temporary vision limits because of low light levels or competing tasks, such as driving, is to support *blind operation* and make the controls accessible without looking at them. This can be more easily done by using physical controls, such as knobs and switches, where you can feel where they are, what their state is, and how their state changes when acted upon. In contrast, touchscreens and other screen-based controls typically do not support blind operation. Touchscreens can be made more accessible by complementing them with voice controls and haptic feedback, such as the brief vibration used on iPhones to give the feel of a button press. Table 9.2 provides more detailed suggestions on making controls more accessible, which we discuss in more detail in the context of making controls discriminable.

> Control design support blind operation, both to benefit those with visual impairments, but also for everyone who might need to use a control while looking elsewhere.

4. **Make discriminable.** Identifying a particular control from an array of controls requires that they are discriminable. As with displays, the lower the proportion of features shared between controls the greater the ability of people to discriminate between them. Here we describe features that make physical controls, such as buttons, switches, knobs, and levers easier to discriminate. Table 9.2 shows primary features to identify and differentiate controls include: location and orientation, shape, size, mode of operation (e.g., slider vs knob), label, and color. Each of these features has certain advantages and disadvantages that depend on the task and task context (See Chapters 4 and 8 for more detail on labels and color coding). As an example, if the task context demands that people wear gloves, then the controller shape may be less effective than location and color. The substantial limits of labeling and color, particularly for low light and blind operation, suggests that neither should be exclusively relied on and that they should be complemented with other ways of coding. Other control features that could be used for coding include the texture of the material used for the control and the resistance of the control to movement.

Figure 9.2 Distinctive shapes and locations make these controls highly discriminable. (Photograph by author: J. D. Lee.)

It is unfortunate that aesthetics in design may sometimes call for an array of uniform controls, with no or minimally visible labels. Design tradeoffs associated with aesthetics and safety are discussed in Chapter 2, but it would seem that such tradeoffs have no place in safety critical systems, such as cars, aircraft, or nuclear power plants. Interestingly, bars have particularly discriminable controls (Figure 9.2).

5. **Exploit redundancy gain.** Just as displays can be more effective if the same message is expressed in multiple ways

	Location	Shape	Size	Operation	Label	Color
Advantages						
Aids visual identification	●	●	●		●	●
Aids non-visual identification (e.g., tactile, kinesthetic)	●	●	●	●		
Aids standardization	●	●	●	●	●	●
Aids identification in low light or colored illumination	●	●	●	●	If lit	If lit
Aids identification of control position		●		●	●	●
Requires little training				●		
Disadvantages						
May require extra space	●	●	●	●	●	
Affects manipulation of control	●	●	●	●		
Limited number of coding categories	●	●	●	●		●
May less effective with gloves		●	●	●		
Must be viewed (not blind operation)					●	●

Table 9.2 Coding of controls for identification and discrimination. (Adapted from MIL-STD 1472 [49]; Ely, Thomson, and Orlansky [548].)

(e.g., the color and position of lights in a traffic signal), redundancy gain can also make controls easier to identify and discriminate. Considered in terms of the advantages and disadvantages of the features for coding controls in Table 9.2, no one code has all the advantages nor avoids all the disadvantages, but combining codes can. For example, a knob that is labeled might not require any training to understand what it controls, but it would not support blind operation; however, it would if it is also differentiated from other knobs by shape and size.

6. **Avoid absolute judgment limits.** Just as absolute judgment limits the distinctions that can be made with color or other variables in displays, the same limits govern how many types of controls can be differentiated by any one type of code, such as color, to no more than five to seven. Combining codes can expand this number. As Table 9.2 shows, only labels avoid the limited number of coding categories associated with the other features. The actuation of controls can also stress the ability of people to judge angle and position accurately and for this reason knobs and selector switches should have *detents*—stops that require extra force to pass— at locations corresponding to the categories of the control setting. This contrasts with a knob that rotates freely and forces people to guess where exactly to position the knob. This is another instance of information theory, where the information (precision and size of movements required) should

The variation and precision required for control should match the variation and precision available in the controller.

match that provided by the controller. If the person is meant to generate discrete inputs the controller should be similarly discrete.

9.3.3 Memory Principles

7. **Knowledge in the world.** Visual displays relieve the burden on memory by placing knowledge in the world rather than forcing people to keep it in their heads. The same applies with controls. The actuation of the control *should be reflected in the control itself,* such that the position of toggle switch indicates its state or when the illumination of a button indicates a system is on. Buttons and levers that return to a set position after people actuate them provide no indication of system status, which forces people to keep that information in working memory if it is not displayed elsewhere. Forcing people to rely on knowledge in the head rather than knowledge in the world increases the chance that they will forget what action they performed, and hence the state of the system they are controlling.

In contrast with physical input devices, software-based controls, such as voice and gesture-based interactions, have few indications of the control opportunities, system state, or control actuation. For some gestures are intuitive, easily learned, such as selecting by touching, pinching to expand, and swiping to reject or accept. Such gestures naturally fit a touch screen, but others are less easily discovered. Gestures and voice lack the codes in Table 9.2 and so place a premium on defining intuitive conventions and providing clear feedback to guide people towards successful control, and issue we return to later in this chapter and in Chapter 10.

8. **Be consistent.** Similar to the principle for visual displays, consistency makes it possible for people to apply skills from one situation to another, reducing errors and response time. Table 9.2 shows that each of the features of control devices can contribute to consistency and standardization. This standardization should be considered for functions within a system, as well as across systems. Our unfortunate driver probably encountered inconsistency in the location of the light control between cars.

9.3.4 Mental Model Principles

In Chapter 8 we discussed the *principle of pictorial realism* in describing the need to align the orientation of a display with the person's mental model, and the *principle of the moving part* in describing the need to align the movement of the displayed system with the person's mental model. *Stimulus-response compatibility* (or display-control compatibility) describes the location or movement of a control response and the location or movement of displayed system [549]. Two principles characterize a compatible (and

Controls that fail to indicate possible commands and whether a control has had its intended effect, rely on limited and effortful knowledge in the head.

hence good) mapping between display and control (or stimulus and response).

9. **Location compatibility.** The control location should be close to (and in fact closest to) the entity being controlled or the display of that entity. Similar to labels for displays, labels that are separated from controls can confuse people as they try to link labels to controls.

10. **Movement compatibility.** The direction of movement of a control should be congruent with the direction both of movement of the displayed indicator and of the system movement itself [550]. A violation of movement compatibility would occur if people need to move a lever to the left to move a display indicator to the right or upward. Population stereotypes describe general expectations regarding expected responses of controls. With a button people in North America expect a system to turn on. With a switch people expect that flipping it up will turn a it on, but not in Europe where the population stereotype is the opposite. With a knob, clockwise is compatible with increasing system variables and displayed indicators moving upwards [550, 487]. However, when the control is to the left or right of the display, the *proximity of movement* leads people to expect displayed element to move in the direction of the control nearest to the display [551, 10]. If the control is on the left of the display, people expect a counterclockwise movement of the control to increase the displayed element. Controls and displays should be arranged so that the effect of the proximity of movement agrees with the expectations of clockwise movement to increase the system variable.

In many workstations, controls will be remote from, and often positioned in different body orientations, the display surface. Here the principle of visual field compatibility [552, 553] asserts that the movement of the control should be in the same direction as the display of the controlled agent if that display were located to the same plane of the control. That is, the frames-of-reference of the display and control axes are rotated into congruence [554, 552].

9.3.5 Response Selection Principles

11. **Avoid accidental activation.** Accidental activation of a control stems from failures of skill-based behavior, where people inadvertently bump or depress a control, and from failures of rule-based behavior, where they intentionally activate a control without sufficient consideration for the situation, an expression of the speed-accuracy tradeoff. Table 9.3 summarizes four general methods to address both of these sources of error: Locate and orient, recess and shield, interlock and sequence operations, and resist, delay and confirm. These methods apply to controls for industrial and military systems

Figure 9.3 A barrier to help people avoid accidentally descending a stairway into the basement during a fire (Photograph by author: J. D. Lee.)

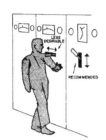

Locate and orient

- Locate and orient the control so that it won't be brushed or bumped when other is used.

- Locate and orient to avoid being included in sequence of routinely activated controls.

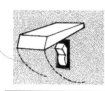

Recess and shield

- Physically protect the control from being brushed or bumped.

- Require an extra step so the control is not activated as part of a routine sequence of actions.

Interlock and sequence operations

- Interlock the control so that it requires actions in at least two directions to activate, such as lateral movement and then longitudinal movement.

- Require a specific sequence of actions, such as a button press that makes the primary movement possible.

- These extra movements make it unlikely that simply bumping the control will activate it.

- They also make it less likely someone will activate the control as part of a routine sequence of actions.

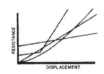

Resist, delay, and confirm

- Require more effort to move the control than would be expected from being brushed or bumped.

- Use resistance from viscous or coulomb friction and spring-loading to distinguish controls.

- Require a confirmation of the primary action, such as the "OK" button before a file is deleted.

Table 9.3 Methods to avoid accidental activation of controls. (Adapted from MIL-STD 1472 [49]; Ely, Thomson, and Orlansky [548].)

as well as computer systems, such as the confirmation dialog box that helps people from accidentally closing a document without first saving it. Methods for avoiding accidental activation tend to force people to consider more evidence and exert more effort to complete a task rather than proceeding according to expectations of what the situations should be rather than what it actually is. This methods are particularly effective in preventing human error that may occur under time-stressed and high-workload circumstances.

The methods in Table 9.3 for protecting against accidental

activation tend to make activating the control more difficult and so the ease and speed of activation must be balanced with the cost of accidental activation. For example, techniques to recess and shield might slow people slightly, but adding an interlock will slow people a lot. The degree of effort and delay should be matched to the consequence of accidental activation. This speed-accuracy trade-off differs from the other principles discussed so far, in that the other 10 principles tend to enhance both speed and accuracy. In fact, the other principles, particularly *Make discriminable*, will reduce accidental activation and enhance both speed and accuracy.

12. **Hick-Hyman Law.** The speed with which an action can be selected is strongly influenced by the number of possible alternative actions that could be selected in that context. This is called the complexity of the decision of what action to select. Each action of a Morse code operator, in which only one of two alternatives is chosen (*dit or dah*) follows a much simpler choice than each action of the typist, who must choose between one of 26 letters. Hence, the Morse code operator can generate a greater number of keystrokes per minute. Correspondingly, users can select an action more rapidly from a computer menu with two options than from the more complex menu with eight options.

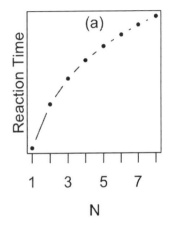

This effect of decision complexity on response selection time is described by the Hick-Hyman law of reaction time (RT):

$$RT = a + b \log_2 N \qquad (9.1)$$

where a represents the movement time, b is the processing speed and N is the number of possible stimulus-response alternatives. As shown in Figure 9.4 [555, 556], when response time is plotted as a function of $\log_2(N)$ (Figure 9.4(b)) rather than N (Figure 9.4(a), the function is linear. This shows a logarithmic increase in RT as the number alternatives (N) increases. The Hick-Hyman law calculates the cost of complexity—every menu option and feature has a cost.

13. **Decision complexity advantage.** The Hick-Hyman law does not imply that systems for people through a long series of simple decisions . In fact, the logarithmic relationship in Figure 9.4(b) suggests it is more efficient to require a smaller number of complex decisions than many simple decision. This is referred to as the *decision complexity advantage* [10]. For example, a typist can convey the same message more rapidly than can the Morse code operator. Although typing keystrokes are made more slowly, there are far fewer of them.

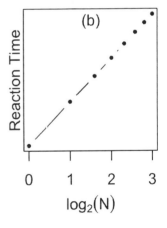

Figure 9.4 Reaction time follows the Hick-Hyman law. (Graph created using R.)

14. **Fitts's Law.** Controls typically require movement of two different sorts: (1) movement is often required for the hands or fingers to reach the control (not unlike the movement

of attention to access information, discussed in Chapter 8), and (2) the control may then be moved in some direction, often to position a cursor. Even in the best of circumstances in which control location and destination are well learned, these movements take time. Fortunately for designers, such movement times (MT) can be relatively well predicted by *Fitts's law* [557, 558]:

$$MT = a + b \, \log_2 \left(\frac{2A}{W} \right) \qquad (9.2)$$

where A = amplitude of the movement and W = width of the target or the desired precision with which the cursor must land. This means that movement time is linearly related to the logarithm of the term, 2A/W, which is the *index of difficulty* the movement. Increasing the index of difficulty increases movement time.

We show three examples of Fitts's law in Figure 9.5, with the index of difficulty calculated to the right. As shown in rows (a) and (b), when the distance to touch the circles doubles, the index of difficulty and therefore movement time increases by a constant amount. Correspondingly, each time the required precision of the movement is doubled—the target width or allowable precision is halved; compare rows (a) and (c). The movement time also increases by a constant amount unless the distance is correspondingly halved; compare rows (b) and (c), showing the same index of difficulty and therefore the same movement time. Smaller targets (reducing W) increases movement time unless they are proportionately moved closer. Another implication of Fitts's law is that if we require a movement of a given amplitude, A, to be made within a shorter time, MT, then the precision of that movement will decrease as shown by an increase in the variability of movement endpoints, represented by W. This characterizes a speed-accuracy tradeoff in pointing movements. The value of W in this case characterizes the distribution of endpoints of the movement: Larger W means larger error.

The mechanisms underlying Fitts's law are quite general, and so the law applies to the physical movement of the hand to a target (e.g., reaching for a key) and to the movement of a cursor to a screen target using a control device (e.g., a mouse to bring a cursor to a particular item in a computer menu [559]). It also applies to movements as coarse as a foot reaching for a pedal [560] and as fine as a manipulation under a microscope [561]. This generality enables designers to predict the costs of different keyboard layouts and target sizes in a wide variety of circumstances [252]. In particular, in comparing rows (b) and (c) of Figure 9.5, the law informs that miniaturized keyboards—reduced distance between keys—will not increase the speed of keyboard use.

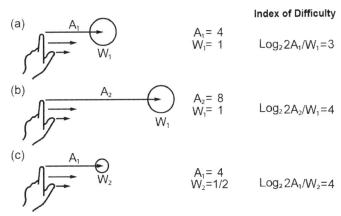

Index of Difficulty

(a) A_1 W_1

$A_1 = 4$
$W_1 = 1$ $\mathrm{Log}_2 2A_1/W_1 = 3$

(b) A_2 W_1

$A_2 = 8$
$W_1 = 1$ $\mathrm{Log}_2 2A_2/W_1 = 4$

(c) A_1 W_2

$A_1 = 4$
$W_2 = 1/2$ $\mathrm{Log}_2 2A_1/W_2 = 4$

Figure 9.5 Movement time (MT) follows Fitts's law. Comparing (a) and (b) shows the doubling of movement amplitude from $A_1 \rightarrow A_2$; comparing (a) to (c) shows halving of target width $W_1 \rightarrow W_2$ (or doubling of target precision); (b) and (c) will have the same MT. Next to each movement is shown the calculation of the index of difficulty of the movement to which MT will be directly proportional.

15. **Provide feedback.** Most actions generate visual feedback that indicates how the system responded to the control input. For example, in a car the speedometer offers visual feedback from the control of the accelerator. However, control design must also be concerned with more direct *transient feedback* of the change in control state. As we learned in Chapter 5, this feedback may be kinesthetic and tactile (e.g., the feel of a button as it is depressed to make contact or the resistance on a lever as it is moved). It may be auditory (e.g., the click of the switch or the beep of the phone keypad), or it may be visual (e.g., the change in position of a switch). Beyond the transient feedback of the change in control state, the control should also provide feedback!persistent regarding its current state (e.g., a light next to a switch to show it is on or even the clear and distinct visual view that a push button has been depressed). This persistent feedback supports knowledge in the world described earlier. Blind operation of controls requires very good tactile or auditory transient and persistent feedback.

Through whatever channel, more feedback of both the current control state and the change in control state is good as long as the feedback is nearly instantaneous—*delayed feedback* can seriously degrade performance. Feedback delayed by as little as 100 ms can be harmful if rapid sequences of control actions are required. Such delays are particularly harmful if the operator is less skilled (and therefore depends more on the feedback) or if the feedback cannot be filtered out by selective attention mechanisms [10]. A good example of such harmful delayed feedback is a voice feedback delay while talking on a radio or telephone. You simply cannot filter out the sound of your voice, and the delay makes it sur-

Control Design Principles

Attention principles
A1 Proximity compatibility
A2 Avoid resource competition

Perception principles
P3 Make accessible
P4 Make discriminable
P5 Exploit redundancy gain
P6 Avoid absolute judgment limits

Memory principles
M7 Knowledge in the world
M8 Be consistent

Mental model principles
MM9 Location compatibility
MM10 Movement compatibility

Response selection principles
R11 Avoid accidental activation
R12 Hick-Hyman Law
R13 Decision complexity advantage
R14 Fitts's Law
R15 Provide feedback

Table 9.4 Control design principles.

prisingly hard to continue talking. For feedback in pressing virtual buttons, such as those on a touchscreen, the feedback delay limits are even more severe: tactile between 5 and 50 ms, audio between 20 and 70 ms, and visual between 30 and 85 ms [562].

9.3.6   Summary of principles

In concluding our discussion of principles for control design, it should be apparent that just as display principles sometimes conflict, so do principles for control design. To show how these conflicts are resolved, we turn to a discussion of various categories of controls. As we encounter each principle in an application, we place a reminder of the principle number in parentheses, for example, (A1) refers to the principle of proximity compatibility, the first principle discussed under attention. The letter refers to the category: attention (A), perception (P), memory (M), mental model (MM), and response selection and execution (R), which are summarized in Table 9.4.

In the following sections we apply these principles to specific applications. With these applications we use the term "guidelines" to distinguish them from the 15 principles; the guidelines are more specific design suggestions derived from the principles.

9.4   Discrete Controls: Buttons and Switches

Our driver in the opening story was troubled, in part, because he simply did not know, or could not find, the right controls to activate the wipers. Many such controls in systems are designed primarily for the purpose of activating or changing the discrete state of some system. In addition to making the controls easily visible, there are several design features that make the activation of such controls less susceptible to errors and delays.

Feedback. Feedback is a critical feature of discrete controls (R15). Some controls offer more feedback channels than others. The toggle switch is very good in this regard. It changes its state in an obvious *visual fashion* and provides an auditory click and a tactile snap (a sudden loss of resistance) as it moves into its new position. The auditory and tactile feedback provide the operator with instant knowledge of the toggle's change in state, while the visual feedback provides continuous information regarding its new state—knowledge in the world (M7). A push button that remains depressed when on has similar features, but the visual feedback may be less obvious, particularly if the spatial difference between the button at the two positions is small.

Care should be taken in the design of other types of discrete controls that the feedback (indicating that the system has received the state change) is obvious. Touch screens do not do this so well; neither do push-button phones or security code entry keys, that lack an auditory cue following each keypress. Computer-based control devices often replace the auditory and tactile state-change feedback with artificial visual feedback (e.g., a light that turns on when the switch is depressed). If such visual feedback is meant to be the only cue to indicate state change (rather than a redundant one), then there will be problems associated both with an increase in the *distance* between the light and the relevant control (this distance should be kept as short as possible; P9) and with the possible electronic failure of the light or with difficulties seeing the light in glare. Hence, feedback lights ideally should be redundant with some other indication of state change (P5, redundancy gain); of course, any visual feedback should be immediate (R15).

Size. Smaller keys are usually problematic from a human factors standpoint. If they are made smaller out of necessity to pack them close together in a miniaturized keyboard, they invite "blunder" errors when the wrong key (or two keys) are inadvertently pressed, an error that is particularly likely for those with large fingers or wearing gloves. If the spacing between keys is not reduced as they are made smaller, however, the time for the fingers to travel between keys increases, following the predictions of Fitts's law (R14).

Discriminability and Labeling. Keypress or control activation errors also occur if the identity of a key is not well specified to the novice or casual user (i.e., one who does not "know" the location by touch). This happened to our driver at the beginning of the chapter. Principles associated with making controls accessible, identifiable, and describable (P3, P4) and exploiting redundancy gain (P5) all help address this problem. These confusions are more likely to occur (a) when large sets of identically appearing controls are unlabeled or poorly labeled and (b) when labels are physically displaced from their associated controls, hence violating P9 and the proximity compatibility principle from Chapter 8.

Fixed and Moving Pointers. Like the moving and fixed pointer for displays, moving pointer for switches are preferred. The switches in Figure 9.6a show that a moving indicator makes check reading easy, where any switch set at anything other than "1" would be immediately apparent. Compare this to the fixed pointer switch (Figure 9.6b), where the scale rotates. Here the advantage is that the selected element always shows in the same location, but when this switch is part of a large array there is no way to quickly assess the state of the group. The moving pointer allows check of the switch position at a glance; the moving scale requires more focal attention, and often good lighting to check its setting.

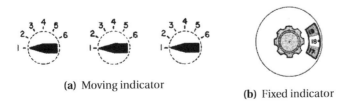

(a) Moving indicator

(b) Fixed indicator

Figure 9.6 Selector switches. (Adapted from Ely, Thomson and Orlansky [548].)

Blind operation. Figure 9.6a shows a switch designed for blind operation. It indicates the setting clearly by its shape, so that the direction can be determined by feel. The switch on the right (Figure 9.6b) does not include these features, making it less able to support blind operation. The setting of the switch can only be determined by looking at the dial and reading the numbers.

9.5 Discrete Controls: Keyboards

Buttons, switches, and knobs do not generally offer a compatible means of inputting or specifying the symbolic, numerical, or verbal information that is involved in system interaction [563]. For this sort of information, keyboards or voice control have generally been the interfaces of choice (see Table 9.1).

9.5.1 Numerical Data Entry

For numerical data entry, numerical keypads or voice remain the most viable alternatives. While voice control is most compatible and natural, it is hampered by certain technological problems that slow the rate of possible input. Numeric keypads, are typically represented in one of three forms: linear array or 3×3 square arrays with either 123 or 789 across the top. The linear array, such as found at the top of the computer keyboard is generally not preferred because of the extensive movement time required to move from key to key. The 3×3 square arrays minimize movement distance (and therefore time). General design guidelines suggest that the layout with 123 on the top row (telephone) is preferable [564], to that with 789 on top (calculator) since it follows the regular reading pattern from top to bottom. However the advantage is probably not great enough to warrant redesign of the many existing "7-8-9" keypads.

Beyond the layout of the keyboard, the details of the keys can have a substantial effect on performance. Flat membrane or glass keypads requiring people to memorize location or visual attention to keep fingers centered. Raised keys with indents provide sufficient discriminability to allow blind operation (P4)[565]. A full keyboard would need raised nibs, typically on the *F* and *J* keys to act as landmarks to orient the fingers.

9.5.2 Text Data Entry

For data entry of linguistic material, the computer keyboard has traditionally been the device of choice. Although some alternatives to the traditional QWERTY layout have been proposed, it is not likely that this design will be changed.

An alternative to dedicated keys that require digit movement is the *chording keyboard* which individual items of information are entered by simultaneously depressing combinations of keys [566, 567]. Chording works by allowing a single complex action to convey a large amount of information and hence benefit from the decision complexity advantage (R13), discussed earlier in this chapter. A single press with a 10-key keyboard can designate any of 2^{10} - 1 (or 1,023) possible actions/meanings. Figure 9.7 shows examples of chording keyboards.

Chording keyboards have three distinct advantages. First, since the hands never need to leave the chord board, there is no requirement for visual feedback to monitor the correct placement of a thumb or finger digit. Consider, for example, how useful this feature would be for entering data in the high-visual-workload environment characteristic of helicopter flight or in a continuous visual inspection task. Second, because there is less finger movement, the chording board is less susceptible to repetitive stress injury or carpal tunnel syndrome (Chapter 13). Finally, after extensive practice, chording keyboards enable faster word transcription than the standard typewriter keyboard because there are less finger movement [566, 568].

Figure 9.7 Several chording keyboards. Letters produced by a combination of simultaneous key presses rather than individual keys. (Author: Dcoetzee, Copyright: CC0 1.0, avaialble on Wikipedia page on Microwriter.)

The primary cost of the chording keyboard is the *extensive training* required to associate the finger combinations with their meaning. In contrast, typewriter keyboards provide knowledge in the world regarding the appropriate key, since each key is labeled on the top and each letter is associated with a unique location in space [16]. For the chording keyboard there is only knowledge in the head, which is more difficult to acquire and may be easier to lose through forgetting. Still, various chording systems have found their way into use; examples are both in postal mail sorting [569] and court transcribing [566], where specialists have invested the necessary training time to speed data input.

Ironically, multi-touch gestures on tablets and smartphones are a type of chording device. The multi-finger gestures invoke commands that would otherwise take many keypresses or menu selections. These devices share the knowledge in the head demands of more traditional chording keyboards, but benefit from some well-learned conventions from interacting with physical systems and from direct visual confirmation of actions. The same may not be true for gesture recognition systems that enable mid-air text entry [570].

Chording keyboards are typically used only by specialists, but multi-touch gestures on smartphones have made aspects of chording commonplace.

9.6 Discrete Controls: Voice Input

Over the last several years, increasingly sophisticated voice recognition technology has made this a viable means of control, although such technology has both costs and benefits [?].

9.6.1 Benefits of Voice Control

Natural decision complexity advantage. Consistent with the decision complexity advantage (R13), chording is efficient because a single action can select one of several hundred items. However, voice control can be even more efficient because a single utterance can represent any one of several thousand. Furthermore, as we know, voice is usually a very "natural" communications channel for symbolic linguistic information and one with which we have had nearly a lifetime's worth of experience. This naturalness may be (and has been) exploited in many control interfaces when the benefits of voice control outweigh the costs.

Reduced resource competition. Particular benefits of voice control may be observed in dual-task situations (A2). When the hands and eyes are busy with other tasks, like driving (which prevents dedicated manual control on a keyboard and the visual feedback necessary to see if the fingers are properly positioned), designs in which the operator can time-share by talking to the interface using separate resources are of considerable value. Some of the greatest successes have been realized, for example, in using voice to enter radio-frequency data in the heavy visual-manual load environment of the helicopter. *Dialing* of cellular phones by voice command while driving is considered a useful application of voice recognition technology [571]. So also is the use of this technology in assisting baggage handlers to code the destination of a bag when the hands are engaged in the "handling" activity. There are also many circumstances in which the combination of voice and manual input for the same task can be beneficial [572]. Such a combination, for example, would allow manual interaction to select objects (a spatial task) and voice to convey symbolic information to the system about the selected object .

9.6.2   Costs of Voice Control

Against these benefits are four distinct costs that limit the applicability of voice control and highlight precautions that should be taken in its implementation. These costs are related closely to the sophistication of the voice recognition technology necessary for computers to translate the complex four-dimensional analog signal that is voice (see Chapter 5) into a categorical vocabulary, which is programmed within the computer-based voice recognition system [573]. More fundamentally, even if voice recognition technology can perfectly transcribe spoken words, that does not mean it can "understand" what the person wants.

Confusion and limited vocabulary size. Because of the demands on computers to resolve differences in sounds that are often subtle even to the human ear, and because of the high degree of variability (from speaker to speaker and occasion to occasion) in the physical way a given phrase is uttered, voice recognition systems are prone to make confusions in classifying similar-sounding utterances (e.g., "cleared to" versus "cleared through"). How such confusions may be dealt with can vary [573]. The recognizing software may simply take its "best guess" and pass it on as a system input. This is what a computer keyboard would do if you hit the wrong letter.

Alternatively, the system may provide feedback if it is uncertain about a particular classification or if an utterance is not even close to anything in the computer's vocabulary. The problem is that if the recognition capabilities of the software are still far from perfect, the repeated occurrences of this feedback will greatly disrupt the smooth flow of voice communications if this feedback is offered in the auditory channel. If the feedback is offered visually, then it may well neutralize the dual-task benefit (i.e., keeping the eyes free). These costs of confusion and misrecognition can be addressed by reducing the vocabulary size and constructing the vocabulary in such a way that acoustically similar items are avoided.

Constraints on speed. Most voice recognition systems do not easily handle the continuous speech of natural conversation. This is because the natural flow of our speech does not necessarily place physical pauses between different words. Hence, the computer does not easily know when to stop "counting syllables" and demarcate the end of a word to look for an association of the sound with a given item in its vocabulary. To guard against these limitations, the speaker may need to speak unnaturally slowly, pausing between each word.

A related point concerns the time required to "train" many voice systems to understand the individual speaker's voice prior to the system's use. This training is required because there are so many physical differences between the way people of different gender, age, and dialect may speak the same word. Hence, the computer can be far more efficient if it can "learn" the pattern of a particular individual (called a *speaker-dependent* system) than it can if it must master the dialect and voice quality of all potential users (*speaker-independent* system). For this reason, speaker-dependent systems usually can handle a larger vocabulary.

Acoustic quality and noise and stress. Two characteristics can greatly degrade the acoustic quality of the voice and hence challenge the computer's ability to recognize it. First, a noisy environment is disruptive, particularly if there is a high degree of spectral overlap between the signal and noise (e.g., recognizing the speaker's message against the chatter of other background conversation). Second, under conditions of stress, one's voice can change substantially in its physical characteristics, particularly increasing the fundamental frequency (the high-pitched "Help, emergency!" [574]. As we will see in Chapter 15, stress appears to occur often under emergency conditions, and hence great caution should be given before designing systems in which voice control must be used as part of emergency procedures.

Voice control, like many technologies to enhance human performance, is prone to fail just when it might be most useful: noisy, high-stress situations.

Compatibility. Finally, we have noted that voice control is less suitable for controlling continuous movement than are most of the available manual devices [575, 572]. Consider, for example, the greater difficulties of trying to steer a car along a curvy road by saying "a little left, now a little more left" than by the more natural manual control of the steering wheel.

Clearly all of these factors—costs, benefits, and associated design considerations (like restricting vocabulary)—play off against each other in a way that makes it hard to say precisely when voice control will be better or worse than manual control. The picture is further complicated because of the continued improvement of computer algorithms have largely addressed two major limitations of voice systems (continuous speech recognition and speaker-dependence). However, even if such systems do successfully address these problems, simpler systems might offer superior performance. For example, even with excellent voice recognition technology, the advantages for voice control over mouse and keyboard data entry are mixed [576]. For isolated words, voice control is faster than typing only when typing speed is less than 45 words/minute, and for numerical data entry, the mouse or keypad are superior.

9.7 Positioning, Tracking, and Continuous Control

A common task is the need to position or point to some entity in space. This may involve moving a cursor to a point on a screen, reaching with a robot arm to contact an object, or moving the setting on a radio dial to a new frequency. Generically, we refer to these spatial tasks as those involving positioning or pointing [564]. A wide range of control devices, such as the mouse, joystick, and thumbpad are available to accomplish such tasks. Before we compare the properties of such devices, however, we consider the important nature of the human performance underlying the positioning task: movement of a controlled entity which we call a *cursor*, to a destination which we call a *target*.

Positioning typically focuses on guiding a cursor to a fixed target either through fairly direct hand movement (the touch screen or light pen) or as mediated by a control device (the trackball, joystick, or mouse). However, much of the world of both work and daily life is characterized by making a cursor or some corresponding system (e.g., vehicle) output follow or "track" a *continuously moving dynamic* target. This may involve tasks as mundane as bringing the fly swatter down on the moving pest or riding the bicycle around the curve, or as complex as guiding an aircraft through a curved flight path in the sky, guiding your viewpoint through a virtual environment, or bringing the temperature of a nuclear reactor up to a target value through a carefully controlled trajectory. These cases and many more are described by the generic task of *tracking* [558, 577]; that is, the task of making a system output (the cursor) correspond in time and space to a time-varying command target input.

Information theory and the associated Hick-Hyman and Fitts's laws are useful for describing discrete control tasks. A different perspective is needed for pointing, positioning, and tracking tasks. For these tasks, *control theory* helps us understand design features that affect how people perform these tasks.

9.7.1 The Tracking Loop: Basic Elements

Figure 9.8 shows the basic elements of a tracking task. Each element receives a time-varying input and produces a corresponding time-varying output. Hence, every signal in the tracking loop is represented as a function of time, $f(t)$. These elements are described in the context of driving a car, although it is important to think about how they generalize to other tracking tasks.

When driving a car, the driver perceives a discrepancy or error between the desired state of the vehicle and its actual state. As an example, the car may have deviated from the center of the lane or may be pointing in a direction away from the road. The driver wishes to reduce this error, $e(t)$. To do so, she applies a force (actually a torque), $f(t)$, to the steering wheel or *control* device. This force in turn produces a rotation, $u(t)$, of the steering wheel itself, called control output. (Note that our frame of reference is the human. Hence, we use the term output from the human rather than the term input to the system.) The relationship between the force applied and the steering wheel control output is defined as the *control feedback dynamics*, which determines how the steering wheel resists the drivers input, and is the source of tactile feedback. A steering wheel with a high degree of resistance might help prevent the accidental input (R11) that nearly caused the driver to crash in the opening vignette.

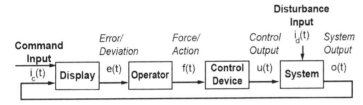

Figure 9.8 Basic elements of the tracking loop.

Movement of the steering wheel or control device, $u(t)$, then causes the vehicle's actual position to move laterally on the highway, or more generally, the controlled system to change its state. This movement is called the *system output, o(t)*. As noted earlier, when presented on a display, the representation of this output position is often called the cursor. The relationship between control output, $u(t)$, and system response, $o(t)$, is defined as the *system dynamics*.

If the driver is successful in the correction applied to the steering wheel, then the discrepancy between vehicle position on the highway, $o(t)$ and the desired or "commanded" position at the center of the lane, $i_c(t)$ is reduced. That is, the error, $e(t)$, is reduced to zero. On a display, the symbol representing the input is called the target. The difference between the output and input signals (between target and cursor) is the error, $e(t)$, which was the starting point of our discussion. A good driver responds in such a way as to keep $o(t) = i(t)$ or, equivalently, $e(t) = 0$.

The system represented in Figure 9.8 is called a *closed-loop* control system, or a negative feedback system because the operator corrects in the opposite direction from (i.e., *negates*) the error. Because errors in tracking stimulate the need for corrective responses, the person need never respond at all as long as there is no error. This might happen while driving on a straight smooth highway on a windless day. However, errors typically arise from one of three sources. *Command inputs, $i_c(t)$*, are changes in the target that must be tracked. For example, if the road curves, it generates an error for a vehicle traveling in a straight line and so requires a corrective response. *Disturbance inputs*, $i_d(t)$, are those applied directly to the system for which the operator must compensate. For example, a wind gust that blows the car off the center of the lane is a disturbance input. So is an accidental movement of the steering wheel by the driver, as happened in the story at the beginning of the chapter. People sometimes intentionally produce the third source of error, *exploratory acion*, to better understand the situation. For example, you might tap the brakes to see if the road is icy or just wet. After tapping the brakes, you might have to null this error by accelerating back up to speed.

Tracking performance is typically measured in terms of *error*, *e(t)*. It may be calculated at each point in time as the absolute deviation and then accumulated and averaged (divided by the number of sample points) over the duration of the tracking trial. This is the *mean absolute error (MAE)*. Sometimes, each error sample may be squared, the squared samples summed, the total divided by the number of samples, and the square root taken. This is the *root mean squared error (RMSE)*, and compared to MAE it emphasizes large deviations more than small ones. Minimizing tracking error is certainly important, but a certain degree of error is unavoidable, and in the case of error that stems from exploratory behavior it is even necessary for robust control.

Now that we have seen the elements of the tracking task, which characterize the human's efforts to make the system output match the command target input, we can ask what characteristics of the human–system interaction make tracking difficult (increased error or increased workload). With this knowledge in mind, the designer can intervene to improve tracking systems. As we will see, some of the problems lie in the tracking system itself, some lie within the human operator's processing limits, and some involve the interaction between the two.

9.7.2 Input and Bandwidth

Drawing a straight line on a piece of paper or driving a car down a straight stretch of road on a windless day are both examples of easy tracking tasks. There is a command target input and a system output (the pencil point or the vehicle position). But the input does not vary; hence, the task is easy. After you get the original course set, there is nothing to do but move forward, and you can drive fast (or draw fast) about as easily as you can drive (or draw) slowly. However, if the target line follows a wavy course, or if the road is curvy, you have to make corrections, and there is uncertainty to process; as a result, both error and workload can increase if you try to move faster. This happens because the frequency of corrections you must make increases with faster movement and your ability to generate a series of rapid responses to uncertain or unpredictable stimuli (wiggles in the line or highway) is limited. Hence, driving too fast on the curvy road, you will begin to deviate more from the center of the lane, and your workload will be higher if you attempt to stay in the center. We refer to the properties of the tracking input, which determine the frequency with which corrections must be issued, as the *bandwidth* of the input. While the frequency of "wiggles" in a command input is one source of bandwidth, so too is the frequency of disturbances from a disturbance input like wind gusts (or drawing a straight line on the paper in a bouncing vehicle).

In tracking tasks, we typically express the bandwidth in terms of the cycles per second (Hz) of the highest input frequency present in the command or disturbance input. It is very hard for people to perform tracking tasks with random-appearing input having a bandwidth above about 1 Hz. In most naturally occurring systems that people are required to track (cars, planes), the bandwidth is much lower, less than 0.5 Hz. High bandwidth inputs keep an operator very busy with visual sampling and motor control, but they do not involve very much cognitive complexity. This complexity, however, is contributed by the order of the system dynamics, to which we now turn.

9.7.3   Control Order

Control order is an important element of system dynamics that determine how the system responds to control inputs. One particularly important feature of this response is *transport lag*, which is the time between initial input to the system and when the system fully responds. Lag increases with system order and, like time delays, makes control more difficult.

Position Control. The *order* the system dynamics refers to whether a change in the position of the control device (by the human operator, *u(t)* in Figure 9.8 leads to a change in the position of system state (*zero-order*), velocity (*first-order*), or acceleration (*second-order*) of the system output. Consider moving a pen across the paper or a pointer across the blackboard, or moving the computer mouse to position a cursor on the screen. In each case, a new position of the control device leads to a new position of the system output. If you hold the control still, the system output will also be still. This is zero-order control (see Figure 9.8).

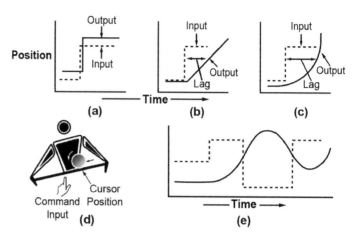

Figure 9.9 Control order. The solid line represents the change in position of a system output in response to a sudden change in position of the input (dashed line), both plotted as a function of time. (a) Response of a zero-order system; (b) response of a first-order system. Note the lag. (c) Response of a second-order system. Note the greater lag in (c) than in (b). (d) A second-order system: Tilt the board so the pop can (the cursor) lines up with the command-input finger. (e) Overcorrection and oscillations typical of control of second-order systems.

Velocity Control System. Now consider the scanner on a typical digital car radio. Depressing the button (a new position) creates a constant rate of change or *velocity* of the frequency setting. In some controls, depressing the button harder or longer leads to a proportionately greater velocity. This is a first-order control or velocity control system. As noted earlier, most pointing-device joysticks use velocity control. The greater the joystick is deflected, the faster will be the cursor motion. An analogous first-order control relation is between the *position* of your steering wheel (input) and the *rate of change* (velocity) of heading of your car (output). As shown in Figure 9.9b, the steering wheel angle (position) brings about a constant rate-of-change of heading (velocity). A greater steering wheel angle leads to a tighter turn (greater rate-of-change of heading). In terms of integral calculus, the order of control system corresponds to the number of time integrals between the input and output; that is, for first-order control or velocity control system,

$$O(t) = \int i(t)dt \qquad (9.3)$$

This relation holds because the integration of position over time produces a velocity.

For zero-order control, there are no (zero) time integrals and the equation becomes:

$$O(t) = i(t) \qquad (9.4)$$

Both zero-order (position) and first-order (velocity) control systems are important in designing manual control devices. Each has its costs and benefits. To some extent, the "which is best?"

question has an "it depends" answer. In part, this depends on the goals. If, on the one hand, accurate positioning is very important (like positioning a cursor at a point on a screen), then position control (with a low gain) has its advantages. On the other hand, if following a moving target or traveling (moving forward) on a path is the goal (matching velocity), then one can see the advantages of first-order velocity control. An important difference is that zero-order control often requires a lot of physical effort. Velocity control can be less effort because you just have to set the system to the appropriate velocity (e.g., rounding a curve defined by the curve radius) and let it go on until system output reaches the desired target (i.e., the new heading coming out of the curve).

Any control device that uses first-order dynamics should have a clearly defined and easily reachable *neutral point* at which no velocity is commanded to the cursor. This is because stopping is a frequent default state. This is the advantage of spring-loaded joysticks for velocity control systems because the natural resting point is set to give zero velocity. It represents a problem when the mouse is configured as a first-order control system, since there is no natural zero point on the mouse tablet. While first-order systems are effort conserving, as shown in Figure 9.9b, first-order systems tend to have a little more lag between when the human commands an output to the device (applies a force) and when the system reaches its desired target position. The amount of lag depends on the gain, which determines how rapid a velocity is produced by a given deflection.

Acceleration Control. Consider the astronaut who must maneuver a spacecraft into a precise position by firing thrust rockets. Because of the inertia of the craft, each rocket thrust produces an acceleration of the craft for as long as the engine is firing. The time course looks similar to that shown in Figure 9.9c. This, in general, is a second-order acceleration control system, described as:

$$O(t) = \int \int i(t)dt \qquad (9.5)$$

To give yourself an intuitive feel for second-order control, try rolling a pop can to a new position or command input, *i*, on a board, as shown in Figure 9.8d. Second-order systems are generally very difficult to control because they are both *sluggish* and *unstable*. The sluggishness can be seen in the greater lag in Figure 9.9c compared to that in zero- and first-order control (Figures 9.9a and 9.9b respectively). Both of these properties require the operator to *anticipate* and *predict* (control based on the future, not the present), and, as we learned in Chapters 6 and 8, this is cognitively demanding and leads to high workload for people.

Because second-order control systems are hard to control, they are not intentionally designed into systems, unlike zero and first-order systems. However, a lot of systems that humans are asked to control have a sluggish acceleration-like response to a position input because of the high mass and inertia of controlled elements in the physical world. As we saw, applying a new position to the thrust control on a spacecraft causes it to accelerate endlessly. Applying a new position to the steering wheel via a fixed lateral rotation causes the car's position, with regard to the center of a straight lane, to accelerate, at least initially. In some chemical or energy conversion processes, application of the input (e.g., added heat) yields a second-order response to the controlled variable. Hence, second-order systems are important to understand because of the things that designers or trainers can do to address their harmful effects (increased tracking error and workload) when humans must control them.

Because of their long lags, second order systems can only be successfully controlled if the tracker anticipates, inputting a control now, for an error that will be predicted to occur in the future. Without such anticipation, unstable behavior will result. As we learned in Chapters 6 and 8, such anticipation is demanding and not always done well.

Sometimes anticipation or prediction can be gained by paying attention to the trend in error. One of the best cues about where things will be in the future is for the tracker to perceive trend information of where they are going right now—that is, attend to the current rate of change. For example, in driving, one of the best clues to where the vehicle will be with regard to the center of the lane is where and how fast it is heading now. This trend information can be perceived by looking down the roadway to see if the direction of heading corresponds with the direction of the road better than it can be by looking at the deviation immediately in front of the car. Predictive information can also be obtained from explicit predictor displays as described in Chapter 8. Finally, as we discuss in Chapter 11, designers often automate the control of higher order systems with lags.

Time delays and transport lags. Higher-order systems, and particularly second-order ones, have a lag (see Figure 9.9b and 9.9c). Lags depend on the order of the controlled system (e.g., inertia), but time delays arise from several other sources. Delays may sometimes occur in systems of lower order as well. When navigating through virtual environments that must be rendered with time-consuming computer graphic routines, there is often a delay between moving the control device and updating the position or viewpoint of the displays (see Chapter 11; [578]). Large robotic arms have substantial lags because of inertia, and control of these arms from earth (if they are on vehicle in outer space) would also have substantial delays because of signal travel time. Time delays produce the same problems of anticipation that we saw with higher-order systems: Delays require anticipation, which is a source of human workload and system error.

Gain. When we discussed input devices, we noted that system gain describes how much output the system provides from a given amount of input. Hence, gain may be formally defined as the ratio $\Delta O/\Delta I$, where Δ is a given change or difference in the relevant quantity. In a high-gain system, a lot of output is produced by a small change of input. A sports car is typically high gain because a small movement of the steering wheel produces a large change in output (change in heading). Note that gain can be applied to

Time delays and lag forces people to anticipate future states, which places substantial demands on people's limited cognitive resources.

any order system, and is used to describe the amount of *change* in position (zero), speed (first), or acceleration (second) produced by a given deflection of the control.

Whether high, low, or medium gain is best is somewhat task-dependent. When system output must travel a long distance (or change by a large amount), high-gain systems are best because the large change can be achieved rapidly and little control effort (for a position control system) or in a rapid time (for a velocity control system). However, when precise positioning is required, high-gain systems present problems of overshooting and under-shooting, or *instability*. Hence, low gain is preferable. As might be expected, gains in the midrange of values are generally best, since they address both issues—reduce effort and maintain stability—to some degree [577]. As a consequence, adjusting gain often involves a speed-accuracy tradeoff that depends on the task. A high gain produces a fast response, but one that might also have large errors.

9.7.4 Stability

Now that we have introduced concepts of lag (due to higher system order or transport delay), gain, and bandwidth, we can discuss briefly one concept that is extremely important in systems that involve tracking: stability. Novice pilots sometimes show unstable altitude control as they oscillate around a desired altitude. Our unfortunate driver in the chapter's beginning story also suffered instability of control. This is an example of unstable behavior known as *closed-loop instability*. It is sometimes called *negative feedback instability* because of the operator's well-intentioned, but ineffective, efforts to reduce the error (i.e., to negate the error). Three factors contribute to closed-loop instability:

1. There is a *delay* somewhere in the control loop in Figure 9.8, either from the system lag, processing delays, or from the human operator's response time.

2. The *gain is too high*. This high gain can represent either the system's gain—too much heading change for a given steering wheel deflection like a sports car—or the human's gain—a tendency to overcorrect if there is an error (our unfortunate driver).

3. The *bandwidth and control order are too high* and the person tries to correct an error too rapidly and does not wait until the system output stabilizes before applying another corrective input. This third factor results when the input bandwidth is high relative to the system lag, and the person tries to correct all of the input "wiggles" (i.e., does not filter out the high-frequency inputs).

Exactly how much of each of these quantities (lag, delay, gain, bandwidth) are responsible for producing the unstable behavior is beyond the scope of this chapter, but there are good models of both the machine and the human that predict the conditions under which this unstable behavior will occur [579, 558, 580].

Human factors engineers can offer five ways to reduce closed-loop instability:

1. Lower the gain (either by system design or by instructing the operator to do so).

2. Reduce the lags and delays (if possible). Reduce delays by reducing the required complexity of graphics in a virtual reality system [581, 582]. Although often impossible, reduce the control order of controlled system to reduce lags.

3. As flight instructors will do with their students, caution the operator to change strategy in such a way that he or she does not try to correct every input but filters out the high-frequency ones, thereby reducing the bandwidth.

4. Change strategy to seek input that can anticipate and predict (like looking farther down the road when driving and attending to heading, or paying more attention to rate-of-change indicators).

5. Change strategy to *open-loop*. This is the final tracking concept we shall now discuss.

9.7.5 Open-Loop Versus Closed-Loop Systems

In all the examples we have described, we have implicitly assumed that the operator is perceiving an error and trying to correct it; that is, the loop depicted in Figure 9.8 is closed. Suppose, however, that the operator did not try to correct the error but just "knew" where the system output needed to be and responded with the precise correction to the control device necessary to produce that goal. Since the operator does not then need to perceive the error and therefore will not be looking at the system output, this is like breaking the loop in Figure 9.8 (i.e., opening the loop). In open-loop behavior the operator is not trying to correct for outputs that may be visible only after they accumulate. As a result, the operator will not fall prey to the evils of closed-loop instability. Of course, open-loop behavior depends on the operator's knowledge of (1) where the target will be and (2) how the system output will respond to his or her control input; that is, a well-developed mental model of the system dynamics [583] (see also Chapter 6). Hence, open-loop behavior is typical only of trackers who are highly skilled in their domain.

A process control operator uses open-loop control when she knows exactly how much the heat needs to be raised in a process to reach a new temperature, tweaks the control by precisely that amount, and walks away. Such behavior also characterizes a skilled baseball hitter who takes one quick look at the fast ball's initial trajectory and knows exactly how to swing the bat to connect. In this case there is no time for closed-loop feedback to guide the response. It also characterizes the skilled computer user who does

Open-loop tracking works only when there is little uncertainty about the future.

not need to wait for screen readout prior to depressing each key in a complex sequence of commands. Of course, such users still receive feedback after the skill is performed, feedback that will be valuable in learning or "fine tuning" the mental model (Chapter 18).

9.8 Pointing Devices

The various categories of control devices that can be used to accomplish these pointing or position tasks may be grouped into four distinct categories. In the first category are direct position controls (light pen and touch screen) in which the position of the human hand (or finger) directly corresponds with the desired location of the cursor. The second category contains indirect position controls—the mouse or touch pad—in which changes in the position of the limb directly correspond to changes in the position of the cursor, but the limb is moved on a surface different from the display cursor surface. Both of these cases correspond to zero-order control. Swiping to scroll on a tablet or mobile phone screen is a hybrid 0 and 1st order system.

The third category contains *indirect velocity controls*, such as the joystick and the cursor keys. Here, typically a movement of control in a given direction yields a velocity of cursor movement in that direction. For cursor keys, this may involve either repeated presses or holding it down for a long period. For joystick movements, the magnitude of deflection typically creates a proportional velocity. Joysticks may be of three sorts: *isotonic*, which can be moved freely and will rest wherever they are positioned; *isometric* (see Chapter 5), which are rigid and produce movement proportional to the force applied; or *spring-loaded*, which offer resistance proportional to both the force applied and the amount of displacement, springing back to the neutral position when pressure is released. The spring-loaded stick, offering both proprioceptive and kinesthetic feedback of movement extent, is typically the most preferred.

Across all display types, there are two important variables that affect usability of controls for pointing (and they are equally relevant for controls for tracking). First, feedback of the current state of the cursor should be salient, visible, and as applied to indirect controls, immediate. Thus, system lags greatly disrupt pointing activity, particularly if this activity is at all repetitive. Second, performance is affected in a more complex way by the system *gain*. Gain may be described by the ratio:

$$\text{Gain} = \frac{\text{Change of cursor}}{\text{Change of control position}}$$

Thus, a high-gain device is one in which a small displacement of the control produces a large movement of the cursor or produces a fast movement in the case of a velocity control device. (Gain is sometimes expressed as the reciprocal of gain, or the control/display ratio; see Figure 9.10) The gain of direct position controls, such as the touch screen and light pen, will obviously be 1.0.

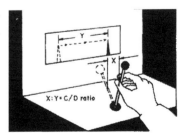

Figure 9.10 Control/Display ratio (1/Gain) for a pointing device (from Ely Thomson, Orlansky [548]).

There is some evidence that the ideal gain for indirect control devices should be in the range of 1.0 to 3.0 [564]. However, two characteristics partially qualify this recommendation. First, humans appear to adapt successfully to a wider range of gains in their control behavior [577]. Second, to elaborate the point made in the previous section, the ideal gain tends to be somewhat task-dependent because of the differing properties of low-gain and high-gain systems. Low-gain systems tend to be effortful, since a lot of control response is required to produce a small cursor movement; however, high-gain systems tend to be imprecise, since it is very easy to overcorrect when trying to position a cursor on a small target.

Selecting the appropriate gain is a matter of a speed-accuracy tradeoff. To the extent that a task requires a lot of repetitive and lengthy movements to large targets, a higher gain is better. This might characterize the actions required in the initial stages of a system layout using a computer-aided design tool where different elements are moved rapidly around the screen. In contrast, to the extent that small, high-precision movements are required, a low-gain system is more suitable. These properties characterize tasks such as uniquely specifying data points in a very dense cluster or performing microsurgery in the operating room, where an overshoot could lead to serious tissue damage. Gain is one of many factors can influence the effectiveness of control devices[564, 584].

9.8.1 Task Performance Dependence

For the most critical tasks involved in pointing (designating targets and "dragging" them to other locations), there is good evidence that the best overall devices are the two direct position controls (touch screen and light pen) and the mouse, as reflected in the speed, accuracy and preference [564, 559], shown in Figure ??.

Analysis using Fitts's law to characterize the range of movement distances and degrees of precision, suggests that the mouse is superior to the direct pointing devices [559]. However, Figure ?? also shows a speed-accuracy tradeoff between the direct position controls, which tend to be very rapid but less accurate, and the mouse, which tends to be slower, but more precise. Problems in accuracy with the direct positioning devices arise from several factors: parallax errors in which the position where the hand or stylus appear to be does not correspond to where it is if the surface is viewed at an angle. There is also the instability of the hand or fingers (particularly on a vertical screen), and in the case of touch screens, the imprecision of the finger area in specifying small targets. In addition to greater accuracy, indirect position devices, like a computer mouse, have another clear advantage over the direct positioning devices. Their gain can be adjusted, depending on the required position accuracy (or effort) of the task.

When pointing and positioning is required for more complex spatial activities, like drawing or handwriting, the advantages for the indirect positioning devices disappear in favor of the most natural feedback offered by the direct positioning devices. The

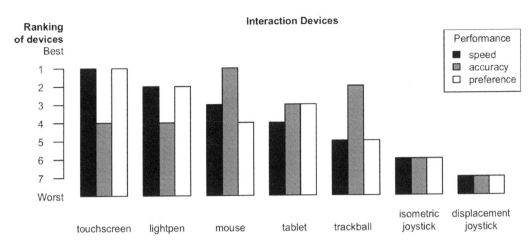

Figure 9.11 A comparison of performance of different control devices, based on speed, accuracy, and user preference. (Adapted from: Baber, C., 1997. Beyond the Desktop. San Diego, CA: Academic Press)

success of tablet devices over the last decade reflects this benefit.

9.8.2 The Work Space Environment

An important property of the broader workspace within which the device is used is the display, which presents target and cursor information [585]. As we have noted, display size (or the physical separation between display elements) influences the extent of device-movement effort necessary to access targets. Greater display size places a greater value on efficient high-gain devices. In contrast, smaller, more precise targets (or smaller displays) place a greater need for precise manipulation and therefore lower gain.

The physical characteristics of the display also influence usability. Vertically mounted displays or those that are distant from the body impose greater costs on direct positioning devices where the hand must move across the display surface. Frequent interaction with keyboard editing creates a greater benefit of devices that are physically integrated with the keyboard (i.e., cursor keys or a thumb touch pad rather than the mouse) or can be used in parallel with it (i.e., voice control). Finally, the available workspace size may constrain the ability to use certain devices. In particular, devices like joysticks or cursor keys that may be less effective in desktop workstations become relatively more advantageous for control in mobile environments, like the vehicle cab or small airplane cockpit, in which there is little horizontal space for a mouse pad. Here the thumb pad or screen swiping, in which repeated movement of the thumb across a small surface moves the cursor proportionately. Finally, the environment itself can have a major impact on usability. A vibrating environment, such as a vehicle cab, greatly undermines performance with direct position control devices.

The preceding discussion should make clear that it is difficult to specify in advance what the best device will be for a particular combination of task, workspace, and environment. It should, how-

ever, be possible to eliminate certain devices from contention in
some circumstances and at the same time to use the factors discussed above to understand why users may encounter difficulties
during early prototype testing. Baber [564] and Proctor and Vu
[584] provide a more detailed treatment of the human factors of
control device differences.

9.9 Displays for Tracking

In contrast with pointing, tracking involves continuous adjustment
with corrective adjustments to keep the tracked object in position.
The source of all information necessary to implement the corrective response is the display (see Chapter 8). For an automobile
driver, the display is the field of view seen through the windshield,
but for an aircraft pilot making an instrument landing, the display
is represented by the instruments depicting pitch, roll, altitude,
and course information.

An important distinction may be drawn between *pursuit* and
compensatory tracking displays, as shown in Figure 9.12. A pursuit display presents an independent representation of movement
of both the target and the cursor against the frame of the display.
Thus, our car driver sees a pursuit display, since movement of the
automobile can be distinguished and viewed independently from
the curvature of the road (the command input; Figure 9.12a). A
compensatory display presents only movement of the error relative
to a fixed reference on the display. The display provides no indication of whether this error arose from a change in system output or
command input [586]. Flight navigation instruments are typically
compensatory displays (Figure 9.12b).

As we noted in Chapter 8, displays may contain *predictive* information regarding the future state of the system, a valuable feature if
the system dynamics are sluggish. The automobile display is a kind
of predictor because the current direction of heading relative to
the vanishing point of the road provides a prediction of the future
lateral deviation. The curvature of the road provides a preview of
the future desired position of the car Figure 9.12a.

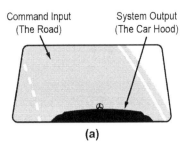
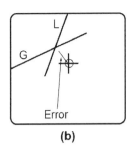

(a) **(b)**

Figure 9.12 Tracking with a pursuit display and a command display.(a) A pursuit display (the automobile); the movement of the car (system output), represented as the position of the hood ornament, can be viewed independently of the movement of the road (command input); (b) a compensatory display (the aircraft instrument landing system). G and L respectively represent the glideslope (commanded vertical input) and localizer (commanded horizontal input). The + is the position of the aircraft. The display will look the same whether the plane moves or the command inputs move.

To summarize the benefits of pursuit displays: They are better if the situation makes it possible and requires that people anticipate disturbances. This benefit, relative to compensatory displays is greater, the greater the input bandwidth. This is the case for driving because curves in the road are easily anticipated disturbances that must be responded to, and this is particularly so for winding roads. In contrast, the compensatory displays are useful when the future error cannot be easily seen, such as anticipating the future state of a second order system.

9.10 Remote Manipulation, Teleoperation, and Telerobotics

There are many circumstances in which continuous and direct human control is desirable but not feasible. Two examples are *remote manipulation*, such as when operators control an underseas explorer or an unmanned air vehicle (UAV), and *hazardous manipulation*, such as is involved in the manipulation of highly radioactive material. This task, sometimes known as *telerobotics or teleoperation* [468, 587, 588], possesses several distinct challenges because of the absence of direct viewing. The goal of the designer of such systems is often to create a sense of "telepresence," that is, a sense that the operator is actually immersed within the environment and is directly controlling the manipulation as an extension of his or her arms and hands. Similar goals of creating a sense of presence have been sought by the designers of virtual reality systems [589, 590]. Yet there are several control features of the situation that prevent this goal from being easily achieved in either telerobotics or virtual reality [591].

9.10.1 Time Delay

Systems often encounter time delays between the manipulation of the control and the availability of visual feedback for the controller. These may be transmission delays or transport lags. For example, the round-trip delay between earth and the moon is five seconds for an operator on earth carrying out remote manipulation on the moon. High-bandwidth display signals that must be transmitted over a low-bandwidth channel also suffer such a delay. Sometimes lags simply result from the inherent sluggishness of high-inertia systems that are being controlled. In still other cases, the delays might result from the time it takes for a computer system to construct and update elaborate graphics imagery as the viewpoint is translated through or rotated within the environment. In all cases, such delays present challenges to effective control.

9.10.2 Depth Perception and Image Quality

Teleoperation normally involves tracking or manipulating in three dimensions. Yet, as we saw in Chapter 4, human depth perception in 3-D displays is often inadequate for precise judgment along the viewing axis of the display. One solution that has proven quite useful is the implementation of stereo. The problem with stereo teleoperation, however, lies in the fact that two cameras must be mounted and two separate dynamic images must be transmitted over what may be a very limited bandwidth channel, for example, a tethered cable connecting a robot on the ocean floor to an operator workstation in the vessel above. Similar constraints on the bandwidth may affect the quality or fuzziness of even a monoscopic image, which could severely hamper the operator's ability to do fine, coordinated movement. It is apparent that the tradeoff between image quality and the speed of image updating grows more severe as the behavior of the controlled robot becomes more dynamic (i.e., its bandwidth increases).

9.10.3 Proprioceptive Feedback

While visual feedback is absolutely critical to remote manipulation tasks, there are many circumstances in which proprioceptive or tactile feedback is also of great importance [587, 592]. This is true because the remote manipulators are often designed so that they can produce extremely great forces, necessary, for example, to move heavy objects or rotate rusted parts. As a consequence, they are capable of doing great damage unless they are very carefully aligned when they come in contact with or apply force to the object of manipulation. Consider, for example, the severe consequences that might result if a remote manipulator accidentally punctured a container of radioactive material by squeezing too hard, or stripped the threads while trying to unscrew a bolt. To prevent such accidents, designers would like to present the same tactile and proprioceptive sensations of touch, feel, pressure, and resistance that we experience as our hands grasp and manipulate objects directly (see Chapter 5). Yet it is extremely challenging to present such feedback effectively and intuitively, particularly when there are substantial delays. Proprioceptive feedback is even more sensitive to delays than is visual feedback. In some cases, visual feedback of the forces applied must be used to replace or augment the more natural tactile feedback.

9.10.4 Design Solutions for Teleoperation

Perhaps the most severe problem in many teleoperator systems is the time delay. As we have seen, the most effective solution is to *reduce the delay*. When the delay is imposed by graphics complexity, it may be feasible to sacrifice some complexity. While this may lower the reality and sense of presence, it is a move that can improve usability [581].

A second effective solution is to *develop predictive displays* that are able to anticipate the future motion and position of the manipulator on the basis of present state and the operator's current control actions and future intentions (see Chapter 8). While such prediction tools have proven to be quite useful, they are only as effective as the quality of the control laws of system dynamics that they embody. Specifically, the prediction horizon should be half the stopping time [593]. Furthermore, the system cannot achieve effective prediction (i.e., preview) of a randomly moving target, and without reliable preview, many of the advantages of prediction are gone.

A third solution is to avoid the delayed feedback problem altogether by implementing a computer model of the system dynamics (without the delay). Creating a *fast-time simulation* allows the operator to implement the required manipulation in "fast time" off line, relying on the now instant feedback from the computer model [594, 468, 578, 588] see Chapter 8). When the operator is satisfied that he or she has created the maneuver effectively, this stored trajectory can be passed on to the real system for execution. This solution has the problem that it places fairly intensive demands on computer power and of course will not be effective if the target environment changes before the planned manipulation was implemented.

Clearly, as we consider designs in which the human plans an action but the computer is assigned responsibility for carrying out those actions, we are crossing the boundary from manual control to automated control, an issue we discuss in depth in Chapter 11. We also note other important aspects of control that are covered in other chapters: process control because of its high levels of automation and its many facets that have little to do with actual control (e.g., monitoring and diagnosis) are also covered in Chapter 11. Finally, many characteristics of telerobotics are similar to those being addressed in the implementation of virtual reality systems, which is discussed again in Chapter 10.

The Dutch Reach (Reaching to open a car door with your right hand can save cyclists. (Photograph of Anna Stewart: by J.D. Lee.))

Photo of Scott Ehardt taken by VinnyR and released by him into the public domain.

The "Dutch Reach" requires the right arm to cross and reach to the left to open the car door. Doing so rotates the body and puts the driver in position to see a cyclist approaching from behind.

Designing controls that enable people to act effectively, depends on considering the role the controls have on perception as much as their effect on action. Just as displays guide action, as in the "Dutch Target" for urinals, controls guide perception, as in the "Dutch Reach" to guide drivers' attention to approaching cyclists.

9.11 Summary

We presented 15 principles to guide design of more effective controls. These principles apply primarily to discrete control devices, such as buttons, knobs, keyboard, and voice control. Good controls minimize the gulf of execution—the gap between the state of the system and the person's goals and intended system state. There is no single "best" way to do this, but consideration of the 15 principles can help to rule out bad control designs. The second half of the chapter considered tracking and continuous control. Instability associated with high gain, long lags, and high bandwidth input shows that system behavior does not follow simply from single responses to displayed information. Instead behavior depends on interactions over time. This chapter also shows how displays and controls begin to combine, particularly with touchscreens and voice interaction. This combination of displays and controls is called the human-computer interface. The next chapter builds on this idea of the human-computer interface and the importance of designing for interactions over time rather than displays and controls for discrete tasks. These interactions include gestures, conversations, and direct brain-computer interface.

Additional Resources

Several useful resources that expand on the content touched on in this chapter include books that address graph design and display design in more detail:

1. **Control design:** This chapter briefly summarized some basic elements of control design, which are covered in much more detail in the following books:

   HFAC. (2012). *Human Engineering Department of Defense Design Criteria Standard*: MIL-STD-1472G.

   Salvendy, G. (2012). *Handbook of Human Factors and Ergonomics*. John Wiley and Songs, Inc.

2. **Control theory:** The concepts of control theory touched on in this chapter have b.

   Jagacinski, R. J., & Flach, J. M. (2003). *Control Theory for Humans: Quantitative approaches to modeling performance*. Mahwah, New Jersey: Lawrence Erlbaum Associates Publishers.

Questions

Questions for 9.1 Types of Controls and Tasks

P9.1 How is the speed-accuracy tradeoff related to the selection and design of control devices?

P9.2 Which of the two attention-related principles—proximity compatibility or avoid resource competition—might guide you to use voice control and why?

P9.3 Describe what it means to support blind operation.

P9.4 Why is it important to support blind operation for even those that have no visual impairment?

P9.5 Table 9.2 describes six features that can make controls more identifiable and discriminable, identify two others not included in that table.

P9.6 What is a benefit of labels that is not shared by any other feature of control devices described in Table9.2?

P9.7 What features would you choose to support blind operation for someone wearing gloves?

P9.8 How does redundancy gain apply to control device design?

P9.9 Describe the implications of absolute judgment for designing the controls to be identifiable and easily manipulated.

P9.10 What principle does a button or lever that returns to its set position violate and what consequence might that have?

Questions for 9.2 Information Theory: Response Selection and Execution

P9.11 Describe how information theory relates to response selection and execution.

P9.12 In terms of information theory, why is a selector switch preferred to a round knob when the task is to select one of three states?

P9.13 When might you use a lever rather than a push button to switch between two states?

Questions for 9.3 Fifteen Principles for Discrete Controls

P9.14 Where would you put a knob relative to the vertical display to avoid a conflict between the population stereotype of clockwise movement to increase a level and the proximity of movement?

P9.15 Many methods to avoid accidental activation increase the effort needed to manipulate a control, forcing the designer to consider a speed-accuracy tradeoff. How would you avoid this speed-accuracy tradeoff?

P9.16 Describe how skill- and rule-based behaviors identify different ways people might accidentally activate a control.

P9.17   What is one purpose of a dead man switch?

P9.18   What does the Hick-Hyman law predict when you provide people with more options in a menu?

P9.19   Given RT=a+b log $_2$(N), calculate the decision complexity advantage for 10 decisions with two alternatives compared to one decision with 20 alternatives. Assume a=1s and b=2 s/bit.

P9.20   How does the decision complexity advantage relate to Morse code, the standard keyboard, and the chording keyboard?

Questions for 9.4 Discrete Controls: Buttons and Switches

P9.21   Given the two selection switches in Figure 9.6, which would you use for a task such as flying that requires your eyes remain on the road.

P9.22   Describe the features of buttons and levers that provide transient and persistent feedback regarding system state.

P9.23   Considering the operation of a stove, describe the role of transient and persistent feedback in the design of controls for a burner.

Questions for 9.5 Discrete Controls: Keyboards

P9.24   Explain why chording keyboards can be useful and give an example of an environment where they had seen success. Do not use courtroom stenographer.

P9.25   Describe the design of transient and persistent feedback to help people avoid confusion regarding the turn signal actuation in a car.

P9.26   How small must the delay be when providing tactile feedback from a virtual button press?

P9.27   Explain why chording keyboards are rarely used.

P9.28   Describe the typical features of a touchscreen or membrane keyboard and why they fail to support blind operation.

Questions for 9.6 Discrete Controls: Voice Input

P9.29   Describe the advantages and disadvantages of using voice control for assembling a shopping list in a kitchen populated by noisy children.

P9.30   What types of tasks does control theory help explain? Give a specific example other than driving.

Questions for 9.7 Positioning, Tracking, and Continuous Control

P9.31 Using Fitts's law, calculate movement time difference for a button 2 ft from a steering wheel with a one inch width when compared to a button that has a 3 inch width. Assume that for this task a=0.75 s and b=0.5 s/bit.

P9.32 What are the consequences in terms of the speed-accuracy tradeoff of increasing gain?

P9.33 Draw a control loop that includes the display, the person, the control device, and system. Adjust the diagram to show open- and closed-loop control.

P9.34 What does "error" mean in the context of closed-loop control?

P9.35 What are three sources of error in closed-loop control and how do they relate to the more general concept of human error?

P9.36 Draw the response of a zero-, first-, and second-order control system to a constant input.

P9.37 Give an example of zero-, first-, and second-order control systems.

P9.38 If people are fundamentally challenged in controlling second-order systems why do we design them?

P9.39 Identify a task on your computer where a high gain for the mouse would be helpful and one where low gain would be helpful.

P9.40 What is the bandwidth of a person?

P9.41 Describe how gain, time delay/lag, and system order affect stability.

P9.42 Is the selection of gain subject to a speed-accuracy tradeoff?

P9.43 Is the selection of time delay and lag subject to a speed-accuracy tradeoff?

P9.44 One strategy to enhance stability is to go open-loop. What are the requirements for open-loop control?

P9.45 What are three strategies an operator might be instructed to adopt to increase the stability of the system (e.g., teaching a novice driver)?

P9.46 What two strategies for enhancing system stability might a design consider?

P9.47 Give an example from the kitchen of open-loop control.

Questions for 9.8 Pointing Devices

P9.48 Considering the features of pointing devices, describe why touchscreens on tablets and smartphones have become so prevalent.

P9.49 What is an advantage of a pursuit display, such as the view out the window as you drive, relative to a compensatory display that shows only the state of the system relative to the goal state?

Questions for 9.10 Remote Manipulation, Teleoperation, and Telerobotics

P9.50 Describe three important approaches for improving performance of teleoperation.

Chapter 10

Human-Computer Interaction

At the end of this chapter you will be able to...

1. understand types of users and tasks and the control interaction styles that support them best

2. link theories of goal-directed behavior and emotions to design

3. apply 15 principles to the design of human-computer interactions

4. understand the five usability criteria for the types of users and tasks

323

Ray Cox, a 33-year-old man, was visiting the East Texas Cancer Center for radiation treatment of a tumor in his shoulder. He had been in several times before and found that the sessions were pretty short and painless. He laid chest-side down on the metal table. The technician rotated the table to the proper position and went down the hall to the control room. She entered commands into a computer keyboard that controlled the radiotherapy accelerator. There was a video camera in the treatment room with a television screen in the control room, but the monitor was not plugged in. The intercom was inoperative. But for Mary Beth this was normal; she had used the controls for the radiation therapy dozens of times, and it was pretty simple.

The Therac-25 radiation therapy machine had two different modes of operation, a high-power x-ray mode using 25-million electron volt capacity and a relatively low-power "electron beam" mode that could deliver about 200 rads to a small spot for cancer treatment. Ray Cox was to have treatment using the electron beam mode. Mary Beth pressed the x key (for the high-power x-ray mode) and then realized that she had meant to enter e for the electron beam mode. She quickly pressed the up arrow key to select the edit function. She then pressed the e key. The screen indicated that she was in the electron beam mode. She pressed the return key to move the cursor to the bottom of the screen. All actions occurred within 8 seconds. When she pressed the b to fire the electron beam, Ray Cox felt an incredible pain as he received 25 million volts in his shoulder. In the control room, the computer screen displayed the message "Malfunction 54." Mary Beth reset the machine and pressed b. Screaming in pain, Ray Cox received a second high-powered proton beam. He died 4 months later of massive radiation poisoning. Similar accidents had happened at other treatment centers because of a flaw in the software. When the edit function was used very quickly to change the x-ray mode to electron beam mode, the machine displayed the correct mode but incorrectly delivered a proton beam of 25,000 rads with 25-million electron volts. (A true story adapted from S. Casey, Set Phasers on Stun and Other True Tales of Design, Technology, and Human Error, 1993).

Computers profoundly affect all aspects of life, whether at work or in the home [595]. They have revolutionized the way people perform office tasks such as writing, communicating with coworkers, analyzing data, maintaining databases, and searching for documents. Computers in the form of smartphones, tablets, wearable devices have also transformed how people keep in touch with friends and family, how to select and find restaurants, and how to keep up with the latest news. Computers are increasingly being used to control manufacturing processes, medical devices, and a variety of other industrial equipment. Computers are becoming so small they can even be implanted in the human body to sense and transmit vital body statistics for medical monitoring.

Because the application of computers is spreading so rapidly, we must assume that most human factors jobs in the future will deal with the design of complex computer software and hardware.

Such work applies to hardware design, functionality of the software, and design of the software interface. Functionality refers to what the user can do with the software and how it supports or replaces human activities. Chapter 11 addresses functionality in more detail by describing how software should be designed when it is used to automate tasks once performed by people.

The *software interface* refers to the combination of displayed information provided by the computer that we see, hear, or touch, and the control mechanisms for entering information into the computer. Historically, this has meant the screen, keyboard, and mouse, but this is rapidly changing as smartphones and tablet computers, internet connected objects, and voice recognition synthesis become common.

Good software interface design must consider the cognitive and perceptual abilities of people, as outlined in Chapters 4, 5, and 6. Interface design also requires the application of display principles, described in Chapter 8, and control principles, described in Chapter 9. Finally, the *human-computer interaction (HCI)* process will affect and/or be affected by other factors such as fatigue, mental workload, stress, and anxiety. Clearly, most of the material in this book is relevant to the design of the software interface to some extent. While we can successfully apply general human factors principles and guidelines to interface design, there is also research and methods that are unique to HCI.

On the hardware side, computer workstations should be designed to maximize task performance and minimize ergonomic problems or hazards, such as cumulative trauma disorders. Chapter 12 discusses design methods for computer workstations and specific hardware components such as keyboards and video display terminals. Chapter 9 discussed various methods for system control with common input devices for computers. Despite the popularity of touchscreens, the feel (e.g., the type of resistance discussed in Chapter 9) of hardware controls remains a critical design element, such as the digital crown on the Apple Watch or the buttons on a smartphone.

HCI differs from simply specifying displays and controls. In this chapter we broaden the discussion from the previous chapters to go beyond displays, controls, and tasks to address interface. *Interface design* considers not just the displays and controls, but the how they combine to support and goal-directed behavior with the system. More broadly, design also considered users' interactions and services that extend over time–sometimes called *interaction design*. We also touch design of the overall experience people have with the hardware, software, and even other people that interact with the user—sometimes called *user experience design* [596]. Such design goes beyond the computer technology to consider elements of customer service and interaction with larger, non-computerized systems, such as all the events during a day at an amusement park.

HCI must also consider how to serve a wide variety of people, performing a wide variety of tasks, with a wide variety of interface technology. *Interaction styles* are ways in which a person and computer system can communicate with each other. In this chapter, we describe how user and task characteristics can be matched to interaction styles (e.g., a command line interface, menu selection), and then we describe underlining theories of interaction, associated design principles, and several emerging application areas.

10.1 Matching Interaction Style to Tasks and Users

HCI affects design for a variety of systems from nuclear power plant control rooms to word processing applications and video games. People using these systems range from highly trained professionals to babies playing with their parents' smartphone. HCI is concerned with design for diverse situations and diverse groups of people. It is therefore important to understand and match interaction styles with the characteristics of the users and their tasks. There are three dimensions that can be used to identify appropriate interaction styles: mandatory versus discretionary use, frequency of use, and task structure.

10.1.1 Understanding User Characteristics and Their Tasks

Mandatory use is where people use a system that is required as part of job, task, or activity. Some software is mandated by the workplace, such as the electronic patient records that hospitals require physicians to use. Other software, such as the Facebook app on a phone is the choice of the person. Such *discretionary* use is where people use a system because they want to, not because they are required to [597]. Discretionary tends to lead users to become experts on a small number of routine tasks, but they may know little regarding anything beyond those tasks With discretionary use, satisfaction, delight, and the overall emotional response, demand greater emphasis than is the case for industrial or military application domains where the focus is on safety and performance.

Some people might do some tasks, such as word processing, eight hours a day, every day. Other tasks, such as making a will, might be done only once or twice in a lifetime. *Frequency of use* describes how often a system is used and it has important implications for the so design for several reasons. For example, people who will be using a software system frequently are more willing to invest initial time in learning; therefore, performance and functionality can take precedence (to some degree) over initial ease of learning [69]. In addition, those who perform tasks frequently will have less trouble remembering keyboard shortcuts from one use to the next. This means that designers can place efficiency of operation over memorability [69]. Frequency of use relates to the degree of expertise, which may range from *novice* to *expert*. Shneiderman [69] describes three common classes of users along this experience scale:

- **Novice users:** People who know the task but have little or no knowledge of the system.

- **Knowledgeable intermittent users:** People who know the task but because of infrequent use may have difficulty remembering the syntactic knowledge of how to carry out their goals.

- **Expert frequent users:** People who have deep knowledge of tasks and related goals, and the actions required to accomplish the goals.

Software for *novice users* tends to focus on use highly restricted vocabulary and simple functionality. Systems built for first-time users are called "walk up and use" systems typical of an electronic check-in system at an airport. Such systems rely heavily on icons, menus, short written instructions, and a *graphical user interface (GUI)*. A GUI consists of buttons, menus, windows, and graphics that enable people to recognize what needs to be done and then do it through intuitive actions. Users select items from menus or groups of icons (recognition memory) rather than recalling text commands, thus reducing the load on long-term memory ("knowledge in the head") or the need to look things up. Rather than typing commands, users directly manipulate objects on the screen with a mouse, touch screen, or thumb pad. In contrast, a *command-line interface* requires users to recall commands and then type them on a keyboard. Because memory for recognition is more reliable than recall, a graphical user interface (GUI) is often more effective than command-line interaction, particularly for novice users (Chapter 6).

Reducing the load on memory is especially critical for *intermittent users*. Such users may have a good idea of how the software works but be unable to recall the specific actions necessary to complete a task. To deal with this, a software interface might have features that accommodate several types of user, as in the case of software that has input either from clicking on buttons or from typed-command entry. However, once people use a GUI, even when they become experienced, they will tend not to switch to the more efficient command line format. For this reason, adaptive interfaces are often desirable, automatically monitoring performance and prompting the user to switch styles as particular tasks become familiar[598]. In Chapter 11 we discuss adaptive automation, which takes this idea a step further by intervening with automatic control when human control performance declines.

Although the classes of novice, intermittent, and expert users provide clear distinctions that can help guide designs, reality is often more complex. Frequently, people may use certain parts of a program frequently and other parts infrequently. This might mean a person is an expert user of the drawing tools of a word processor, an intermittent user of the automatic table of contents function, and a novice user of the mail merge function. In addition, expertise may refer to experience with the software or with a particular domain. A secretary with 20 years of experience may be an expert in document production, but a novice with a particular word processor. These distinctions demonstrate the potential danger in using the simple categories of expert, intermittent, and novice users to guide software design. A more sophisticated approach requires a deep understanding of the specific types of expertise of the likely users. This understanding can be summarized with the concept of personas described in Chapter 3 [21].

Task structure is the third dimension that influences the choice of interaction style, and it describes the degree to which a task has clearly defined steps that must be completed in a particular sequence. For example, an electronic check-in system at an airport. In contrast, other tasks, such as working with a drawing program to create a figure, or a visualization tool to explore complex data, has little structure regarding the order a person might use specific drawing tools.

Whether a system caters to mandatory or discretionary use, whether it serves infrequent use by novices or frequent use by experts, or whether it supports structured or unstructured tasks should guide the choice of an interaction style.

10.2   Interaction Styles

Interaction styles define the ways in which a person and computer system can communicate with each other. Most interactions with computers depend on various manual input methods (e.g., keyboards, track boards, or finger swipes) and viewing text or graphic displays on a monitor, but voice conversational interfaces are becoming more common and even direct brain-computer interfaces are becoming possible [599]. Although there is a great deal of dynamic interaction, designers still must focus heavily on the components and arrangement of *static* screen design, that is, what each screen looks like as a display panel. Most current screen layout and design focuses on two types of elements, output interface elements: (information given by computer) and input interface elements (dialog boxes, buttons, slider switches, or other input elements that may be displayed directly on the screen). For more information on display elements see Chapter 8, and for input control elements see Chapter 9.

Given that computers are information-processing systems, people engage in a dialog with computers, which consists of iteratively giving and receiving information. Computers are not yet technologically sophisticated enough to use unrestricted human natural language [600], so the interface must be restricted to a dialog that both computer and user can understand.

There are currently several basic dialog styles that are used for most software interfaces:

- **Command lines:** At prompt, user types in commands with limited, specific syntax.

- **Function keys:** Commands are given by pressing special keys or combinations of keys.

- **Menu selection:** Provides users with a list of items from which to choose one of many.

- **Fill-in forms:** Provides blank spaces for users to enter alpha or numeric information.

- **Question and answer:** Provides one question at a time, and user types answer in field.

- **Direct manipulation:** Users perform actions directly on visible objects.

- **Multi-touch and gesture:** Users interact through finger, hand, and even body movement.

- **3-D navigation:** Users move through virtual space (Jacco chapter)

- **Conversational:** A computer understands a restricted set of spoken messages, which includes interactions with voice-based telephone systems, chat bots and virtual agents (Jacco chapter), such as the Microsoft's Cortana, Amazon's Alexa, and Apple's Siri.

While it is sometimes difficult to distinguish perfectly between these dialog styles, it is still convenient to categorize them as such for design purposes. Some interaction styles are suited to specific applications or types of task, and several dialog styles are frequently combined in one application. Each style has somewhat different human factors design considerations and associated guidelines. Shneiderman [69] describes such guidelines in great depth, a few of which are included in the following discussion.

Command line. At a line prompt, such as">", the user types in commands that require use of a very specific and limited syntax (such as UNIX, C++, or Java), and unlike menus, a command line does does not require much screen space. Command languages are appropriate for users who have a positive attitude toward computer use, high motivation, medium- to high-level typing skills, high computer literacy, and high task-application experience. Designers who are creating a command language should strive to:

- Make the syntax as natural and easy as possible.

- Make the syntax consistent.

- Avoid arbitrary use of punctuation.

- Use simple, consistent abbreviations.

```
This exam has a total of \numquestions~questions.

\textbf{Select only the one best answer for each question}.
Good luck!
\vspace{5mm}
\begin{questions}


\question
The system design cycle can be simplified into three phases of:
\begin{choices}
\choice Understand the user, design, implement
\choice Use common sense to design, evaluate with the user, implement
\CorrectChoice Understand the user, design, evaluate
\end{choices}


\question
Which of the following correctly describes the content of a typical accident investigation report, which are similar to the case studies in this course?
\begin{choices}
\choice    Overview of the facts, identify those people at fault, recommendations to avoid future accidents.
\CorrectChoice Overview of the facts, interpretation to identify contributing causes, recommendations to avoid future accidents.
\choice   Overview of the facts, identify those people at fault, identify those legally responsible.
\choice Overview of the facts, identify those people at fault, human error tendencies underlying the mishap.
\end{choices}
```

Figure 10.1 Latex is a command language for creating documents that requires people to remember commands that format text, which Microsoft Word shows as menu or icons.

This interface requires extensive training and practice. The limitations of command line languages are that the commands and sequence of commands need to be remembered. Hence, it requires the user to retrieve information from long-term memory and does demand manual typing input. The interface also has low tolerance for errors (lower case a is different from upper case A), and thus, the error rates will be high. Further, error messages and online assistance is limited given the many diverse possibilities.

Function keys. In this dialog style, users press special keys or combinations of keys to provide a particular command. An example is pressing and holding the control button and then pressing the "B" key to change a highlighted section of text to boldface type. For users who perform a task frequently, want application speed, and have low-level typing skills, function keys are extremely useful. Like command languages, the arbitrary nature of function keys place substantial demands on long-term memory. Designers should consider the following guidelines:

- Reserve the use of function keys for generic, high-frequency, important functions.
- Arrange in groups of three to four and base arrangement on semantic relationships or task flow.
- Place frequently used keys within easy reach of home row keys.
- Place keys with serious consequences in hard to reach positions and not next to other function keys.
- Minimize the use of "qualifier" keys (alt, ctrl, command, etc.) that must be pressed on the keyboard in conjunction with another key.

Menus. Menus have become very familiar to anyone who uses the Apple or Windows operating systems. Menus provide a list of actions to choose from, and they vary from menus that are permanently displayed to pull-down or multiple hierarchical menus. Menus should be used as a dialog style for users with little computer experience, poor typing skills, and perhaps even low motivation.

One approach to menu design is to rely on simple guidelines. For example, a series of studies have found that each menu should be limited to between four and six items to reduce search time (see Chapter 4) [120]. The number can be increased by grouping menu items into categories and separating them with a dividing line. Menus that have a large number of options can be designed to have few levels with many items per level ("broad and shallow") or to have many levels with few items per level ("narrow and deep"). In general, usability is higher with broad and shallow menus. Shneiderman [69] provides the following guidelines (among others):

- Gray out of inactive menu items.

- Create logical, distinctive, and mutually exclusive semantic categories.

- Menu choice labels should be brief and consistent in grammatical style.

- Use existing standards for desktop applications, such as: File, Edit, View.

Menu Selection interfaces are great for novice and intermittent users. They afford exploration and structure decision making. Unfortunately, menu designs can be more complex than these simple guidelines suggest. Even with a relatively simple set of menu items, the number of possible ways to organize the menu design options can explode substantially as the number of menu items increase [601]. Hence, a disadvantage of menus is that if there are too many menu options, it can lead to information overload. A comprehensible menu structure enables users to select correct menu options more quickly. One way to make menus comprehensible is the *card-sorting* technique. With this approach, representative users are given cards with menu terms written on them and are then asked to sort cards into groups. This simple process can reveal how people think about terms that might not be otherwise obvious.

Fill-in forms. Fill-in forms are like paper forms: They have labeled spaces, termed fields, for users to fill in alphabetical or numeric information. Like menus, they are good for users who have little system experience. Fill-in forms are useful because they are easy to use, and a "form" is a familiar concept to most people. Wizards that lead people through a series of steps are a multi-screen version of fill-in forms. They require minimal training, and provide convenient, guided assistance. The disadvantage of fill-in forms is that they do tend to consume screen space, requires manual keystrokes, and some handling of typing errors.

Like menus, fill-in forms should be designed to reflect the content and structure of the task itself. An example is a form filled out by patients visiting a doctor's office. The form could look very similar to the traditional paper forms, asking for information about the patient's name, address, medical history, insurance, and reason for the visit. Fill-in forms should be designed according to the following basic guidelines:

- Organize groups of items according to the task structure.

- Use white space and separate logical groups.

- Support forward and backward movement.

- Keep related and interdependent items on the same screen.

- Indicate whether fields are optional.

- Prompts or reminders of incomplete entries should be brief and unambiguous.

Figure 10.2 shows the fill-in form interface for the TSA Pre application. Related elements, such as the person's name and contact information are grouped, the "back" and "next" buttons clearly indicate how to navigate, and the red asterisk highlights required fields.

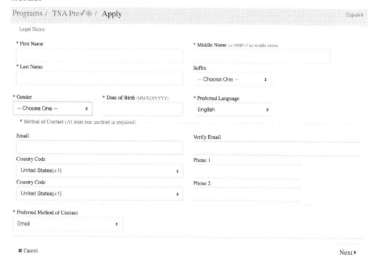

Figure 10.2 Form interface for the Transportation Security Administration (TSA) Pre application. From: universalenroll.dhs.gov

Question and Answer. In this interaction style, the computer displays one question at a time, and the user types an answer in the field provided. This method is suitable when short data entry is required such as observed in wizards and installation software. It is appropriate for discretionary use with tasks that occur infrequently, and have a clear structure. This interaction style does not expect the user to remember many things, is self-explanatory, and requires little to no training. The disadvantage to the user is that there is no opportunity to deviate and hence, the user has little system control. The answer must be valid and there is often limited support for correcting errors. Question and answer methods must be designed so that the intent of the question and the required response is clear:

- Use visual cues and white space to clearly distinguish prompts, questions, input area, and instructions.

- State questions in clear and simple language.

- Provide flexible navigation.

- Minimize typing requirements.

Direct manipulation. Direct manipulation means performing actions directly "on visible objects" on the screen [300, 602]. It usually involves a pointing device that indicates the object to be manipulated and acted upon. An example is using a mouse to position the cursor to a file title or icon, clicking and holding the mouse button down, dragging the file to a trash can icon by moving the mouse, and dropping the file in the trash can by letting up on the mouse key. Direct manipulation dialog styles are popular because they can map well onto a user's mental model of the task, are easy to remember and learn, and do not require typing skills. Direct manipulation is a good choice for discretionary users who have a negative to moderate attitude toward computers, low-level typing skills, and moderate to high task experience. Direct manipulation interface design requires a strong understanding of the task being performed and creativity to generate ideas for metaphors or other means of making the direct manipulation interface comprehensible to the user. Shneiderman [69] provides the following design guidelines, among others:

- Minimize semantic distance between user goals and required input actions.

- Choose a consistent icon design scheme.

- Design icons to be concrete, familiar, and conceptually distinct.

- Accompany the icons with names if possible.

The disadvantage of direct manipulation is that they may be more difficult to program. They may not be suitable for small graphic displays, and depending on the resolution and the screen space, the manipulation of screen objects may not be directly identical to real-world objects.

Multi-touch and gesture. Multi-touch interactions use input from multiple fingers on a touchscreen, which contrasts with simple pointing and selecting with touchscreens discussed in Chapter 9 [603]. Multi-touch enables multi-finger gestures such as pinching and spreading to zoom in and out of images. Motion tracking systems make it possible to use gestures produced by hands, arms, and even the whole body as input [604, 605]. Gestures promise a natural and intuitive interaction style, but their success depends on several considerations [606]:

- Provide context to signal what gestures are can be used.

- Make gestures easy to perform.

- Use appropriate metaphors, such as spread to zoom.

- Minimize fatigue and stress with repetitive use.

- Ensure gestures are easily differentiated by the computer.

3-D Navigation. Computers now make it possible to create and interact with virtual worlds, particularly in the context of games, but also for exploring actual physical spaces without the need to physically travel to the location. These spaces require unique interaction styles to navigate. Some navigation methods include physical movement where moving through the physical world produces movement through the virtual world. People can also navigate by manipulating the viewpoint and by steering. Rather than exerting direct control movement can be enacted automatically with the person specifying a target endpoint and the system automatically moves to the endpoint. Similarly, route planning achieves a similar result, but the person exerts more control by specifying waypoints. Specific design considerations from Bowman et. al, [607] include:

- Use "Automatic" interactions, such as specifying the target endpoint, when the focus is on efficiency.

- Use "Natural" interactions, such as physical movement, when a replication of the physical world is important.

- Constrain input with physical or virtual constraints to help guide user input.

Conversational. Natural language or conversation interaction style an increasing number of application. In this style, users speak or write using natural language. Because it is a natural rather than artificial style for people, natural language can be thought of as the"interface of choice" for the types of interactions that occur between people. Many early, but promising applications have emerged in the form of virtual assistants such as Google Now, Microsoft Cortana, Apple Siri, Amazon Alexa, Facebook M.

As with the discussion in Chapter 9 concerning voice as an input method, natural language processing is improving, but formidable challenges remain [600]. Some of these are technical, such as processing sound more efficiently to extract words and meaning. Recognizing spoken commands, as discussed in Chapter 9 on control design, is not the same as understanding the meaning of a spoken sentence. Even if they understand meaning, natural language interfaces may never be a perfect communication channel because this interaction style involves designing a personality as much as it does effective natural language processing. Some specific considerations for design include[608, 609, 610]:

- Use the context of interaction to guide interpretation of meaning.

- Pay particular attention to error recover and conversation repair.

- Use the context of interaction to minimize interruption of ongoing activity.

- Build the dialog around a plan. Plan-based theories of communication suggest that the speaker's speech act is part of a plan and that it is the listener's job to identify and respond appropriately to this plan [16].

- Consider communication as a collaborative act that requires both the person and the technology to work together to obtain a mutual understanding. This encourages clarifications and confirmations [48].

Summary of Interaction Techniques No dialog style is best for all applications. Table 10.1 provides a rough guide to matching the characteristics of the dialog style to those of the user and the tasks. For example, certain tasks are better performed through direct manipulation than natural language. Consider the frustration of guiding a computer using a command language to tie your shoe compared to simply tying the shoe manually. More realistically, it is feasible to control the volume of a car stereo with a conversational interface, but the continuous control involved in volume adjustment makes the standard knob more appropriate. Also the interactions styles are not mutually exclusive. Most modern applications support several interaction styles—you can copy and paste text using the menu, function keys, and direct manipulation drag and drop. Figure 10.3 shows how the Apple operating system integrates command line to launch an application by typing its name, complementing the graphical interface method of clicking on an icon to launch an application. Selecting and combining interaction styles is more complicated than indicated by Table 10.1. To address this complexity we describe two theoretical perspectives on interaction design and 15 associated principles.

Figure 10.3 A command language embedded in a graphical user interface.

10.3   Theories Underlying Interface and Interaction Design

Guidance for interface and interaction design falls into several categories: high-level theories, general principles, specific guidelines, and methods for evaluation and testing. In this section, we review a few of the more commonly used theories and models. Such theories provide a general framework for designers to conceptualize their problem and discuss issues, using a language that is application independent. The theories described below can help designers develop an overall idea of user capabilities, including a

Interaction Style	Features	Use Discretionary	Task Frequent	Structured
Command line	Fast and powerful, but vulnerable to error and requires learning	○	●	◑
Function keys	Fast and simple, but requires learning	○	●	◑
Menus	Avoids typing and typos, but slows experts and consumes screen space	●	◑	◑
Fill-in forms	Simple, but inflexible and consumes screen space	●	◑	●
Question/answer	Simple, but inflexible	●	◑	●
Direct manipulation	Intuitive and easy to learn, but challenging to program and requires a good metaphor	●	◑	◑
Gesture	Sometimes intuitive, but doesn't support discovery	●	◑	◑
3-D Navigation	Intuitive, but can be challenging to control	●	◑	○
Conversational	Accessible, but unreliable and requires dialog for clarification and context	●	◑	●

Table 10.1 Interaction styles, their characteristics and associated use and task characteristics. ●: recommended applications, ◑: acceptable applications, ○: not recommended.

description of the kinds of cognitive activity taking place during during interactions and how to support these interactions.

10.3.1 Goal-directed behavior

One theory that has been useful in guiding user-oriented interface design is Norman's [611] seven stages of action. It consists of two "bridges" and seven steps (Figure 10.4). A user starts with goals, needs to understand what to do to accomplish those goals, how to do it. These steps bridge the *of execution,* which is the mismatch between the user's intentions and the actions supported by the software. This gulf can be narrowed by controls designed according to the control principles discussed in Chapter 9. The user then processes and evaluates feedback on whether and how well those goals are achieved. These steps bridge the *gulf of evaluation,* which is the mismatch between the user's expectations and the system state. This gulf can be narrowed by providing interpretable displays, following the principles of display design discussed in Chapter 8.

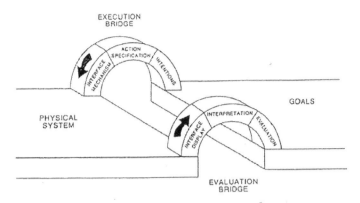

Figure 10.4 Bridging the gulf of execution and evaluation. (Source: Norman, D., 1986. Cognitive engineering. In D. A. Norman, S. W. Draper [eds.], *User-Centered System Design: new perspectives on human-computer interaction*. Copyright ©1986. Reprinted with permission from Taylor and Francis Group LLC Books provided by Copyright Clearance Center.)

The person first establishes a goal, such as sending an email to a friend. If the person feels that this goal is something that he or she might be able to accomplish using the system, the user forms an intention to carry out actions required to accomplish the goal. Next, the user identifies the action sequence necessary to carry out the goal; the execution bridge. It is at this point that a user may first encounter difficulties. Users must translate their goals and intentions into the desired system events and states and then determine what input actions or physical manipulations are required. The discrepancy between psychological variables and system variables and states may be difficult to bridge. Closing this gap is particularly important for novices who use a system infrequently. For situations where people "walk up and use" the system, it must be very clear how they should begin the interaction. Supporting the first step of the interaction is critical because these users are likely to walk away and use another system. This is particularly true for Web sites and Web-based applications.

Even if the user successfully identifies needed input actions, the input device may make them difficult to carry out physically. For example, the "hot" portion of a small square to be clicked using a mouse might be so small that it is difficult to be accurate. Norman notes that the entire sequence must move the user over the gulf of execution (see Figure 10.4). A well-designed interface makes that translation easy or apparent to the user, allowing him or her to bridge the gulf. A poorly designed interface results in the user not having adequate knowledge and/or the physical ability to make the translation and therefore be unsuccessful in task performance.

Once the actions have been executed, people must compare the system events and states with their original goals and intentions; evaluation bridge. This means perceiving system display components, interpreting their meaning with respect to system events and current state, and comparing this interpretation with the goals. The process moves the user over the gulf of evaluation. If the system displays have been designed well (Chapter 8), it will be relatively easy for the user to identify the system events and states and compare them with original goals. As a simple example, consider a user who is trying to write a friend via email. This user has composed a letter and is now ready to send it. The goal is to "send letter," and the user clicks on the button marked "send." This is a relatively straightforward mapping, allowing easy translation of goal into action. However, after the button is pressed, the button comes up and the screen looks like it did before the user clicked on it. This makes evaluation difficult because the user does not know what system events occurred (i.e., did the letter get sent?). System design will support the user by making two things clear—what actions are needed to carry out user goals and theories regarding what events and states resulted from user input. The seven steps needed to bridge the gulfs of execution and evaluation provide a useful way of organizing the large number of more specific design guidelines and principles.

Bridging the gulfs of execution and evaluation often depends on the mental model of the user, which can best be described as a set of expectancies regarding what human actions are necessary to accomplish certain steps and what computer actions will result. As described in Chapter 6, an effective mental model is one that is relatively complete and accurate, and supports the required tasks and subtasks. It allows the user to correctly predict the results of various actions or system inputs. As a consequence, a good mental model will help prevent errors and improve performance, particularly in situations that the user has not encountered before. The development of effective mental models can be facilitated by system designers.

One way to promote an accurate *mental model* is by developing a clearly defined *conceptual model*. A conceptual model is "the general conceptual framework through which the functionality is presented" [612]. Often the success of a system hinges on the quality of the original conceptual model that designers used to create the system. For example, the success of the cut-and-paste feature in many programs is due to the simple but functional conceptual model of this component (cut and paste). Several ways a conceptual model can be made clear to the person include [612]:

- **Make invisible parts and processes visible.** For example, clicking on an icon that depicts a file and dragging it to a trash can icon makes an invisible action (getting rid of a file) visible to the user.

- **Provide feedback.** When an input command is given, the system can report to the user what is happening (e.g., loading application, opening file, searching, etc.).

- **Build in consistency.** People are used to organizing their knowledge according to patterns and rules. If a small number of patterns or rules are built into the interface, it will convey a simple yet powerful conceptual model of the system.

(a) With skeumorphic features   (b) Without skeumorphic features

Figure 10.5 Skeuomorphic design (a) and revised version without skeuo-morphic features (b). (Screenshots using Apple iPhone.)

- **Present functionality through a familiar metaphor.** Designers can make the interface look and act similar to a system with which the user is familiar. This approach uses a metaphor from the world with which the user is supposedly familiar.

Metaphors are a particularly useful approach for helping users develop an effective mental model. A metaphor is the relationship between objects and events in a software system and those taken from a non-computer domain , which supports the transfer of knowledge [613]. Metaphors provide knowledge about possible actions, how to accomplish tasks, and so forth. *Skeuomorphs* are a specific way that metaphors can be conveyed, and are features of the physical implementation that are retained in the software implementation. Skeuomorphic features include the texture of leather and the stitching that might be found on a physical notepad. These features can provide clues of how the software works by relating it to their physical analogs and can be more engaging than minimalist designs. These features can also stand in the way of realizing the benefits of computer-based versions—Why burden people with turning pages of an electronic book? Skeuomorphic features also clutter the interface and waste space. Figure 10.5 shows the old, skeuomorphic version of the iOS and the new version. Notice the stitching and leather texture at the top of the screen.

In summary, users will invariably develop a mental model of the system. Designers must try to make this mental model as accurate as possible. This can be done by making the conceptual model of the system as explicit and can sometimes be aided by real-world metaphors, but can adhering too closely to surface features of the real-world can limit the potential of software systems.

10.3.2   Affect, Emotion, and Aesthetics

The emotional and affective elements of software design are becoming increasingly important as computers become more complex and ubiquitous. Affective computing is the study and development of systems and devices that can recognize, interpret, process, and simulate human emotion. Affective computing suggests that computers may be more readily accepted if the computer can sense and respond to user's emotional states [614]. One potential outcome of affective computing is that future computers will sense your emotional state and change the way they respond when they sense you are becoming frustrated. Humans respond socially to technology and react to computers similarly to how they might respond to human collaborators [615]. For example, the similarity attraction hypothesis in social psychology predicts that people with similar personality characteristics will be attracted to each other. This finding also predicts user acceptance of software [616]. Software that displays personality characteristics similar to that of the user are more readily accepted. Considering affect in system design is not just about designing for pleasure. Designers should consider how to create unpleasant emotional responses to signal dangerous situations [617]. Overall, affective computing provides a way to enhance acceptance and minimize frustration. What is beautiful tends to work better [618].

Emotion is important to making appropriate decisions and not just in reducing frustration and increasing the pleasure of computer users. Norman, Ortony, and Russell [619] argue that affect complements cognition in guiding effective decisions. A specific example is the role of trust in Internet-based interactions. People who do not trust an Internet service are unlikely to purchase items or provide personal information. In many cases, trust depends on surface features of the interface that have no obvious link to the true capabilities [620]. Credibility depends heavily on "real-world feel," which is defined by factors such as speed of response, listing a physical address, and including photos of the organization. Visual design factors of the interface, such as cool colors and a balanced layout, can also induce trust [621]. Similarly, trusted Websites tend to be text-based, use empty space as a structural element, have strictly structured grouping, and use real photographs [622]. These results show that trust tends to increase when information is displayed in a way that provides concrete details in a clear and consistent manner. Chapter 11 expands on this discussion to describe the role of trust in guiding reliance on automation. Understanding how emotions, such as trust, affect response to software can help people use technology more appropriately.

10.4   Fifteen Principles for HCI Design

As in previous chapters, the high-level theories can be related to design principles that can be more easily applied. Nielson [56] recognized that some usability guidelines might be more predictive of common user difficulties than others. To assess this possibility, he conducted a study evaluating how well each of 101 different usability guidelines explained usability problems in a sample of 11 projects. Besides generating the predictive ability of each individual heuristic, Nielson performed a factor analysis and successfully identified a small number of usability factors that "structured" or clustered the individual guidelines and that accounted for most of the usability problems. We extend and update ten principles. Some of these principles build on information processing stages that describe principles in the preceding chapters, others reflect considerations from macrocognition (Chapter 7), and several are specific to goal-directed behavior and the role of emotions described in this chapter.

10.4.1 Attention Principles

1. **Anticipate needs.** People should not be required to search for information or remember it from one screen to the next. Provide necessary information and tools for a particular task and context. Forcing people to look for information pulls their attention away from their goal and makes errors, such as those associated with interruptions, more likely. Task analysis techniques described in Chapter 2 help designers anticipate user needs, and direct attention appropriately.

2. **Highlight changes.** The system should clearly identify changes in the systems' status. A blinking cursor on a computer screen directs a user quickly to the location of where they were last typing. Other ways to highlight important changes include color coded lists, or highlighted the last set of commands.

3. **Limit interruptions and distractions.** Attention is limited and shifting one activity to another means performance on one or both will suffer. Notifications and alerts that catch users' attention will compromise responses to the primary task and, as discussed in Chapter 6, if the interruption draws attention away from the ongoing task for more than a few seconds it is likely to cause errors and even lead people forget to return to the primary task. Increasing the complexity of the primary task, the complexity of the interrupting task, the duration of the interruption, and the similarity of the primary tasks and the interruption all compromise performance of the primary task. Giving people control over when to respond the interruption and timing the interruption of arrive after the end of a sub-task can mitigate mitigate the negative effect [623, 624]. Because of these effects many people have found it helpful to turn off email and text message notification for period of the day.

4. **Mimimize information access cost.** Sometimes both the physical and cognitive effort required to retrieve information can be excessive; via several mouse clicks or key presses,

and because of the need to remember the appropriate sequence of activities [224, 625]. These excessive physical and attention demands can both inhibit the retrieval of such information, or interfere with other concurrent cognitive activities, such as remembering the information that needs to be entered in the location that is being retrieve

10.4.2 Perception Principles

Figure 10.6 Gesture control for a hand dryer lacks a signifier for the affordance and so must be indicated with a label.(Photograph by author: J. D. Lee.)

5. **Make system structure and affordances visible.** Affordances describe a relationship between a property of the environment and the person that enables certain actions [300, 103]. In natural environments, affordances are direclty perceived and help to make many of our daily interactions with the physical world so intuitive, such as walking down a flight of stairs or knowing what piece of furniture to sit on. Affordances are not a natural feature of computers and so must be designed to be visible and some technologies, such as gesture and multi-touch present challenges in creating affordances that indicate what actions are possible. Indicators of what actions are possible are termed signifiers, which indicate an affordance. A door affords passage, and the handles signify how to open it. The capabilities of a systems should be visible in the interface—discoverable. Graphical user interfaces signify possible actions with the icons, buttons, and other interface elements, to make possible actions easily discoverable. Gesture interfaces often lack these signifiers and so make possible actions less discoverable [626].

10.4.3 Memory Principles

6. **Support recognition rather than forcing recall.** Minimize memory load by presenting options as part of the interface rather than require that they be remembered. Graphical user interfaces comply with this principle, but command lines do not.

7. **Be consistent.** Consistency means that same type of information should be located in the same place on different screens, the same actions always accomplish the same task, and so forth [627]. Functions should be logically grouped and consistent from screen to screen. This internal consistency ensures that design elements are repeated in a consistent manner throughout the interface. External consistency refers to existing standards for platform on which it will run. For example, a Microsoft Windows application must be designed to be consistent with standardized Windows icons, groupings, colors, dialog methods, and so on. This consistency acts like a mental model—the user's mental model of "Windows" allows the user to interact with the new application in an easier and quicker fashion than if the application's interface components were entirely new.

10.4.4 Mental Model Principles

8. **Match system to real world.** This the idea that the software interface should use concepts, ideas, and metaphors that are well known to the user and map naturally onto the user's tasks and mental goals. Familiar objects, characteristics, and actions cannot be used unless the designer has a sound knowledge of what these things are in the user's existing world. This helps ensure that interface elements have clear meaning. People will avoid and even ignore things they cannot understand. Avoid designing interface elements that make people wonder what to do with it (see Figure 6). Such information is gained through performing a task analysis, as discussed in Chapter 3. This is not to say that the interface should only reflect the user's task as the user currently performs it. Computers can provide new and powerful tools for task performance that move beyond previous methods [56].

9. **Make credible and trustable.** Just trust and credibility influence how people come to rely on other people, trust and credibility affect how people engage with technology. If people find a website credible and trustworthy they are more likely to provide their credit card and buy something. Some features that enhance trust and credibility include the absence of typographical errors, broken links, and other errors. Connection to the real people and the ability to contact them. Also the alignment between the purpose and process used by the software and those of the user [628, 189]. Trust becomes a key element in human-automation interaction, discussed in the following chapter.

10. **Consider aesthetics and simplicity.** Eliminate irrelevant elements because they compete with the relevant. Similarly, design with a maximum of three typefaces and three sizes. The same holds for color. Because of the prevalence of color-blindness, one should always design the interface so that it can be understood in black and white [69]. Color should as redundant coding. A common suggestion is that websites include only four colors: primary, secondary, accent, and background or text color.

10.4.5 Response Selection Principles

11. **Choose appropriate defaults.** As noted later, many human-computer interfaces can be personalized for the individual user. Hence, there are many functions available to meet the demands of different environments. Facing a complex interface or complex choice of multiple options can be overwhelming and confusing to users. Designers must remember that users may not have the knowledge and understanding of the system in the way that designers do. An interface that presents lots of information and lots of options will actually appear more difficult. Hence, the values set for the system

defaults have become increasingly important. Many users do not notice defaults, but they have a tremendous impact on how we use systems. Many users do not change the default settings on their TV, refrigerator, car, or even our mobile device. Hence, it is important to make sure default values are as useful and practical as possible as the majority of people will never change them.

12. **Simplify and structure task sequences.** Reduce complex actions into smaller, simpler steps. Users are more inclined to perform a complex action if it's broken down into smaller steps. Filling out long, complicated forms is tiresome, overwhelming and difficult to double-check. However, if you split the information into multiple steps or screens, with a progress bar, it becomes more manageable. 10 seconds corresponds roughly to a single task and (Johnson, 2010).

10.4.6 Interaction Principles

13. **Make the system state visible.** Provide prompt feedback. Feedback should occur within 50 ms for a virtual button press. It should occur within 100 ms for people to feel a cause and effect relationship, such as clicking on a menu and the sub-menu appearing. Delays of 1000 ms (1 s) is the largest expected gap in a conversation, and delays longer than 1 second should be indicated by the computer. Ten seconds corresponds roughly to the span of attention to a single task and for delays longer than this should be indicated and completion announced with an auditory cue so that the user to start another task (Johnson, 2010).

14. **Support flexibility, efficiency, and personalization.** The goal is to have the software match the needs of the user. For example, software can provide shortcuts or accelerators for frequently performed tasks. These include facilities such as function or command keys that capture a command directly from screens where they are likely to be most needed, using system defaults (Greenberg, 1993; [55]. In other words, they are any technique that can be used to shorten or automate tasks that users perform frequently or repeatedly in the same fashion.

People often benefit from interfaces that can be tailored for individual preferences. For example, different workers on the same general information system may wish to configure their computer windows differently.. This personalized feature provides a functional aspect by accounting for individual direction finding skills. Other personalized options relate to the use of the interface while performing a complex task or for multi-tasking. Personalization options (as in social media sites) might include selection of background color and font sizes.

15. **Make robust to errors and exploration.** Avoid error-prone situations, make it more difficult to commit severe errors with confirmation buttons and other techniques described in Chapter 9 to avoid accidental activations. As operators continue to engage in a system, clear and consistent feedback is needed to ensure that the user can recognize the error and then respond appropriately. The system should provide information so that they can learn from these errors. This can be in the form of messages, help screens, etc.

Because errors cannot always be prevented and because they often accompany exploration, design should minimize the negative consequences of errors or to help users recover from their errors [55]. Such error-tolerant systems rely on a number of methods. First, systems can provide "undo" facilities as discussed previously. Second, the system can monitor inputs (such as "delete file") and verify that the user actually understands the consequence of the command. Third, a clear and precise error message can be provided, prompting the user to (1) recognize that he or she has made an error, (2) successfully diagnose the nature of the error, and (3) determine what must be done to correct the error. Error messages should be clearly worded and avoid obscure codes, be specific rather than vague or general, should constructively help the user solve the problem, and should be polite so as not to intimidate the user (e.g., "ILLEGAL USER ACTION") [69].

The accident described in the beginning of this chapter occurred because (1) the software system had a bug that went undetected, (2) there was not good error prevention, and (3) there was not good error recognition and recovery. As an example, when the operator saw the message "Malfunction 54," she assumed the system had failed to deliver the electron beam, so she reset the machine and tried again.

10.4.7 Summary of principles

Because interface and interaction design involves both displays and controls it is clear that the 15 principles for discussed here need to be considered with principles of display and control. Because human-computer interaction is such a broad domain we apply these principles to a diverse sample of HCI application: interactive visualization, website and application design, tangible and wearable technology, and computers in cars. As we encounter each principle in an application, we place a reminder of the principle number in parentheses, for example, (A1) refers to the first attention-related principle: "anticipate needs". The letter refers to the category: attention (A), perception (P), memory (M), mental model (MM), response (R), and interaction (I), which are summarized in Table 10.2.

Interaction Design Principles

Attention principles
A1 Anticipate needs
A2 Highlight changes
A3 Limit interruptions and distractions
A4 Minimize information access cost

Perception principles
P5 Make system structure and affordances visible

Memory principles
M6 Support recognition rather than force recall
M7 Be consistent

Mental model principles
MM8 Match system to real world
MM9 Make credible and trustable
MM10 Consider aesthetics and simplicity

Response selection principles
R11 Choose appropriate defaults
R12 Simplify and structure task sequences

Interaction principles
I13 Make the system state visible
I14 Support flexibility and personalization
I15 Make robust to errors and exploration

Table 10.2 Interaction design principles.

10.5   Interactive Visualization

In Chapter 8, we described principles for data visualization and graph design. At that time, we briefly pointed the substantial advantages of being able to interact with such displays to"explore" the data. Interactive data visualization becomes important when there are two many data points to render on the screen, two many dimensions to show at once, or multiple relationships that cannot be captured by a single image. Here we consider interactive data visualization in detail.

When data cannot be rendered in the space of a single screen, people often "visit" different parts of the data by "traveling", as when "zooming" or "panning". Such travel can be carried out in a variety of ways. Direct travel involves "flying" through the data base with a joystick or other control device, much like the tracking task we discussed in Chapter 9. There is a danger in providing too much interactivity."Flying" in 3D space is not a skill natural to human evolution, and if three axes of travel are added to three axes of viewing orientation, the 6-axis control problem can become complex (and possibly unstable). As we discussed in Chapter 9, stability of such flight depends upon avoiding a control gain that is too high (leading to overshoots in getting to one's goal) or too low (leading to long delays in traveling). There is little conclusive empirical evidence that one approach is superior to the other [546]; however "Flying" through a data best supports data where there is a clear 3-D analog for the data (MM8), such when the data are tied to the latitude and longitude of a map. For these reasons, there are advantages for discrete point and click systems when three (or higher) dimensional data are involved.

Indirect travel can be accomplished by a point-and-click system where targeted areas of the data are expanded to provide "details on demand"[69, 629]. Interactive visualizations succeed when these interactions minimize access costs (A4). Typical interactions associated with indirect travel through the data include:

- **Identifying:** Get numeric value or label for a data point.
- **Selecting:** Select a set of points in a graph.
- **Filtering:** Apply a Boolean operation to subset the data.
- **Mapping:** Change the axis of the graph, also including panning and zooming.
- **Linking:** Highlight points across graphs.

Identification is often enabled by showing a small pop-up window with the details of the selected data point. Linking highlights a selected point in one graph and a point that it corresponds to in another window. Mapping can be as simple as zooming in on a selected subset of points by having the selected subset of data fill the screen. Mapping can also involve placing the selected points in a different coordinate system. We use the interactive visualization in Figure 10.7 to make these concepts more concrete.

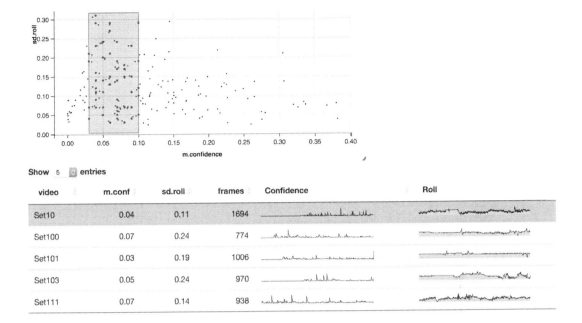

Figure 10.7 Brushed points to select a subset of data. (Created using RShiny (https://shiny.rstudio.com).)

One simple form of *selecting* data that has proven successful is "brushing", where a subset of data is selected by drawing a box or lasso around a plot of data [630]. For example, Figure 10.7 depicts data from a study of driver distraction. Each point in the upper scatter plot is an aggregate measure of the drivers' head positions over a drive. Data were sampled at 10 Hz, so each data point represents thousands of individual elements of raw data that have been aggregated over time. The shaded rectangle covering part of the scatterplot shows points that have been "brushed". Dragging the cursor across the screen changes this brushed area and selects different points, naturally highlighting changes in the selection (A2).

Large data sets generally require *filtering* to limit the amount rendered or to focus on a subset that meets certain criteria. Such filtering may involve Boolean queries of the data. In Figure 10.7 this might be accomplished by selecting a subset of the drives by setting a desired range of a summary variable in the table, such as filtering to include only those drives where the head tracking confidence is greater than .80.

Finally, *linking* makes it possible to identify a associate a datapoint along one dimension with the same datapoint plotted along a different dimension. Linking connects data mapped to one coordinate system to the same data in another coordinate system. In Figure 10.8, the shaded row in the table on the top of the figure links to the graphs below. Also, the large dot on the timeline links that datapoint with the same datapoint in the scatter plot.

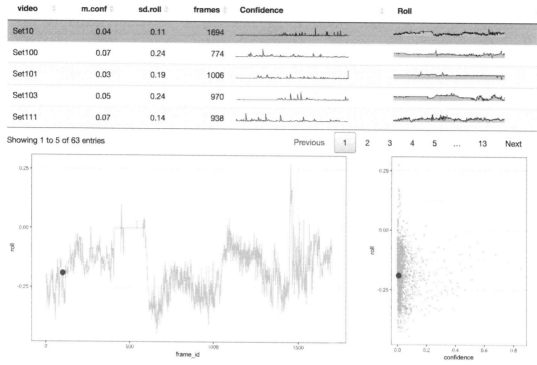

Figure 10.8 A timeline and scatter plot linked to a row of summary data, and a linked point highlighted by the large dot on the timeline and the large dot in the scatter plot. (Created using RShiny (https://shiny.rstudio.com).)

10.6 Website and Application Design

Although the computer was originally designed as a traditional computational device, one of its greatest emerging potentials is in information handling. Currently, computers can make vast amounts of information available to people far more efficiently than was previously possible. Large library databases can be accessed from a computer or event a smartphone, eliminating trips to the library and long searches through card catalogs. As suggested in Chapter 7 (Decision Making), diagnostic tests and troubleshooting steps can be called up in a few key presses instead of requiring the maintenance technician to page through large and cumbersome maintenance manuals containing out-of-date information or missing pages. Physicians can rapidly access the records of hundreds of cases of a rare disease to assess the success of various treatments. All of these uses require people to interact with the computer to search for information. Supporting information search and retrieval is a critical emphasis in human factors and software interface design. Some of the issues involved in designing information systems for websites and application that should be considered include:

- **Navigation:** A good website or app will be able to guide users through workflows and make it easy for people to take action

(R12). A simple rule of thumb is the "three click rule" which means that users should be able to find the information they seek within three clickable buttons.

- **Grid layouts:** Present information in a grid format using sections, columns and boxes. The layout should line up and feel balanced, which leads to a less cluttered feel (MM10).

- **Readability:** Eye tracking studies have shown that people scan computer screens in a "F" pattern (Nielson, 2006), or two horizontal stripes followed by a vertical stripe. Given that users will most likely not read all text on a webpage thoroughly, the most important information should be stated earlier with subheadings and bullet points that can easily scan on the left side of your content in the final stem of the F-pattern (A4).

- **Legibility:** Sans Serif fonts such as Arial, Helvetica, and Verdana are easier to read online or in handheld device.

Figure 10.9 Amazon Dash. By Alexander Klink (Own work) [CC BY 4.0 (http://creativecommons.org/licenses/by/4.0)], via Wikimedia Commons.

Figure 10.10 Apple Watch. By Tscott3714 (Own work) [CC BY-SA 4.0 (http://creativecommons.org/licenses/by-sa/4.0)], via Wikimedia Commons.

10.7   Tangible and Wearable Technology

HCI is moving beyond desktop applications in the workplace, where the main interactions occur through the screen, keyboard, and mouse. Increasingly, people interact with computers in a way that parallels how they interact with the rest of the world: with hands, ears or mouth, but with even the entire body [631]. *Tangible interfaces* blend the physical world with computer technology [632, 633]. The Amazon Dash Button has moved some activities, such as ordering detergent, from the computer into the world. Just put the button on your washing machine, press it when you run out and Amazon ships the detergent.

Interactions can be integrated into clothing and accessories that incorporate practical functions and features. The Fitbit, Android Wear, and Apple Watch are examples of wearable technology that tracks various activities (movements, sleep patterns, heart rate) with a physical band embedded with electronics and sensors. These devices can also exchange information with the user more directly than in possible with a smartphone or computer, such as haptically with a tap on the wrist. Consequently it is even more important to limit interruptions distractions these alerts might represent (A3).

Special purpose wearable devices, such as the Disney magic band, enhance the customer experience by providing seamless access to parks, hotel rooms, and fast pass through rides. The band can even be used to buy food and merchandise while at a Disney location). More importantly, the Magic band identifies customers to Disney staff so that the staff can provide a personalized experience. "The goal was to create a system that would essentially replace the time spent fiddling with payments and tickets for moments of personal interactions with visitors." [634] Such wearable technology

Figure 10.11 FitBit. Image Courtesy: US CPSC (www.flickr.com/photos/uscpsc/13104103473) Licensed under the Creative Commons Attribution 2.0 Generic, via Flickr.

could bring similar benefits to other service industries, such as for caring for patients at hospitals.

Wearable technologies promise to improve many facets of our lives including exercising, sleeping, shopping, education, and healthcare. There are many considerations for designers to ensure that these wearable technologies will enhance the safety, productivity, and satisfaction of users within a very constrained space. Specific considerations include:

- **Comfort:** Wearable technologies need to be comfortable for users while they perform everyday activities. Effective technology considers the context of the user, the environment, and the person's lifestyle (MM10).

- **Prioritization of information:** Given the constrained spaces of wearable technologies it is critical to identify what information should be filtered and what information pushed forward (A3).

- **Accessible with a single glance:** Information should be simplified to its essential elements so that people can assimilate it with a single glance. Paging through multiple screens should be very limited (A4).

- **Use non-visual cues (tactile, haptic, auditory):**. Other non-visual cues can be useful to ensure that the user does not need to look at the technology.

- **Simplifying complex tasks:** Determine how best to collapse an entire payment process into a series of swipes and taps, as Amazon did with the Dash Button, eliminate them entirely (R12).

- **Connectivity:** Ensure that people can activate and deactivate the detection of their devices near appropriate scanners. Limitations of connectivity (either technical or those imposed by the person) and should link to changes in the world that are easily perceived and understood (MM8).

10.8   Computers in Cars

Although computers have been incorporated in low-level vehicle control since the 1970's, but over the last 10 years they have become an increasingly visible part of the driving experience. Computerization of cars includes both entertainment and vehicle control. For vehicle control, computer systems aim to increase safety and reduce the burden of many vehicle operations such speed maintenance with cruise control and car following with adaptive cruise control. Computers are even automating complex activities such as as parallel parking and emergency braking. In some cases it might be unclear to the driver who is actually driving the car. We will return to the topic of vehicle automation in Chapter 11.

Beyond controlling the car, computers also provide drivers with a great variety of information and entertainment. Dashboard displays have always provided speed, miles driven, and gas tank status. These systems now include various features and functions linked to the driver's smart phone that provide information on navigation (current location status and time to destination), vehicle status (gas, engine, climate, radio), driver status (fatigued, distracted, drunk), as well as restaurant reviews, connectivity to the internet, phone call dialing, and even video chats. One of the central design challenges to design such systems to avoid distraction and ensure safety, while also satisfying drivers' information needs [223]. Unlike information and entertainment systems in other situations, a poorly designed interaction can kill. It is important to help facilitate the drivers' awareness of the environment, and with more automated cars, the driver needs to know when to take control. Other sources provide many specific guidelines [95], but some important considerations include:

- **Limit the amount of visual information on displays:** This can be done by anticipating needs and presenting information without requiring the driver request it (A1) defaults and timing. In can also be done by limiting the access cost (A4).

- **Limit distractions**. Limiting the amount of visual information helps limit distraction (A2), but distraction can also be caused by alerts the pull attention away from safety critical driving information, such as the alert for an incoming text message.

- **Simplify interactions**. As with wearable, computers reducing the number of options, steps, and screens in an interaction is essential. In driving, it can be a life and death issue (R12). Systems that demand glances longer than two seconds pose an unacceptable risk

- **Reduce errors** Errors can lead to frustration and long glances away from the road, even for systems that use a speech interface. Creating conversational interactions that are robust to error in particularly important for in-vehicle systems (I15).

- **Make the role of the driver clear** Increasingly automated vehicles can encourage drivers to delegate control of the car to the automation and disengage from driving, even when the automation is incapable of safely controlling the car. To make the role of the driver clear, the state of the automation should be indicated in multiple redundant ways and should be the most salient feature of the vehicle controls and displays (I13). It should also highlight changes in that would make it unable to operate properly (A2).

10.9 Evaluation Criteria for HCI

Human factors engineers strive to maximize the safety, performance, and satisfaction of products, experiences, and environments. These goals all apply to interface and interaction design, and these general goals are often addressed in terms of the more narrow concept of software *usability*. Usability generally describes how intuitive and easy it is to use a design. Beyond this general definition many specific dimensions have been used to define what makes something "easy to useâĂŹ". One common approach describes usability with five dimensions [55]:

- **Learnability:** How easy is it people to use the design for the first time?

- **Efficiency:** After extended and frequent use of the design how fast they perform tasks?

- **Memorability:** After having not uses the design for an extended period, how easily can people return to the previous level of performance?

- **Errors:** How often do people make errors when interacting with the design, what is the consequence, how can people recover, and how can they learn from the errors?

- **Satisfaction:** How pleasant and delightful are interactions with the design?

Failing to create usable systems often has large financial consequences, and with consumer products usability can mean the difference between success or failure. When software controls life-critical systems, such as air traffic control systems, power utilities, ship navigation, and medical instruments (such as a device for delivering radiation treatment), it can easily become a matter of life and death. In fact, poor software design of the radiation therapy machine, mentioned at the start of the chapter, contributed to the death of the patient [635]. As discussed in Chapter 3, this variety of systems has implications for the relative emphasis on safety, performance, and satisfaction, and there are similar implications for the relative importance of different usability criteria.

Careful design can substantially reduce learning time, enhance speed, help people remember how to use the system, reduce error rates, and boost satisfaction. However the focus emphasis differs according to the types of users and tasks discussed at the start of this chapter. For high frequency of use or mandatory use, designers should emphasize efficiency [69]. However, low frequency of use or discretionary use, ease of learning and memorability should have priority over efficiency. In addition, for safety-critical systems or hazardous equipment, error rates may the most important of the five usability criteria; that is, longer training periods are acceptable but that should produce fast, efficient, and error-free performance. Although designers might lower the priority for ease of learning, it is still generally the case that design should strive to address all five of the usability criteria.

Although the expert or mandatory user differs from the novice or discretionary user in terms of which criteria are considered most important (efficiency, accuracy for the expert, learnability and memorability for the novice), an important challenge is to have a single product satisfy all five criteria for both populations.

Another aspect of usability testing mentioned in Chapter 3 merits emphasis here: the selection of people included in the tests is critical. The people should be representative of the full population of intended users. Testing on engineers and designers who helped to create the product will tend to overestimate the usability of the product. Generally, designers should aspire to *universal design*, where the product works for everyone, independent of age and cognitive and physical abilities[565]. This means usability tests should oversample from populations with functional impairments that the design is intended to support.

10.10   Summary

This chapter highlights the need to design interactions and experiences, not just displays and controls. This means creating systems that works effectively for users by considering how to integrate information and control over time. As HCI has expanded beyond the desktop to include wearable and automotive applications a full evaluation of cognitive, physical (see Chapters 12-15), and social considerations of users and their environment become more relevant to HCI (see Chapters 16-18). This expansion also emphasizes the need to consider universal design, where the emphasis is on making systems usable for people of all ages, and all perceptual and motor abilities. Such universal design is a general theme through all of human factors design addressed throughout this book, but sometimes neglected in HCI. As software becomes more "intelligent" and able to adapt to usersâĂŹ needs based on the particular context it can start completing tasks that might otherwise be performed by a person. Instead of simply executing command from the user the software makes decisions with some degree of autonomy, such as putting email messages in a junk folder. Such technology is often termed automation and is becoming prevalent in military systems, aircraft, cars, and robots. Chapter 11 describes techniques to address design challenges with such systems.

Additional resources

A variety of books and journals are written exclusively on this topic (e.g., *Human-Computer Interaction* and *International Journal of Human-Computer Interaction*), and annual meetings result in proceedings reflecting the cutting-edge views and work, such as Computer-Human Interaction (CHI). Two recent handbooks cover a range of HCI issues, including computers in the car, the home, and the hospital.

1. **HCI handbooks:** Two important handbooks that cover the concepts of this chapter in more detail include:

Jacko, J. A. (Ed.). (2012). *Human Computer Interaction Handbook: Fundamentals, evolving technologies, and emerging applications.* CRC Press. Boca Raton, FL.

Shneiderman, B., Plaisant, C., Cohen, M., Jacobs, S., Elmqvist, N., & Diakopoulos, N. (2016). *Designing the User Interface: Strategies for Effective Human-Computer Interaction* (Sixth edition). New York: Pearson.

Questions

Questions for 10.1 Matching Interaction Style to Tasks and Users

P10.1   How does interface design differ from design of displays and design of controls?

P10.2   Is design of hardware, such as buttons, and other input devices part of a human-computer interface (HCI)?

P10.3   What is interaction design and how does it relate to user experience design?

P10.4   How do the three dimensions of user and task characteristics—mandatory versus discretionary use, frequency of use, and task structure–inform interaction design?

Questions for 10.2 Interaction Styles

P10.5   What interaction style is best suited to a situation where the use of the system is mandated and the system is frequently used?

P10.6   What interaction style is best suited to a situation where the use of the system is not mandatory but the system is still frequently used?

P10.7   What interaction style is best suited for people who use the system occasionally and where the interaction is highly structured (e.g., purchasing a ticket for a commuter train)?

P10.8   What interaction style is well-suited to occasional, discretionary use for unstructured tasks (e.g., word processing)?

Questions for 10.3 Theories Underlying Interface and Interaction Design

P10.9   Describe why metaphors can be useful in bridging the gulfs of evaluation and execution.

P10.10   Skeuomorphic features are controversial elements that might be part of a design that uses a metaphor. Describe what they are and why they are controversial.

P10.11   Describe two ways emotions might be considered in a design.

P10.12   What method can be used to address the principle, Anticipate needs?

P10.13   Describe a situation where the principle, Highlight changes, would be particularly important.

P10.14   According to the principle, Limit interruptions and distractions, how should notifications and other features that might interrupt people be designed?

P10.15   What is an example of an affordance in a Windows or Apple operating system?

P10.16   Give an example of supporting recognition rather than forcing recall.

P10.17   Why does match system to real world help give clear meaning to interface elements?

P10.18   What is a skeuomorph, and what is its role in match system to real world?

Questions for 10.4 Fifteen Principles for HCI Design

P10.19   Describe two ways you might inadvertently undermine trust in a website you developed to sell used textbooks?

P10.20   What is the maximum number of colors you should use on a website (not including colors in pictures that might be included)?

P10.21   Why is it important to carefully choose defaults?

P10.22   Considered in terms of the idea of task structure discussed at the start of the chapter, describe why it is important to enable user control and freedom.

P10.23   If you are simplifying a long task sequence into to manageable chunks, how long should each set of steps take to complete?

P10.24   At what length of delay should the system indicate that it is "thinkingâĂŹ'?

Questions for 10.6 Website and Application Design

P10.25   Why is robustness to errors or error tolerance important for helping people learn how to use the system?

P10.26   What is the F-pattern and what are its implications for web-site design?

Questions for 10.7 Tangible and Wearable Technology

P10.27   How might the characteristics of wearable technology associated with Disney's Magic band enhance healthcare services?

Questions for 10.8 Computers in Cars

P10.28   What does interaction design for cars need to consider that would not be a priority for desktop computers?

Questions for 10.9 Evaluation Criteria for HCI

P10.29   What are the five criteria for usability evaluation?

P10.30   What usability criteria are most important for discretionary use?

P10.31   What usability criteria are most important for intermittent use?

P10.32   What is universal design?

P10.33   How does automation relate to HCI and interaction design?

Chapter 11

Human-Automation Interaction

At the end of this chapter you will be able to...

1. identify types of tasks and the types of automation that support them best

2. describe three good reasons for implementing automation and one bad reason

3. identify situations when automation can help and hinder productivity, safety, and satisfaction

4. apply principles for human-centered automation

The pilots of the commercial airlines transport were flying high over the Pacific, allowing their autopilot to direct the aircraft on the long, routine flight. Gradually, one of the engines began to lose power, causing the plane to tend to veer toward the right. As it did, however, the autopilot appropriately steered the plane back to the left, thereby continuing to direct a straight flightpath. Eventually, as the engine continued to lose power, the autopilot could no longer apply the necessary countercorrection. As in a tug-of-war when one side finally loses its resistance and is rapidly pulled across the line, so the autopilot eventually "failed." The plane suddenly rolled, dipped, and lost its airworthiness, falling over 30,000 feet out of the sky before the pilots finally regained control just a few thousand heart-stopping feet above the ocean (National Transportation Safety Board, 1986; [144]). Why did this happen? In analyzing this incident, investigators concluded that the autopilot had so perfectly handled its chores during the long routine flights that the flight crew had been lulled into a sense of complacency, not monitoring and supervising its operations as closely as they should have. Had they done so, they would have noted early on the gradual loss of engine power (and the resulting need for greater autopilot compensation), an event they clearly would have detected had they been steering the plane themselves.

Automation is a machine (usually a computer) that performs a task that is otherwise performed by a person, or that has never been performed before. This contrasts with the discussion of human-computer interaction in Chapter 10 where the person is typically fully in control and the computer does relatively little on behalf of the person. Automation often shifts the person from direct control to supervisory control, where the person's role is to manage the automation. As the aircraft example illustrates, automation is somewhat of a mixed blessing. When it works well, it usually works very well—so well that we sometimes trust it more than we should. Yet on the rare occasions when it does fail, those failures may often be more catastrophic, less forgiving, or at least more frustrating than any potential corresponding failures of a person in the same circumstance. Sometimes, of course, these failures are relatively trivial and benign—like my copier, which keeps insisting that I have placed the book in an orientation that I do not want (when that's exactly what I do want). At other times, however, as with the aircraft incident and a host of aircraft crashes that have been attributed to automation problems, the consequences are severe [144, 636, 637, 638, 639]. Although automation is most visible when it controls aircraft and cars, automation also plays a role in the financial system and even in heating and cooling our homes. Automation can cause planes to crash and it can also cause stock markets to do the same [640].

If the serious consequences of automation resulted merely from failures of software or hardware components, then this would not be a concern for human factors engineers. However, the system problems with automation are distinctly and inexorably linked to human issues of attention, perception and cognition. These issues arise in managing the automated system in its normal operating state, when the system that the automation is serving has failed or been disrupted, or when the automated component itself has failed [187, 641]. The performance of most highly automated systems depends on the interaction of people with the technology. Before addressing these problems, we first consider why we automate and describe some of the different types of automation. After discussing the various human performance problems with automation and providing some solutions, we discuss automation issues in industrial process control, manufacturing, vehicles, and robots.

11.1 Why Automate?

There are several reasons that designers develop machines to replace or aid human performance and they can be roughly placed into four categories.

1. **Impossible or hazardous.** Some processes are automated because it is either dangerous or impossible for humans to perform the equivalent tasks. A clear example was provided in Chapter 9, with teleoperation. The use of robotic handling of hazardous material (or material in hazardous environments) will remove the need for humans in an otherwise unsafe situation. There are also many circumstances in which automation can serve the needs of special populations whose disabilities may leave them unable to carry out certain skills without assistance [565]. Examples include automatic guidance systems for the quadriplegic or automatic readers for the visually impaired. In many situations, automation enables people to do what would otherwise be impossible and helps enhance mobility and productivity.

2. **Difficult or unpleasant.** Other processes, while not impossible, may be *very challenging* for the unaided human operator, such that humans carry out the functions poorly. Of course, the border between "impossible" as described above in the category above and "difficult" is somewhat fuzzy. For example, a calculator "automatically" multiplies digits that can be multiplied in the head. But the latter is generally more effortful and error producing. Robotic assembly cells automate highly repetitive and fatiguing human operations. Workers can do these things but often at a cost to their health, morale, and overall safety. Autopilots on aircraft provide more precise flight control and can also relieve pilots of the fatiguing task of continuous control over long-haul flights, and train engineers can benefit from automation that avoids overspeed derailments, which often occur during long boring drives .

Chapter 7 describes expert systems that can replace humans in routine situations where it is important to generate very consistent decisions. As another example, we learned in

Chapter 4 and 13 that humans are not very good at *vigilant monitoring*. Hence, automation is effective in monitoring for relatively rare events, and the general class of warning and alert systems, like the light that appears when your oil pressure or fuel level is low in the car. Of course, sometimes automation can impose more vigilant monitoring tasks on the human, as we saw in the airplane incident [642, 183]. This is one of the many "ironies of automation" [643]. Ideally, automation makes difficult and unpleasant tasks easier.

Figure 11.1 Automation makes it possible for a car to drive itself under in some situations. (By DimiTVP (Own work) [CC BY-SA 4.0 (http://creativecommons.org /licenses/by-sa/4.0)], via Wikimedia Commons.)

3. **Extend human capability.** Sometimes automated functions may not replace but may simply *aid humans* in doing things in otherwise difficult circumstances. For example, we saw in Chapter 6 that human working memory is vulnerable to forgetting. Automated aids that supplement memory are useful. The decision aids discussed in Chapter 7 and the predictive displays discussed in Chapters 6 and 8, are examples of automation that relieves the human operator of some cognitively demanding mental operations. Automated planning aids have a similar status [467]. Automation is particularly useful in extending people's *multitasking* capabilities. For example, pilots report that autopilots can be quite useful in temporarily relieving them from duties of aircraft control when other task demands temporarily make their workload extremely high. In many situations, automation should help enhance productivity and extend rather than replace the human role in a system.

4. **Technically possible.** Finally, sometimes functions are automated simply because the technology is there and inexpensive, even though it may provide little or no value to the person using it. Many of us have gone through painfully long negotiations with automated "phone menus" to get answers that would have taken us only a few seconds with a human operator on the other end of the line. But it is probable that the company has found that a computer operator is quite a bit cheaper. Some automation maximizes the gain for the company rather than the customer [644]. Many household appliances and vehicles have several automated features that provide only minimal advantages that may even present costs and, because of their increased complexity and dependence on electrical power, are considerably more vulnerable to failure than the manually operated systems they replaced. It is unfortunate when the purported "technological sophistication" of these features are marketed, because they often have no real usability advantages. Automation should focus on supporting system performance and humans' tasks rather than showcasing technical sophistication.

For which of these reasons are people designing driverless cars? Maybe all four?

11.2   Types of Automation and Types of Tasks

Automation and the types of tasks it supports are very diverse ranging from home thermostats and stock trading to cars and power plants. Here we discuss three dimensions of automation: levels and stages, number of automation modes and elements, and the span of control. When considering automation-interaction, tasks differ on two particularly important dimensions: timescale and consequence, where timescale refers to how quickly the system changes, and consequence describes how bad it can get when things go wrong. These dimensions of automation and tasks are important because they indicate likely problems and point towards useful design principles.

11.2.1   Types of Automation

One way of representing what automation does is in terms of the *stages* of human information processing as presented in Chapter 6. There are four stages of human information processing that automation can replace (or augment). The amount of cognitive or motor work that automation replaces is defined by the level of automation. The four stages and the different levels for each combine to define types of automation [645, 646, 647].

1. **Information acquisition, selection, and filtering (Stage 1).** Automation replaces many of the cognitive processes of human selective attention, discussed in Chapter 6. Examples include warning systems and alerts that guide attention to inspect parts of the environment that automation deems to be worthy of further scrutiny [194]. Automatic highlighting tools, such as the spell-checker that redlines misspelled words, is another example of attention-directing automation. In the medical domain, stage 1 automation can highlight important symptoms related to a cancerous growth. In transportation, automatic target-cueing devices can help direct drivers to a sharp curve or a car ahead [648, 649]. Finally, more "aggressive" examples of stage 1 automation may be seen in intelligent warning systems, which may filter or delete altogether information assumed to be unworthy of operator attention.

2. **Information integration (Stage 2).** Automation replaces (or assists) many of the cognitive processes of perception and working memory, described in Chapters 6 and 7, to provide the operator with a situation assessment, inference, diagnosis, or easy-to-interpret "picture" of the task-relevant information. Examples at lower levels may configure visual graphics in a way that makes perceptual data easier to integrate (Chapter 8). Examples at higher levels are automatic pattern recognizers, predictor displays, diagnostic expert systems as in in medicine ([440]; Chapter 7). Many intelligent

warning systems that guide attention (stage 1) also include sophisticated integration logic necessary to infer the existence of a problem or dangerous condition ([650]; stage 2). So too would be the statistics package that automatically computes a "significance level" from the raw data, providing an inference that an effect is (or is not) "true" within the population.

3. **Action selection and choice (Stage 3).** As described in Chapter 7, diagnosis is quite distinct from choice, and in Chapter 4, sensitivity is quite different from the response criterion. In both cases, the latter entity explicitly considers the *value* of potential outcomes. In the same manner, automated aids that diagnose a situation at stage 2 are quite distinct from those that recommend a particular course of action. In recommending action, the automated agent must explicitly or implicitly assume a certain set of values for the operator who depends on its advice. An example of stage 3 automation is the airborne traffic alert and collision avoidance system (TCAS), which explicitly (and strongly) advises the pilot of a vertical maneuver to take to avoid colliding with another aircraft. In this case, the values are shared between pilot and automation (avoid collision); but there are other circumstances where value sharing might not be so obvious, as when an automation medical decision aid recommends one form of treatment over another for a terminally ill patient.

4. **Control and action execution (Stage 4).** Automation may replace different levels of the human's action or control functions. As we learned in Chapter 9, control usually depends on the perception of desired input information, and therefore control automation also includes the automation of certain perceptual functions. (These functions usually involve *sensing* position and trend rather than *categorizing* information). Autopilots in aircraft, cruise control in driving, robots in industrial processing, exploration and health care [588] are examples of control automation, as are automatic trading algorithms in the stock market. More mundane examples of stage 4 automation include electric can openers and automatic car windows. As with robots, full automation at stage 3 often, not necessarily, involves automation at all stages. Such as system can therefore be defined as one having a high degree of autonomy, such as unmanned air vehicles that can operate for hours without immediate supervision of people. No systems are truly autonomous, but involve oversight of people who are increasingly distanced from the direct control, an issue we return to at the end of the chapter.

We noted that levels of automation characterized the amount of "work" done by the automation (and therefore, workload relieved from the human). It turns out that it is at stages 3 and 4, where the levels of automation take on critical importance. Table 11.1, adapted from Sheridan [468], summarizes eight levels of

Table 11.1 Levels of automation range from complete manual control (level 1) to complete automatic control (level 8). (Adapted from Sheridan [468].)

Levels	Automation's role
1.	Offers no aid; **human in complete control**.
2.	**Suggests multiple alternatives**; filters and highlights what it considers to be the best alternatives.
3.	**Selects an alternative**, one set of information, or a way to do the task and suggests it to the person
4.	**Carries out the action if the person approves.**
5.	Provides the person with **limited time to veto the action** before it carries out the action
6.	Carries out an action and then **informs the person.**
7.	Carries out an action and **informs the person only if asked.**
8.	Selects method, executes task, and **ignores the human** (i.e., the human has no veto power and is not informed).

automation that apply particularly to stage-3 and stage-4 automation, characterizing distribution of authority between human and automation in choosing a course of action.

The importance of both stages and levels emerges under circumstances when automation may be *imperfect* or *unreliable* [641]. Automation at different stages and levels have different consequences for human performance. People are unlikely to catch the mistakes made by high levels of late stage automation. Because of this, designers should exercise great caution before imposing high levels of stage 3 automation in safety critical systems, an issue we return to toward the end of this chapter.

Another way of representing automation is in terms of the *span of control*. Span of control describes the number of things controlled and choices made by the automation on behalf of the person [651, 652]. With a low span of control, few choices are made by the automation and it is easy for people to see whether the automation achieved their goal and did it in a way they expected. With a high span of control the automation makes many choices, making it hard to know whether the final outcome was as intended. A single rule that automatically highlights from a friend based on her address represents low span of control. In contrast, an algorithm that combines data from your email reading history and that of other people to highlight important email messages represents high span of control. Another example of high span of control automation would be automation that translates a single command of an operator into the control and coordination of a team of drone aircraft.

Automation also differs according to the *number of interacting elements*. Although much of the discussion in this chapter describes interaction with a single element of automation, many systems are much more complex and the automation has many modes. A modern aircraft might have many modes in the flight management system. Also many jobs require people to manage many separate automated systems, some of which interact and

Stages and levels, span of control, as well as the number of interacting elements are critical characteristics of automation.

some of which might be independent, such as drone operators who oversee flight control automation, information acquisition automation, and path planning automation to name a few, each of which might have multiple modes.

11.2.2 Types of Tasks

Two aspects of tasks are particularly important for designing automation: time span and consequence. *Time span* can range from milliseconds to months. In your car, automation at the millisecond level modulates the brakes to keep you from skidding when you brake hard. At the time span of minutes, automation tells you when to turn as it guides you to your destination. At the time span of months, automation illuminates the "check engine" light to get you to take your car in for maintenance. Operation that operates at a short time span typically requires a very fast response from the person if the automation were to fail, whereas automation that operates at a very long time span requires sustained attention over a long period. The demands of very short and very long time span tasks are at odds with human cognitive processes described in Chapter 4, and make it harder for people to detect and intervene to address problems with automation.

The *consequence* of tasks describes the cost failing. This cost might be counted in money or lives. High-consequence tasks include flying aircraft and driving cars; whereas low-consequence tasks include vacuuming a room. Consequence roughly corresponds to the goals of safety, performance, and satisfaction: high-consequence systems are those where safety is central. Consequence corresponds to the worst that can happen with a system if it fails. A system where satisfaction is central the worst might be that it disappoints or frustrates the person. A car is a high-consequence system because, even though satisfaction and enjoyment of the entertainment system is important, a failure to control the car can result in death.

As a general rule, different types of automation combine with different types of tasks place different demands on people and to generate different problems, which we address in the following section.

11.3 Problems with Automation

Whatever the reason for choosing automation, and no matter which kind of function (or combination of human functions) are being "replaced," the history of human interaction with such systems has revealed certain problems [641, 187, 653]. In discussing these problems, however, it is important to stress that they must be balanced against the number of very real *benefits of automation*. There is little doubt that the ground proximity warning system in aircraft, for example, has helped save many lives by alerting pilots to possible crashes they might otherwise have failed to note [654]. Autopilots have contributed substantially to fuel savings; robots have allowed workers to be removed from unsafe and hazardous jobs. They have greatly improved industrial productivity although unfortunately this has often been at the cost of worker unemployment; and computers have radically improved the efficiency of many human communications, computations, and information-retrieval processes. Still, there is room for improvement, and the direction of those improvements can be best formulated by understanding the nature of the remaining or emerging problems that result when humans interact with automated systems.

11.3.1 Automation Reliability

To the extent that automation can be said to be reliable, it does what the human operator expects it to do. Cruise control holds the car at a set speed, a copier faithfully reproduces the number of pages requested, and so forth. However, what is important for human interaction is not the reliability per se but the perceived reliability. There are at least four reasons why automation may be perceived as unreliable.

1. **It may be unreliable.** A component may fail or may contain design flaws. In this regard, automated systems typically are more complex and have more components than their manually operated counterparts and therefore contain more components that could go wrong at any given time [635], as well as working components that are incorrectly signaled to have failed. The nature of these alarm false alarms in warning systems was addressed in Chapter 5.

2. **The automation does not operate or perform well in certain situations.** All automation has a limited operating range within which designers assume it will be used. Using automation for purposes not anticipated by designers leads to lower reliability. For example, cruise control is designed to maintain a constant speed on a level highway. It does not use the brakes to slow the car, so cruise control will fail to maintain the set speed when traveling down a steep hill.

3. **The human operator may incorrectly "set up" the automation.** By one estimate, keypress errors occur 4% of the time [257], and so configuration errors can be common. Nurses sometimes make errors when they program systems that allow patients to administer periodic doses of painkillers intravenously. If the nurses enter the wrong drug concentration, the system will faithfully do what it was told to do and give the patient an overdose [655]. Thus, automation is often described as "dumb and dutiful", blindly operating the way it was set up to do.

4. **It appears to be acting erroneously to the operator.** There are circumstances when the automated system does exactly what it is supposed to do, but the logic behind the system is sufficiently complex and poorly understood by the human

operator—a poor mental model. This is a particular problem for automation with a large span of control. Sarter and Woods [656, 657] observed that these *automation induced surprises* appear relatively frequently with the complex flight management systems in modern aircraft. The automation triggers certain actions, like an abrupt change in air speed or altitude, for reasons that may not be clear to the pilot. If pilots perceive these events to be failures and try to intervene inappropriately, disaster can result [658, 638, 639].

The term, *unreliable automation,* has a certain negative connotation. However, it is important to realize that automation is often asked to do tasks, such as weather forecasting or prediction of aircraft trajectory or enemy intent, that are simply impossible to do perfectly given the uncertain nature of the dynamic world in which we exist. [466]. Hence, it may be better to label such automation as "imperfect" rather than "unreliable." To the extent that such imperfections are well known and understood by the operator, even automation as low as 70 percent reliable can still be of value, particularly under high workload and low consequence situations [189, 659]. The value that can be realized from imperfect automation relates directly to the concept of trust.

11.3.2   Trust: Calibration and Mistrust

The concept of *perceived automation reliability* is critical to understanding the human performance issues because of the relation between reliability and *trust*; and the corresponding relationship between trust, a cognitive state, and automation dependence or use, a behavioral measure. Trust in another person is related to the extent to which he or she believes that the other will carry out actions that are expected and consistent with his or her goals. Trust has a similar function in a human's belief in the actions of an automated component [660, 661, 662, 190, 189]. Trust in automation is "... the attitude that an agent will help achieve an individual's goals in a situation characterized by uncertainty and vulnerability." [189, p. 54]

Ideally, when dealing with any entity, whether a friend, a salesperson, a witness in a court proceeding, or an automated device, *trust should be well calibrated.* This means our trust in the agent, whether human or computer, should be in direct proportion to its reliability. Mistrust occurs when trust fails correspond to the reliability of the automation. As reliability decreases, our trust should go down, and we should be prepared to act ourselves and be receptive to sources of advice or information other than those provided by the unreliable agent.

Although this relation between reliability, trust and human cognition holds true to some extent [663, 661, 664], there is also some evidence that human trust in automation is not always *well calibrated:* Sometimes it is too low (*distrust*), sometimes too high (*overtrust*) [187]. Distrust is a type of mistrust where the person fails to trust the automation as much as is appropriate, leading to disuse. For example, in some circumstances people prefer manual control to automatic control of a computer, even when both are performing at precisely the same level of accuracy [665]. A similar effect is seen with automation that enhances perception, where people are biased to rely on themselves rather than the automation [648]. Distrust of alarm systems with high false alarm rates, the so called "cry wolf effect" is a common syndrome across many applications as discussed in Chapter 5 [183, 666]. Distrust in automation may also result from a failure to understand how the automated algorithms produce an output, whether that output is a perceptual categorization, a diagnosis, a decision, or a control action. This can be a particularly important problem for decision-making aids, as described in Chapter 7.

The consequences of distrust are not necessarily severe, but they may lead to inefficiency when distrust leads people to reject the good assistance that automation can offer. For example, a pilot who mistrusts a flight management system and prefers to fly the plane by hand may become more fatigued and may fly less efficient routes. Often "doing things by hand" rather than using a computer can lead to lower performance that may be less accurate, when the computer-based automation is highly reliable. However, over trust of false-alarm prone automated warning systems can lead people to ignore legitimate alarms [667, 668, 183].

11.3.3 Overtrust, Complacency, and Out-of-the-Loop Behavior

In contrast to distrust, *overtrust* of automation, sometimes referred to as *complacency*, or automation bias that occurs when people trust the automation more than is warranted and can have severe negative consequences if the automation is less than fully reliable [669, 187]. We saw at the beginning of the chapter the incident involving the airline pilot who trusted his automation too much, became complacent in monitoring its activity, and nearly met disaster. The cause of complacency is probably an inevitable consequence of the human tendency to let experience guide our expectations. As we learned in the earlier chapters, our expectations are affected by our *top-down processing* of information that signals that the automation should work since it worked in previous situations. Here top-down processing can overrule bottom-up processing.

One consequence of expectations that the automation will control the system well can be described as *out-of-the-loop behavior*. Chapter 9 described the control loop in which perception of the state relative to the goal guides corrective actions. People are *in-the-loop behavior* to the extent that they are actively attending and controlling the process. Monitoring the system is sometimes described as being *on-the-loop behavior*, which requires sustained attention that is hard to maintain. This effort leads people to move out of the loop and delegate responsibility to the automation. This delegation is particularly likely when people are working in systems that have many control loops demanding attention. The limits of attention described in Chapter 6 make it impossible for someone to be in-the-loop with more than a few activities. In such cases, it is only in hindsight that someone's monitoring can be judged complacent.

It is likely that many people using a particular system may never encounter failures, and hence their perception of the reliability of the automation is that it is perfect (rather than the high, but still less than 100 percent, that characterize all operations of the system in question). If the operator perceives the device to be perfectly reliable, a natural tendency would be to cease monitoring its operation or at to least monitor it far less vigilantly than is appropriate [643, 670]. This situation is exacerbated by the fact that, as we learned in Chapter 3, Displays are often designed in ways that do not integrate important information for monitoring tasks [127, 671].

Of course, the real problem with complacency, the failure to monitor adequately, only surfaces in the infrequent circumstances when something does fail (or is perceived to fail) and the human must (or feels a need to) intervene. Automation then has three distinct implications for human intervention related to detection, situation awareness, and skill loss [641].

1. **Detection.** The complacent operator will likely be slower to detect a real failure [672, 673]. As noted in Chapters 4 and 13, detection in circumstances in which events are rare (the automation is reliable) is generally poor, since this imposes a vigilance monitoring task. Indeed, the more reliable the automation, the rarer the "signal events" become, and the poorer is their detection [674]. A large component of this detection failure is simply the failure to look at (scan) those areas of the workplace where the failure would be evident [641] such as, in a recent tragic example, the roadway ahead in a self-driving car.

Perception while controlling is very different from perception while monitoring.

2. **Situation awareness.** People are better aware of the dynamic state of processes in which they are active participants, selecting and executing its actions, than when they are passive monitors of someone (or something) else carrying out those processes, a phenomenon known as the generation effect [675, 676]. Perception that guides action (an is guided by action) is very different than perception when monitoring [677]. Hence, independent of their ability to detect a failure in an automated system, they are less likely to intervene correctly and appropriately if they are out of the and do not fully understand the system's momentary state [656]. Such was also the case with the 2008 financial disaster, when complex automatic trading algorithms functioned in ways that were not intended. With cruise control, the driver may remove her foot from the accelerator and become less aware of how the accelerator pedal moves to maintain a constant speed. Thus, she may be slower to put her foot on the brake when the car begins to accelerate down a hill because cruise control does not apply the brakes. The issue of situation awareness can be particularly problematic if the system is designed with

poor feedback regarding the ongoing state of the automated process.

3. **Skill loss.** A final implication of being out of the loop has less to do with the response to failures than with the long-term consequences [678]. Wiener [679] described deskilling as the gradual loss of skills an operator may experience by virtue of not having been an active perceiver, decision maker, or controller during the time that automation assumed respon- sibility for the task. Such skill loss has two implications. First, it may make the operator less self-confident in his or her own performance and hence more likely to continue to use automation even when it may not be appropriate to do so [680]. Second, skill loss may further degrade the operator's ability to intervene appropriately should the system fail, an operation often referred to as "return to manual" or "manual takeover" [681] which is of critical importance in self-driving cars. Imagine your calculator failing in the middle of a math or engineering exam, when you have not done unaided arith- metic for several years. The relation between trust and skill loss is shown in Figure 16.1. The figure also makes clear the distinction between automation trust and automation de- pendence or use. Several features affect trust, only one of which is reliability.

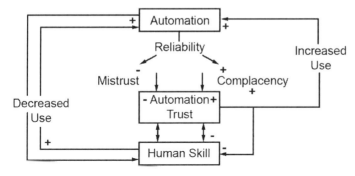

Figure 11.2 Elements of automation reliability and human trust. The + and indicate the direction of effects. For example, increased (+) automa- tion reliability leads to increased (+) trust in automation, which in turn leads to increased (+) use and a decrease (-) in human skill.

11.3.4 Workload and Situation Awareness

Automation is often introduced with the goal of reducing opera- tor workload (see Chapter 6). For example, an automated lane keeping and headway maintenance in driving may reduce driv- ing workload [682, 683] and free mental resources to drive more safely. Alerting automation in aircraft is vital to reduce the load of continuous monitoring so the pilot has perceptual and cogni- tive resources available to devote to other aspects of safe flying. However, in practice, sometimes the workload is reduced by au- tomation in environments when workload is already too low and loss of arousal rather than high workload is the most important problem (e.g., driving on a freeway at night). In fact, as we will see in Chapter 15,it is incorrect to think that vigilance tasks are low workload because they require attention if the person is to respond to events in a timely manner [73]. In addition, sometimes automation directly undermines situation awareness, because the operator is not actively involved in choosing and executing actions as we discussed above. There is a correlation between situation awareness and workload; as automation level moves up the scale in Table 11.1 both workload and situation awareness tend to go down [676, 647].

Sometimes automation has the undesirable effect of both reducing workload during already low-workload periods and increasing it during high-workload periods. This problem of *clumsy automation* is where automation makes easy tasks easier and hard tasks harder. For example, a flight management system tends to make the low-workload phases of flight, such as straight and level flight or a routine climb, easier, but it tends to make high-workload phases, such as the maneuvers in preparation for landing, more difficult, as pilots have to share their time between landing procedures, communication, and programming the flight management system.

Another aspect of clumsy automation is that the circumstances in which some automated devices fail are often the same circumstances that are most challenging to human: automation tends to fail when it is most needed by the human operator. Such was the case with the failed engine in our opening story. These circumstances may also occur with decision aids that are programmed to handle ordinary problems but must "throw up their hands" at complex ones. It is, of course, in these very circumstances that the automated system may hand off the problem to its human counterpart. But now, the human, who is out of the loop and may have lost situation awareness, will be suddenly asked to handle the most difficult, challenging problems, hardly a fair request.

11.3.5 Mode Confusion and Managing Multiple Elements of Automation

Many automation systems operate in multiple modes, so that a single human action may accomplish different system responses depending on the mode setting. The simple calculator is an example in which a mode setting can change the operation of a key press from squaring to computing the square root. This can present generates *mode!errors* if the person forgets which mode is in operation, like typing "&&$*@" instead of the intended "77482". But more complex and intelligent automation may itself automatically change modes. As described above, cruise control may disengage on a hill, or in trains, the positive train control system, designed to prevent derailment at high speeds, may disengage when the track is not equipped to support it. At even higher levels of complexity one part of automation may change the modes of another (and not just disengage it). This becomes a major causes of the "automation surprises" that have confronted pilots in highly automated aircraft: "what did it do, and why did it do that?" Such confusion will often lead the human supervisor to intervene, and sometimes make matters worse, leading to fatal crashes in aviation.

11.3.6 Loss of Human Cooperation

In non-automated, multi-person systems, there are many circumstances in which subtle communications, achieved by nonverbal means or voice inflection, provide valuable sources of information (Chapter 18; [684]). The air traffic controller can often tell if a pilot is in trouble by the sound of the voice. Sometimes automation may eliminate valuable information channels that may depict urgency, stress, or even a sense of calm. For example, in the digital datalink system [685, 686, 687], which is replacing some air-to-ground radio communications with digital messages that are typed in and appear on a display panel, such relevant information will be gone. In addition, the "party line" effect, where pilots hear conversations between ATC and other pilots shares information and supports a shared situation awareness that datalink might not. There may also be circumstances in which negotiation between humans, necessary to solve nonroutine problems, may be eliminated by automation. Many of us have undoubtedly been frustrated when trying to interact with an uncaring, automated phone menu in to get a question answered that was not foreseen by those who developed the automated logic.

11.3.7 Job Satisfaction

We have primarily addressed performance problems associated with automated systems, but the issue of job satisfaction goes well beyond performance to consider the morale of the worker who is replaced by automation, and indeed damage to world wide employment numbers [688]. In reconsidering the reasons to automate, we can imagine that automation that improves safety or unburdens the human operator will be well received. But automation introduced merely because the technology is available or that increases job efficiency may not be appreciated. Many operators are highly skilled and proud of their craft. Replacement by robot or computer can eliminate the opportunity to develop and demonstrate skill. If the unhappy, demoralized operator then is asked to remain in a potential position of resuming control, should automation fail, an unpleasant situation could result.

11.3.8 Training and Certification

Errors can occur when people lack the training to understand the automation. As increasingly sophisticated automation eliminates many physical tasks, complex tasks may appear to become easy, leading to less emphasis on training. On ships, the misunderstanding of new radar and collision avoidance systems has contributed to collisions[689]. One contribution to these accidents is training and certification that fails to reflect the demands of the automation. An analysis of the exam used the by the U.S. Coast Guard to certify radar operators indicated that 75 percent of the items assess skills that have been automated and are not required by the new technology [690]. The new technology makes it possible to monitor a greater number of ships, enhancing the need for interpretive skills such as understanding the rules of the road and the limits of the automation. These very skills are underrepresented on the test. Further, the knowledge and skills may degrade because they are used only in rare but critical instances. Automation design should carefully assess the effect of automation on the training and certification requirements [690]. Chapter 18 describes how to identify and provide appropriate training.

11.4 Allocating Functions between People and Automation

How can automation be designed to avoid these problems? One approach is a systematic allocation of functions to the human and to the automation based on the relative capabilities of each. We can allocate functions depending on whether the automation or the human generally performs a function better. This process begins with a task and function analysis, described in Chapter 2. Functions are then considered in terms of the demands they place on the human and automation. A list of human and automation capabilities guides the decision to automate each function. Table 11.2 lists the relative capabilities originally developed by Fitts [691] and adapted from Sheridan [468] and Fuld [692].

> Automation provides consistency and precision, and people provide flexibility and capacity.

Many functions can be accomplished by either a person or technology, and designers must identify appropriate functions for each. To do this, designers must first evaluate the basic functions that must be performed by the human-machine system to support or accomplish the activities identified in the task analysis. Designers then determine whether each function is to be performed by the system (automatic), the person (manual), or some combination. This process is termed function allocation and is an important, sometimes critical, step in human factors engineering [693, 694].

Function allocation is sometimes complex. There are numerous reasons for allocating functions to either machine or person. In 1951, Paul Fitts provided a list—the Fitts's list— of those functions performed more capably by humans and those performed more capably by machines [691]. This may be referred to as "HABA-MABA" (humans better at–machines better at). Although technology has advanced tremendously, people still remain more flexible and adaptable and machines provide precise and consistent responses. Many such lists have been published since that time, and some researchers have suggested that allocation simply be made by assigning a function to the more "capable" system component. Given this traditional view, where function is simply allocated to the most capable system component (either human or machine),

Table 11.2 "Fitts's List" showing the relative benefits of automation and humans.)

Humans are better at...	Automation is better at...
Detecting small amounts of visual, auditory, or chemical signals (e.g., evaluating wine or perfume)	Monitoring processes (e.g., warnings).
Detecting a wide range of stimuli (e.g., integrating visual, auditory, and olfactory cues in cooking).	Detecting signals beyond human capability (e.g., measuring high temperatures, sensing infrared light and x-rays).
Perceiving patterns and making generalizations (e.g., "seeing the big picture")	Ignoring extraneous factors (e.g., a calculator doesn't get nervous during an exam).
Detecting signals in high levels of background noise (e.g., detecting a ship on a cluttered radar display)	Responding quickly and applying great force smoothly and precisely (e.g., autopilots, automatic torque application.
Improvising and using flexible procedures (e.g., engineering problem solving, such as on the Apollo 13 moon mission as described in Chapter 7).	Repeating the same procedure in precisely the same manner many times (e.g., robots on assembly lines).
Storing information for long periods and recalling appropriate parts (e.g., recognizing a friend after many years)	Storing large amounts of information briefly and erasing it completely (e.g., updating predictions in a dynamic environment).
Reasoning inductively (e.g., extracting meaningful relationships from data)	Reasoning deductively (e.g., analyzing probable causes from fault trees).
Exercising judgment (e.g., choosing between a job and graduate school)	Performing many complex operations at once (e.g., data integration for complex displays, such as in vessel tracking).

we might ultimately see a world where humans only needed for emergency use, compensating for situations the automation can not address. Such an approach to function allocation strategy is known as the *leftover approach*, where people perform the functions left after all that can be automated is automated.

As machines become more capable, human factors specialists have come to realize that functional allocation is more complicated than simply assigning each function to the component (human or machine) that is most capable in some absolute sense. There are other important factors, including whether the human would simply want to perform the function. Also, as discussed above, there is the notion that assignment is not "either-or", but can fall along the graded continuum of stages and levels (see Table 11.1. Most importantly, functions should be shared between the person and the automation so that the person is left with a coherent set of tasks that he or she can understand and respond to when the inherent flexibility of the person is needed. Several researchers have written guidelines for performing function allocation in a myriad of domains [695, 696].

Automation for ships provides a concrete example. An important function in maritime navigation involves tracking the position and velocity of surrounding ships using the radar signals. This "vessel tracking" function involves many complex operations to determine the relative velocity and location of the ships and to estimate their future locations. Fitts's original list of the relative benefits of humans and automation is shown in Table 11.2. In this list, automation is better suited for tasks that require repetition in precisely the same manner (drilling holes, stamping operations) and when there are many complex operations to be performed at once (inspection over large areas). In contrast, the course selection function involves considerable judgment regarding how to interpret the rules of the road. Because humans are better at exercising judgment, these types of tasks should be allocated to the human.

Applying the information in Table 11.2 to determine an appropriate allocation of function is a starting point rather than a simple procedure that can completely guide a design. One reason is that there are many *interconnections between functions*. In the maritime navigation example, the function of vessel tracking interacts with the function of course selection. Course selection involves substantial judgment, and may be best for the human. However, the mariner's ability to choose an appropriate course depends on the vessel-tracking function, which is performed by the automation. Although vessel tracking and course selection can be described as separate functions, the automation must be designed to support them as an integrated whole. In general, you should not fractionate functions between human and automation but strive to give the human a coherent job.

Autonomy in the World (Photo by J.D. Lee.)
A skeptical dog considers an "autonomous" system that has lost its way.

Theoretically, Roomba operates autonomously after being setup to clean the house. Every day it will vacuum at 10:00 and return to its base to recharge. Is there a role for a person? What could go wrong?

Rude interactions: Sensor limits and algorithm choices lead to rude collisions with furniture, pets, and people.

Entrapment in furniture: Rare combinations of furniture heights trap the Roomba.

Cord entanglement: Power cords trap Roomba and so require a person to organize the environment to facilitate autonomy.

Getting lost: On way back to charging station, Roomba loses its way and requires a person to help it home.

Clock reset: Getting lost can completely discharge the battery and erase settings, causing Roomba to start cleaning at odd hours.

Vacuuming dog poop. Despite being ill-equipped to vacuum some things, Roomba tries anyway, succeeding only in spreading the mess.

Devices produced by the makers of Roomba are on the battlefield. How autonomous are they? How autonomous should they be? What could go wrong?

The need to design coherent activities rather than just allocate functions points to the general trap designers can fall into: the substitution myth [657]. The substitution myth reflects the tendency of engineers to see functions as independent and humans and machine roles to be interchangeable. Rather than simply substituting an reliable machine for an unreliable person, rarely has the desired effect and often leads to unanticipated demands on the person and new ways the system can fail, as highlighted in the sidebar.

Any cookbook approach that uses comparisons like those in Table 11.2 will be only partially successful at best; however, Table 11.2 contains some general considerations that can improve design. Human memory tends to organize large amounts of related information in a network of associations that can support effective judgments requiring the consideration of many factors. People understand context and can see overall patterns, but are less effective with details. For these reasons it is important to leave the "big picture" to the human and the details to the automation [468].

11.5   Fifteen Principles of Human-Centered Automation

Perhaps the most important limit of the function allocation approach is that automation design is not an either/or allocation between the automation or the human. It is often more productive to think of how automation can support and complement the human in adapting to the demands of the system. Ideally, the automation design should focus on creating a human-automation partnership by incorporating the principles of *human-centered automation* [144]. Of course, human-centered automation might mean keeping the human more closely in touch with the process being automated; giving the human more authority over the automation; choosing a level of human involvement that leads to the best performance; or enhancing the worker's satisfaction with the workplace. In fact, these characteristics are important human factors considerations, even if they are not always completely compatible with each other. We present 15 human-centered automation principles that help harmonize the relationship between the human, system, and automation. As in the previous chapters, we divide these principles into categories: mental model (MM), attention (A), perception (P), interaction (I), response (R), organizational (O). Organizational considerations relate to the problems of automation undermining teamwork and job satisfaction, issues discussed in more detail in Chapter 18.

11.5.1 Mental model principles

1. **Define and communicate the purpose of automation (MM1):** We began this chapter describing reasons for implementing automation: remove people from hazardous situations, perform difficult or unpleasant tasks, or extend human capability. The specific purpose of the automation should be communicated to the user because automation designed for one of these purposes might not serve another purpose well. With vehicle automation, this might occur when designers create automation intended to reduce the demand of steering, but drivers see the purpose of the automation as allowing them to neglect steering and attend to other activities. Communicating the designers' purpose of the automation to the users can help calibrate trust and avoid misuse.

   Defining the purpose of automation ensure it satisfies a need, and avoids indifference towards the people it should support.

2. **Define and communicate the operating domain (MM2):** The *operating domain* is the range of situations the automation is designed for. All automation has limits and will fail in situations that exceed these limits: the operating domain defines these limits. The operating domain can be designed by structuring the environment can make the automation simpler and more reliable. For example, industrial robots have the potential to injure people, and placing these robots in structured environment that limits human contact ensures safety. This strategy only works if the boundaries of the operating domain are clearly defined are communicated to people. This communication can take the form of locking out functions or signaling limits on displays. Stating these limits in manuals is typically ineffective. More effective is to design the operating domain in a in a way that matches

   For any safety-critical system, failing to define the operating domain can have severe consequences.

people's current mental models. As an example, the operating domain of a partially self-driving could be defined as only freeways. This corresponds to how people already think about limited access highways relative to other roads.

3. **Design the role of the person and automation (MM3):** Blind application of the Fitts's list can leave the person with a set of unrelated tasks that lack connection, which can create unexpected multitasking, monitoring, and prospective memory burdens. Simply allocating tasks the automation cannot perform to the person invites human error and frustration. Designing roles helps solve this problem. Role is defined by a purpose, responsibilities, authority, and activities . Well-designed roles align authority with responsibility. This alignment is violated when a high level of automation leaves the person responsible for control, but gives the person little authority for accessing information or for guiding the automation [697].

4. **Simplify the mode structure (MM4):** The problems of mode confusion often stem from overly complex networks of modules of automation[698]. Training or supporting people with a well-designed overview display are not the most effective ways to reduce mode confusion. A better solution is to simplify the number of modes and possible transitions between modes, recognizing that each additional mode introduces potential for additional confusion[699]. Many new cars have adaptive cruise control and conventional cruise control, and each has several modes. Because adaptive cruise control brakes for a slowing vehicle ahead and cruise control does not, confusing the modes could be dangerous. Rather than creating a display to communicate this distinction, it might be possible to re-think the automation design and simplify it by reducing the number of modes [700].

5. **Make trustable and polite (MM5):** Automation should be considerate of the people it works with. In part, this means it should be polite and follow etiquette of social norms that smooth interactions between people [701]. As an example, automation should avoid interrupting people unless justified by the importance of the interrupting activity. As we saw in Chapter 7, interruptions undermine safety and performance, but also satisfaction. Similarly this means the automation should avoid contributing to workload peaks by helping people plan their interactions with the automation [702]. Avoiding interruption requires that the automation either has subtle signals that can be ignored or that it has a sophisticated model of the person's activities, role, and conventions.

Because people tend to respond to automation in social fashion and attribute intentions to automation behavior [615], the behavior of automation might be interpreted as rude, aggressive, or uncertain. This behavior should be designed

Like putting makeup on a corpse, there is a limit to what interface can do to fix overly complex automation.

Design for appropriate trust: Too much trust causes overreliance and too little trust leads to poor acceptance.

to communicate the intended feeling. For example, tuning algorithms can make automation trustable by conveying confidence when the automation is well within its operating domain, and conveying uncertainty as it approaches the edge of its operating domain.

11.5.2 Attention principles

6. **Signal inability to satisfy role (A6):** Automation fails for a variety of reasons, such as violating the boundaries of the operating domain, software bugs, and sensor failures. Because people are very poorly equipped to monitor and detect these rare events, an important responsibility of the automation is to detect failures and alert the person [702, 703]. This alert must be delivered in advance of the failure so that the person has time to understand the situation and compensate for the automation. Automation should actively direct people's attention to situations that might undermine the role of the automation [144]. Ironically, the algorithms to detect failure in the automation may be substantially more challenging to create than the automation itself.

11.5.3 Perception principles

7. **Transparency—Keep the person informed (P7):** However much authority automation assumes in a task, it is important for the operator to be informed of what the automation is doing and why, using good displays. Bisantz and Seeong [662] refer to this as automation "transparency". People should have the "big picture," knowing what the automation is doing now, why it is doing it, and what it will do next [144]. A well-designed display provides a window into the automation. As a positive example, the pilot should be able to see the amount of thrust delivered by an engine as well as the amount of compensation that the autopilot might have to make to keep the plane flying straight, and how this compensation will change in the future as the thrust is reduced.

Transparency can enhance performance, particularly for imperfect automation that requires a person to compensate for its limits [190, 704]. Transparency typically means more information in the display and more clutter, but that does not need to be the case if properly designed [705]. Of course, merely presenting information is not sufficient to guarantee that it will be understood. Coherent and integrated displays (Chapter 8) are necessary for the information to understood. Being able to see the state of the automation is often not enough, the person should also be able to adjust, manipulate and direct in order to understand its behavior, addressed in principle I11 below.

For greatest benefit, transparency should be paired with design elements that make the automation directable.

The importance of transparency depends on the nature of automation. It is most critical for automation that has a broad span of control. Automation with a broad span of control takes input from people and acts on this input over many minutes and many elements of the system to produce a response. In this situation, it is not easy for a person to simply observe the state of the system to know whether the automation is acting properly. Hence, transparency is particularly critical for automation with a broad span of control.

11.5.4 Response selection principles

8. **Avoid accidental activation and deactivation (R8):** Accidental activation and deactivation occurs when people accidentally turn on or off, or otherwise change the setting of the automation. This might occur by bumping a button. Chapter 9 describes a range of strategies to avoid these accidents, such as requiring the button be pressed and held rather than simply pressed. In general, techniques that are most effective at preventing accidental activation have the negative consequence of slowing the person and adding effort. This additional effort needs to be weighted against the consequence of changing the state of automation by accident. High-consequence situations should be paired with those mechanisms that are most effective even if they add effort. Importantly, the consequences for inadvertent activation and deactivation could be very different and require very different protective mechanisms.

11.5.5 Interaction principles

9. **Keep the person in the loop (I9):** Earlier in this chapter we discussed two tradeoffs. First, automation at later stages and higher levels can be more efficient and support better performance when it works well, but may be more problematic when it fails. Sebok and Wickens [641] refer to this as the "lumberjack analogy": the higher the tree, the harder it falls. Second, more automation (later stages and higher levels) can decrease workload, but often at the cost of reduced situation awareness. These tradeoffs mean that performance for routine situations increases with increasing degrees of automation, but, performance for unusual situations, such as when the automation fails, can be catastrophically poor [647].

How then do we do we keep the human operator sufficiently "in the loop", so that situation awareness does not degrade leading to challenges in managing automation failures, without defeating the very purpose of automation in the first place? Is there a "sweet spot" in the tradeoff function that can preserve the best qualities of high and low degrees of automation?

There is indeed some evidence that there is a "sweet spot". Retain relatively high levels of automation at earlier stages of attention, perception and information integration assistance [676, 706], but resist the temptation to invoke high levels of decision automation [707]. That is, require the human to actively choose the action, even if the automation makes a recommendation, or preferably a small set of preferred actions. That is, only levels 2 or 3 in Table 11.1. Forcing the operators to choose should also encourage them to consider the situation underlying that choice, and hence preserve better awareness of the state of affairs should things go wrong; and of course this choice will require the person to practice the decision making skills, that might otherwise degrade with higher levels of automation. For higher levels of automation, this might also mean designing tasks to engage the operator in a meaningful manner [702].

When choosing lower levels of automation for high-risk decision aiding, it is important to realize the tempering effect of time pressure and the demands for multi-tasking. There is no doubt that if a decision must be made in a time-critical situation, later stages of automation (choice recommendation or execution) can usually be done faster by automation than by human operators. Hence the need for time-critical responses may temper the desirability for low levels of stage 3 automation. This is particularly true when the person has other tasks to attend to. In this situation, it may be more important to make it easy for the operator to re-enter the loop rather than always keeping the operator in the loop.

> Sometimes the purpose of the automation is to enable people shift attention to other tasks and be "out of the loop."

10. **Support smooth re-entry into the loop (I10):** The purpose of some automation is to enable people to be "out of the loop" for one activity to do other activities. In this situation, design should focus on attracting the person's attention at the appropriate time and coordinating a smooth and timely—bumpless—transfer of control [708]. This coordination requires that the automation determine when the person has begun actively controlling the system and the automation must also signal the person that he or she is now in control. A critical design tradeoff involves balancing smooth re-entry with the extra effort imposed by features designed to prevent accidental de-activation of the automation (principle R8).

11. **Make automation directable (I11):** Often the behavior of automation is not perfectly matched to the current activity or the person's preferences. Directable automation enables people to shift the behavior of the automation without disengaging it. With adaptive cruise control, the driver can set the car following distance to comply with preference and traffic conditions [702]. Directable automation avoids unnecessary reversion to manual control, but it can also enhance trust

> Giving people control–preserving agency–can be critical for acceptance and success, even if the automation can function without the person's input.

and acceptance by giving people a sense of control. A more general strategy is to allow people to create policies to guide automation. Policies allow people to define the boundary of the operating domain to reflect the current assessment of the automation's capacity in the current situation [703].

12. **Make the automation flexible and adaptable (I12):** Making automation directable concerns adjusting its behavior within a level, and making automation adaptable concerns choosing between levels of automation. The amount of automation needed for any task is likely to vary from person to person and within a person to vary over time. Hence, flexible or *adaptable automation* is where person can change the level of automation over time, which is often preferable over one that is fixed and rigid. Flexible automation simply means that different levels are possible. One driver may choose to use cruise control, the other may not. The importance of flexible automation parallels the flexible and adaptive decision-making process of experts. As discussed in Chapter 7, decision aids that support that flexibility tend to succeed, and those that do not tend to fail. This is particularly true in situations that are not completely predictable. Flexibility seems to be a useful way to enhance safety, performance, and satisfaction. However, adaptable automation adds modes and the benefits of flexibility need to be tempered by the need to simplify the automation (Principle MM4).

13. **Consider adaptive automation (I13):** *Adaptive automation*, goes one step further than flexible automation by automatically changing the level of the automation based on characteristics of the environment, user, and task [709, 710, 711, 712, 713]. With adaptive automation, the level of automation increases as either the workload imposed on the operator increases or the operator's capacity decreases (e.g., because of fatigue). For example, when psychophysiological (e.g., heart rate) measures indicate a high workload, the degree of automation can be increased [714]. An essential element of adaptive automation concerns how to monitor the person to detect and prevent errors[144]. Adaptive automation is becoming more prevalent because powerful machine learning techniques make it possible to detect the state of the person [715, 716], such as Microsoft's openly available system for identifying the emotional state of a person from a video stream.

While such systems have proven effective [709, 712] in environments such as aircraft flight decks in which workload can vary over time, they should be implemented only with great caution because of several potential pitfalls. First, because such systems are adaptive closed-loop systems, they may fall prey to problems of negative feedback, closed-loop instability, as discussed in Chapter 9. Second, humans do not always easily deal with rapidly changing system configurations. Remember that consistency is an important feature in design (Chapter 8). Finally, as Rouse [709] has noted, computers may be good at taking control (e.g., on the basis of measuring degraded performance by the human in the loop), but are not always good at giving back control to the human.

11.5.6 Organizational Principles

14. **Keep people trained (O14):** Training for automation should include: training of automation management, intervention and failure recovery, and manual control. Automation can make complex tasks seem simple when manual interactions are automated. At the same time, automation often changes the task so that operators must perform more abstract reasoning and judgment in addition to understanding the limits and capabilities of the automation [688]. These demands of automation management strongly argue that training is needed so that the operator uses the automation appropriately and benefits from its potential [690]. The operator should have substantial training in exploring the automation's various functions and features in an interactive fashion, rather than simply reading a description in a manual [717].

If the automated system was to fail and require human intervention, it is essential for the human to develop skills to carry out the intervention and subsequent control. This is particularly important because people often need to intervene during extreme situations where they have the least experience but the control requirements are most demanding. For partially automated functions where manual control is needed in a more routine situation, training may be needed to minimize problems that can occur due to skill loss.

Because training is typically part of broader organizational design considerations, such as selection and team configuration, we discuss training in more detail in Chapter 17.

15. **Consider organizational consequences (O15):** A worker's acceptance and appreciation of automation can be greatly influenced by the management philosophy [718]. If the workers view the automation as being "imposed" because it can do the job better or cheaper than they can, the human workers' attitude toward the automation will probably be poor [719]. However, if automation is introduced as an aid to improve human-system performance and a philosophy can be imparted in which the human remains the master and automation the servant, then the attitude will likely remain more accepting and cooperative [144]. This can be accompanied by good training of what the automation does and how it does its task. Under such circumstances, a more favorable attitude can also lead to better understanding of automation, better appreciation of its strengths, and more effective utilization of its features. Indeed, studies of the introduction

of automation into organizations show that *management is often responsible for making automation successful* [720]. Providing prior automation exposure is also key for altering user acceptance of automation [721].

More broadly, the introduction of automation often has substantial effects on people not directly considered in its design. As an example, Roomba and other household robots, affect pets and small children even though these devices were certainly not designed to serve them. Although the unintended consequences of automation can be difficult to anticipate, design should broadly consider the operating environment and the people who might be affected even if they are not the primary beneficiaries or users of the technology. This is an example of systems thinking discussed in Chapter 2.

11.5.7 Summary of principles

As with design principles offered in previous chapters, the conflict and the art of design reconciles these conflicts. Part of this reconciliation involves understanding what principles are most applicable the particular automation being design. Because automation design involves elements of displays, controls, and interaction design, it is clear that the principles discussed here need to be considered with principles discussed in the previous three chapters.

11.6 Increasingly Autonomous Technology

Automation plays a particularly critical role *process control*— situations when a small number of operators must control and supervise a very complex set of remote processes, whose remoteness, complexity, or high level of hazard prevents much "hands on" control, of the sort described in Chapter 9. Automation here is not optional, it is a necessity [468, 588]. Examples of such systems include the production of continuous quantities, such as energy, in chemical process control, the production of discrete quantities, in the area of manufacturing control [722, 723], and the control of remotely operated vehicles and robots, in the area of robotics control. In these cases, the human supervisor/controller is challenged by some or all of several factors with major human factors implications: the remoteness of the entity controlled from the operator, the complexity (multiple-interacting elements) of the system, the sluggishness of the system, following operator inputs, and the high level of risk involved, should there be a system failure. More detailed treatments of the human factors of these systems are available elsewhere [724, 468]. In this section, we only highlight a few key trends that are human factors relevance.

For process control, such as involved in the manufacturing of chemicals, energy production, or other continuous commodities, the systems are so complex that high levels of automation must be implemented. Thus the key human factors question is how to support the supervisor in times of failures and fault management,

Automation Design Principles

Mental model principles
MM1. Define and communicate the purpose of automation
MM2. Define and communicate the operating domain
MM3. Design the role of the person and automation
MM4. Simplify the mode structure
MM5. Make trustable and polite

Attention principles
A6. Signal inability to satisfy role

Perception principles
P7. Transparency–keep the person informed

Response selection principles
R8 Avoid accidental activation and deactivation

Interaction principles
I9. Keep the person in the loop
I10. Support smooth re-entry into the loop
I11. Make automation directable
I12. Make automation flexible and adaptable
I13. Consider adaptive automation

Organizational principles
O14. Keep people trained
O15. Consider organizational consequences

Table 11.3 Automation design principles.

so that disasters such as Three Mile Island [391] and Chernobyl [725] do not occur because of poor diagnosis and decision making. Tools for such support were suggested in Chapter 7, where the importance of decision support for knowledge-based behavior was emphasized, and in Chapter 8, where the concepts of predictor displays and of ecological interfaces were introduced. Such interfaces have two important features: (1) they are highly graphical, often using configural displays to represent the constraints on the system, in ways that these constraints can be easily perceived, without requiring heavy cognitive effort. (2) they allow the supervisor to think flexibly at different levels of abstraction [726, 513], ranging from physical concerns like a broken pipe or pump, to abstract concerns, like the loss of energy, or the balance between safety and productivity. In some regards, many aspects of air traffic control mimic those of process control [727].

Automation in the form *robotics in manufacturing* is desirable because of the repetitious, fatiguing, and often hazardous mechanical operations involved, and is sometimes a necessity because of the heavy forces often required. Here a critical emerging issue is that of agile manufacturing, in which manufacturers can respond quickly to the need for high-quality customized products [728]. In this situation, decision authority is often transferred from the traditional role of management to that of operators empowered to make important decisions. In this situation, automation needs to support an integrated process of design, planning and manufacturing, and integrate information so that employees can make decisions that consider a broad range of process considerations. (See Chapter 7 for a description of how decision aids might support this process.)

A second use of robots is in unmanned air and ground vehicles, such as the *unmanned air vehicles (UAVs)* that provide surveillance for military operations, or ground vehicles that can operate in cleaning up hazardous waste sites. Here a major challenge is the control-display relationships with remote operators. How can the resolution of a visual display, necessary to understand a complex environment, be provided with a short enough delay, so that control can be continuous and relatively stable [729]. If this is not possible, because of bandwidth limitations on the remote communications channels, a greater degree of autonomy is needed. An important point is are *increasingly autonomous* as opposed to *autonomous* systems. Increasing autonomy means that the person is not removed from control, but simply distanced from moment control. UAVs and similar examples of highly automated systems are not autonomous, but simply involve human control at a greater distance and level of abstraction.

Beyond automation associated with process control and robotics, automation in the form of *machine learning and artificial intelligence* has a much more personal influence on most of our lives. Based on large volumes of data algorithms can produce medical diagnoses, guide parole decisions, trade stocks, decide who to fire, and guide our understanding of the world [640, 39, 365]. Contrary to automation associated with process control robotics, the purpose of the automation is not always clear. With such automation, the goals of the developers are not always aligned with the people using the services. With social networking sites that provide news stories, the goal might not be to provide useful information, but instead maximize the number of page views and associated advertising revenue. This could prioritize news stories without regard for their link to reality. This would seem to be a clear violation of Dieter Rams suggestion discussed in Chapter 2: "Indifference towards people and the reality in which they live is actually the one and only cardinal sin in design."[39]

Automation that aggregates personal data to make decisions to hire, fire or even people would see to offer the appeal of greater precision and objectivity than [644]. Unfortunately, these systems often cover the biases of designers and data sources with the appearance of objectivity. More troubling, is that these algorithms are often proprietary and the companies that design them have a clear motivation to avoid transparency. With algorithms that trading of stocks and other securities, the timescale that the operate at (nanoseconds) and complex interactions with the market make it very difficult to protect against potentially damaging market fluctuations. One such flash crash occurred on May 6 2010, when the US stock market dropped more than 5%. Most agree that high-frequency trading algorithms contributed to the event, but the precise cause is still not understood. The broad societal consequences of such systems confronts designers with similar ethical challenges as those associated with choice architecture in Chapter 7.

11.7 Summary

Automation has greatly improved safety, comfort, and satisfaction in many applications; however, it has also led to many problems. Careful design that considers the role of the person can help avoid these problems. In this chapter, we described automation classes and levels and used them to show how function allocation and human-centered approaches can improve human-automation performance. In many situations automation supports human decision making, and Chapter 7 discusses these issues in more detail. Although the domains of process control, manufacturing, and hortatory control already depend on automation, the challenges of creating useful automation will become more important in other domains as automation becomes more capable—entering our homes, cars, and even our bodies.

Automation is sometimes introduced to replace the human and avoid the difficulties of human-centered design; however, as systems become more automated, the need for careful consideration of human factors in design becomes more important, not less. In particular, requirements for decision support (Chapter 7), good displays (Chapter 8), and training (Chapter 17) become more critical as automation becomes more common.

Automation does not remove the human role, it only distances people from the controlled process.

Additional resources

Readable accounts of automation challenges:

1. Degani, A. (2004). *Taming HAL: Designing interfaces beyond 2001.* New York: Springer.

2. Lewis, M. (2014). *Flash Boys: A Wall Street revolt.* New York: Norton.

3. O'Neil. (2016). *Weapons of Math Destruction: How big data increases inequality and threatens democracy.* New York: Crown.

Automation handbooks: Two handbooks that cover the concepts of this chapter in more detail include:

1. Nof, S. Y. (Ed.). (2009). *Springer Handbook of Automation.* Berlin: Springer.

2. Lee, J. D., & Kirlik, A. (2013). *The Oxford Handbook of Cognitive Engineering.* New York: Oxford University Press.

Questions

Questions for 11.1 Why Automate?

P11.1   Name one good reason to automate and a bad one.

P11.2   What are the reasons for automating driving and creating a driverless car.

P11.3   What is a consequence of automating a system simply because it is feasible.

Questions for 11.2 Types of Automation and Types of Tasks

P11.4   What are the four stages that are one way to define types of automation?

P11.5   What are the levels of automation?

P11.6   What combination of levels and stages of automation that is vulnerable to people neglecting imperfect automation?

P11.7   What combination of levels and stages of automation that is vulnerable to people neglecting imperfect automation?

P11.8   Beyond the levels and stages of automation, describe two other ways of describing types of automation.

P11.9   Give and example of automation with a low span of control and an example of automation with a high span of control.

P11.10   Why is the number of interacting elements an important way of describing types of automation?

P11.11   Why does the time span of tasks matter in describing how well people might manage various types of automation?

P11.12   How might task consequence guide the choice of levels and stages in automation design?

Questions for 11.3 Problems with Automation

P11.13   Why is the distinction between actual and perceived reliability important?

P11.14   Why does automation not need to be perfect to be useful?

P11.15   Why is too little trust (distrust) in automation?

P11.16   What is the cry wolf effect?

P11.17   What does it mean to be out of the loop when interacting with automation?

P11.18   Explain how being out of the loop contributes to skill loss how skill loss contributes to an operator's reluctance to intervene.

P11.19   what is meant by clumsy automation and what contributes to it?

P11.20   Give an example of a mode on your computer and how it might lead to confusion and errors.

P11.21 Describe how automation might undermine human cooperation in a manufacturing environment.

P11.22 What are two ways automation can undermine cooperation between people?

P11.23 Why might training be more important rather than less important when automation is introduced?

Questions for 11.4 Allocating Functions between People and Automation

P11.24 What is the left over approach to function allocation, any why might it cause problems for the people managing the automation?

P11.25 What is the substitution myth and why is important to keep in mind when allocating functions to people and automation?

P11.26 What general statement can be made about how the roles of people and automation given their general strengths and limits.

Questions for 11.5 Fifteen Principles of Human-Centered Automation

P11.27 What are some important dimensions in the design of an operating domain for a n automatic lawn watering system?

P11.28 Which automation principles would be most useful in guiding the design of a Roomba and why?

P11.29 Why is a good interface useful, but insufficient for some cases of complex automation?

P11.30 Why is it important, but difficult for automation to signal to the person when it is not able to satisfy its role.

P11.31 How might you make the Roomba more polite, particularly considering the algorithms that control its motion?

P11.32 What is the difference between adaptable and adaptive automation? Given this difference which is more likely to confuse people and which is more likely to increase workload?

P11.33 Training to prevent skill loss is one of three types of training that automation demands. What are the other two?

P11.34 What are the organizational consequences that might go addressed in the design of vehicle automation.

Questions for 11.6 Increasingly Autonomous Technology

P11.35 What ethical issues arise with increasingly autonomous systems, particularly those systems responsible for support parole and hiring decisions?

P11.36 Why is it inaccurate to say that any unmanned aerial vehicle (UAV) or other increasingly autonomous system is actually autonomous?

P11.37 Identify an important issue and design principle associated with each of the application areas at the end of the chapter.

Bibliography

[1] F. Taylor. *The Principles of Scientific Management*. Harper and Brothers, 1911.

[2] P. M. Fitts and R. E. Jones. Analysis of factors contributing to 460 âĂŻpilot errorâĂŻ experiences in operating aircraft controls (Report No. TSEAA-694-12). Technical report, Aero Medical Laboratory, Air Materiel Command, US Air Force., Dayton, OH:, 1947.

[3] J. McNish and S. Silcoff. The Inside Story of How the iPhone Crippled BlackBerry, 2015.

[4] J. Smetzer, C. Baker, F. D. Byrne, and M. R. Cohen. Shaping systems for better behavioral choices: Lessons learned from a fatal medication error. *Joint Commission journal on quality and patient safety*, 36(4):152–163, 2010.

[5] K. J. Vicente. *The Human Factor*. Routeledge, New York, 2004.

[6] H. W. Hendrick. Good ergonomics is good economics. *Proceedings of the Human and Ergonomics Society 40th Annual Meeting*, pages 1–15, 1996.

[7] H. R. Booher. *MANPRINT: An Approach to Systems Integration*. Van Nostrand Reinhold, New York, 1990.

[8] H. W. Hendrick. Determining the cost-benefits of ergonomics projects and factors that lead to their success. *Applied Ergonomics*, 34(5):419–427, 2003.

[9] A. Lange. Lillian Gilbreth's Kitchen Practical: How it reinvented the modern kitchen., 2012.

[10] C. D. Wickens, J. G. Hollands, S. Banbury, and R. Parasuraman. *Engineering Psychology and Human Performance*. Routledge, Taylor and Francis Group, New York, fourth edition, 2013.

[11] J. Rasmussen, A M Pejtersen, and L P Goodstein. *Cognitive Systems Engineering*. John Wiley, New York, 1994.

[12] J. D. Lee and A. Kirlik. *The Oxford Handbook of Cognitive Engineering*. Oxford University Press, New York, 2013.

[13] D. Meister. Conceptual Aspects of Human Factors. *John Hopkins University Press*, Baltimore,, 1989.

[14] M. S. Sanders and E. J. McCormick. *Human Factors in Engineering and Design*. McGraw-Hill, New York, seventh edition, 1993.

[15] R. W. Proctor and T. Van Zandt. *Human Factors in Simple and Complex Systems*. Taylor and Francis, Boca Raton, FL, second edition, 2008.

[16] D. A. Norman. *The Design of Everyday Things: Revised and Expanded Edition*. Basic Books, second edition, 2013.

[17] J. R. Wilson and S. Sharples. *Evaluation of Human Work*. Taylor and Francis, Boca Raton, FL, fourth edition, 2015.

[18] D. B. Chaffin, G. B. Andersson, J., and B. J. Martin. *Occupational Biomechanics*. John Wiley and Sons, New York, fourth edition, 2006.

[19] G. Salvendy. *Handbook of Human Factors and Ergonomics*. John Wiley and Sons, New York, 2013.

[20] D. A. Boehm-Davis, F. T. Durso, and J. D. Lee. *APA Handbook of Human System Integration*. APA press, 2015.

[21] A. Cooper. *The Inmates Are Running the Asylum: Why High Tech Products Drive Us Crazy and How to Restore the Sanity*. SAMS, Indianapolis, IN, 1999.

[22] H. R. Hartson and P. S. Pyla. *The UX Book: Process and guidelines for ensuring a quality user experience.* Morgan Kaufmann, Waltham, MA, 2012.

[23] N. R. Tague. *The Quality Toolbox.* ASQ Quality Press, 2004, second edition, 2004.

[24] L. Rising and N. S. Janoff. The Scrum software development process for small teams. *IEEE Software*, pages 26–32, 2000.

[25] D. A. Norman and S W Draper. *User Centered System Design: New Perspectives on Human-Computer Interaction.* Lawrence Erlbaum, Hillsdale, New Jersey, 1986.

[26] ISO. ISO 9241-210 Ergonomics of human-system interaction–Part 210: Human-centred design for interactive systems. Technical report, ISO, 2010.

[27] K. E. Fletcher, S. Saint, and R. S. Mangrulkar. Balancing continuity of care with residents' limited work hours: Defining the implications. *Academic medicine : journal of the Association of American Medical Colleges*, 80(1):39–43, 2005.

[28] D. D. Woods, E. S. Patterson, J. Corban, and J. Watts. Bridging the gap between user-centered intentions and actual design practice. In *Proceedings of the Human Factors and Ergonomics Society 40th Annual Meeting*, volume 2, pages 967–971, Santa Monica, CA, 1996. Human Factors and Ergonomics Society.

[29] T. K. Landauer. *The Trouble with Computers: Usefulness, usability, and productivity.* MIT Press, Cambridge, MA, 1995.

[30] J. Nielsen. Iterative user-interface design. *Computer*, 26(11):32–41, 1993.

[31] S. Lanoue. IDEO's 6 Step Human-Centered Design Process: How to Make Things People Want | UserTesting Blog.

[32] H. Beyer and K. Holtzblatt. Contextual design. *Interactions*, 6(1):32–42, 1999.

[33] B. W. Crandall and R. R. Hoffman. Cognitive task analysis. In J. D. Lee and A. Kirlik, editors, *The Oxford Handbook of Cognitive Engineering*, pages 229–239. Oxford University Press, New York, 2013.

[34] N. A. Stanton, M. S. Young, and C. Harvey. *Guide to Methodology in Ergonomics: Designing for human use.* CRC Press, Boca Raton, FL, 2014.

[35] J. Flanagan. The critical incident technique. *Psychological Bulletin*, 51(4):327–359, 1954.

[36] D. W. Stewart and P. N. Shamdasani. *Focus Groups: Theory and Practice.* Sage, Thousand Oaks, CA, 2015.

[37] S. Caplan. Using focus group methodology for ergonomic design. *Ergonomics*, 33(5):527–533, may 1990.

[38] T. Both. The Wallet Project, 2016.

[39] S. Lovell and K. Kemp. *Dieter Rams: As Little Design as Possible.* Phaidon, 2011.

[40] B. Tognazzini. First Principles of Interaction Design (Revised and Expanded), 2014.

[41] J. Nielsen. 10 Heuristics for User Interface Design, 1995.

[42] P. McAlindon, K. Stanney, and N. C. Silver. A comparative analysis of typing errors between the Keybowl and the QWERTY keyboard. In *Proceedings of the Human Factors and Ergonomics Society Annual Meeting'*, pages 635–639. Sage, 1995.

[43] J. R. Hauser and D. Clausing. The house of quality. *Harvard Business Review*, pages 63–73, 1988.

[44] S. G. Bailey. Iterative methodology and designer training in human-computer interface design. In *In Proceedings of the INTERACT'93 and CHI'93 Conference on Human factors in Computing Systems*, pages 198–205. ACM, 1993.

[45] J. Nielsen. How Many Test Users in a Usability Study?, 2012.

[46] B. Kirwan and L. K. Ainsworth. *A Guide to Task Analysis.* Taylor & Francis, Washington, D.C., 1992.

[47] B. Hanington and B. Martin. *Universal Methods of Design: 100 Ways to Research Complex Problems, Develop Innovative Ideas, and Design Effective Solutions.* Rockport, Beverly, MA, 2012.

[48] K. R. Boff and J. E. Lincoln. Guidelines for alerting signals. *Engineering Data Compendium: Human Perception and Performance*, 3:2388–2389, 1988.

[49] HFAC. Human Engineering Department of Defense Design Criteria Standard: MIL-STD-1472G. Technical report, Department of Defense, 2012.

[50] P. Reed, P. Billingsley, E. Williams, A. Lund, E. Bergman, and D. Gardner-Bonneau. Software ergonomics comes of age: The ANSI/HFES-200 Standard. *Proceedings of the Human Factors and Ergonomics Society Annual Meeting*, 40,:323–327, 1996.

[51] H. P. Van Cott and R. G. Kinkade. Human Engineering Guide to Equipment Design (revised edition). *McGraw-Hill*, Inc.:New York, 1972.

[52] D. L. Strayer and W. A. Johnston. Driven to distraction: Dual-task studies of simulated driving and conversing on a cellular telephone. *Psychological Science*, 12(6):462–466, 2001.

[53] R. Rosenthal. *Meta-analytic Procedures for Social Research*. Sage, Newbury Park, CA, 1991.

[54] H. Cooper, L. V. Hedges, and J. C. Valentine. *The Handbook of Research Synthesis and Meta-Analysis*. Russell Sage Foundation, New York, 2009.

[55] J. Nielsen. *Usability Engineering*. Academic Press, San Francisco, CA, 1993.

[56] J. Nielson. Heuristic evaluation. In *Usability Inspection Methods*, pages 25–64. Wiley, New York, 1994.

[57] M. H. Blackmon, M. Kitajima, and P. G. Polson. Repairing usability problems identified by the cognitive walkthrough for the web. *Proceedings of the conference on Human factors in computing systems - CHI '03*, 5:497–504, 2003.

[58] L. M. Holson. Putting a bolder face on Google, 2009.

[59] G. Keppel and T. D. Wickens. *Design and Analysis: A Researcher's Handbook*. Engelwood Cliffs, fourth edition, 2004.

[60] J. W. Creswell and V. Clark. *Designing and Conducting Mixed Methods*. Sage, second edition, 2011.

[61] Z. Solomon, M. Mikulincer, and S. E. Hobfoll. Objective versus subjective measurement of stress and social support: Combat-related reactions. *Journal of Consulting and Clinical Psychology*, 55(4):577–583, 1987.

[62] A. D. Andre and C. D. Wickens. When users want what's not best for them. *Ergonomics in Design*, October:10–14, 1995.

[63] G. Cumming. The new statistics: Why and how. *Psychological Science*, 25(1):7–29, nov 2014.

[64] G. Cumming. Replication and p intervals: p values predict the future only vaguely, but confidence intervals do much better. *Perspectives on Psychological Science*, 3(4):286–300, jul 2008.

[65] T. Eissenberg, S. Panicker, S. Berenbaum, N. Epley, M. Fendrich, R. Kelso, L. Penner, and M. Simmerling. IRBs and Psychological Science: Ensuring a collaborative relationship. *http://www.apa.org/research/responsible/irhs-psych-science.aspx IRBs*, pages 1–10, 2004.

[66] J. G. Adair. The Hawthorne Effect: A reconsideration of the methodological artifact. *Journal of Applied Psychology*, 69(2):334–345, 1984.

[67] J. Johnson. *Designing with the mind in mind: SImple guide to understanding user interface design rules*. Morgan Kaufmann, Burlington, MA, 2010.

[68] S. C. Seow. *Designing and engineering time: the psychology of time perception in software*. Addison-Wesley Professional, 2008.

[69] B. Shneiderman, C. Plaisant, M. S. Cohen, S. Jacobs, N. Elmqvist, and N. Diakopoulos. *Designing the User Interface: Strategies for effective human-computer interaction*. Pearson, New York, sixth edit edition, 2016.

[70] K. A. Ericsson, M. J. Prietula, and E. T. Cokely. The making of an expert. *Harvard Business Review*, 85(7/8):114–121, 2007.

[71] P. Lally, Van Jaarsveld, H.W. W. Potts, and J. Wardle. How habits formed: Modelling habit formation in the real world, 2010.

[72] P. McCauley, L. V. Kalachev, D. J. Mollicone, S. Banks, D. F. Dinges, and H. P. A. Van Dongen. Dynamic circadian modulation in a biomathematical model for the effects of sleep and sleep loss on waking neurobehavioral performance. *Sleep*, 36(12):1987–97, dec 2013.

[73] J. S. Warm and R. Parasuraman. Vigilance requires hard mental work and is stressful. *Human Factors*, 50(3):433–441, 2016.

[74] Statistic Brain. Attention Span Statistics, 2016.

[75] D. Bordwell. *The way Hollywood tells it: Story and style in modern movies*. University of California Press, 2006.

[76] N. Cowan. Evolving conceptions of memory storage, selective attention, and their mutual constraints within the human-informaiton processing system, 1988.

[77] A. Newell and S. K. Card. The prospects for psychological science in human-computer interaction. *Human Computer Interaction*, 1:209–242, 1985.

[78] R. B. Miller. Response time in man-computer conversational transactions. *Fall Joint Computer Conference*, pages 267–277, 1968.

[79] F. F.-H. Nah. A study on tolerable waiting time: how long are Web users willing to wait? *Behaviour & Information Technology*, 23(3):153–163, 2004.

[80] R. Flesch. A new readability yardstick. *Journal of Applied Psychology*, 32(3):221–233, 1948.

[81] M. Heldner and J. Edlund. Pauses, gaps and overlaps in conversations. *Journal of Phonetics*, 38(4):555–568, 2010.

[82] M. Green. "How long does it take to stop?" Methodological analysis of driver perception-brake times. *Transportation Human Factors*, 2(3):195–216, sep 2000.

[83] Jeffrey W Muttart. Development and evaluation of driver response time predictors based upon meta analysis. *Society of Automotive Engineering*, 2003.

[84] J. Nielsen. Powers of 10: Time Scales in User Experience, 2009.

[85] G. Wyszecki and W. S. Stiles. *Color Science*. Wiley, New York, 1982.

[86] H. Widdel and D. L. Post, editors. *Color in Electronic Displays*. Springer Science and Business Media, New York, 2013.

[87] P. R. Boyce. *The Human Factors of Lighting*. CRC Press, Boca Raton, FL, third edition, 2014.

[88] J. Theeuwes, J. Alferdinck, and M. Perel. Relation between glare and driving performance. *Human Factors*, 44(1):95–107, 2002.

[89] W. K. E. Osterhaus. Office lighting: a review of 80 years of standards and recommendations. In *Proceedings of the 1993 IEEE Industry Applications Society Annual Meeting*, pages 2365–2374, 1993.

[90] D. DiLaura, K. W. Houser, R. G. Misrtrick, and R. G. Steffy. *The Lighting Handbook 10th Edition: Reference and Application*. Illuminating Engineering Society of North America 120, 2011.

[91] A. A. Kruithof. Tubular luminescence lamps for general illumination. *Philips Technical Review*, 6(3):65–96, 1941.

[92] R. P. O'Shea. Thumb's rule tested: Visual angle of thumb's width is about two deg. *Perception*, 20:415–418, 1991.

[93] C. Ware. *Information Visualization: Perception for design*. Elsivier, Boston, MA, 2013.

[94] G. E. Legge and C. A. Bigelow. Does print size matter for reading? A review of findings from vision science and typography. *Journal of Vision*, 11(5):1–22, 2011.

[95] J. L. Campbell, C. M. Richard, J. L Brown, and M. McCallum. Crash Warning System Interfaces: Human Factors insights and lessons learned. Technical report, NHTSA, Washington D.C., 2007.

[96] A. Degani. On the Typography of Flight-Deck Documentation. Technical report, NASA, Moffet Field, CA, 1992.

[97] FAA Human Factors Research and Engineering Group. Baseline Requirements for Color Use in Air Traffic Control Displays (DOT/FAA/HF-STD-002). Technical report, Federal Aviation Administration, Washington, DC, 2007.

[98] D. E. Broadbent and M. H. Broadbent. Priming and the passive/active model of word recognition. *Attention and Performance*, VIII:419–433., 1980.

[99] M. Perea and E. Rosa. Does "whole-word shape" play a role in visual word recognition? *Perception & psychophysics*, 64(5):785–794, 2002.

[100] B Shneiderman. *Designing the user interface*. Addison-Wesley, Reading, MA, 1987.

[101] P. F. Waller. The older driver. *Human Factors*, 33(5):499–505, 1991.

[102] D. Shinar and F. Schieber. Visual requirements for safety and mobility of older drivers. *Human Factors*, 33(5):507–519, 1991.

[103] J. J. Gibson. *The ecological approach to visual perception*. Houghton-Mifflin, New York, 1979.

[104] F. H. Previc and W. R. Ercoline, editors. *Spatial Disorientation in Aviation*. American Institute for Aeronautics and Astronautics, Reston, VA, 2004.

[105] H. W. Leibowitz. Perceptually induced misperception of risk: A common factor in transportation accidents. In L. P. Lipsitt and L. L. Mitnick, editors, *Self-regulatory Behavior and Risk Taking: Causes and consequences*, pages 219–229. Ablex, Norwood, NJ, 1991.

[106] J. E. Cutting, P. M. Vishton, and P. A. Braren. How we avoid collisions with stationary and moving obstacles. *Psychological Review*, 102(4):627–651, 1995.

[107] D. A. Kleffner and V. S. Ramachandran. On the perception of shape from shading. *Perception & Psychophysics*, 52(1):18–36, 1992.

[108] D. Regan and A. Vincent. Visual processing of looming and time to contact throughout the visual field. *Vision Research*, 35(13):1845–1857, 1995.

[109] D. O'Hare and S. N. Roscoe. *Flightdeck performance: The human factor*. Iowa State Press, Ames, IA, 1990.

[110] R. E. Eberts and A. G. MacMillan. Misperception of small cars. *Trends in ergonomics/human factors II*, pages 33–39, 1985.

[111] G. G. Denton. The influence of visual pattern on perceived speed. *Perception*, 9:393–402, 1980.

[112] S. Godley, T. J. Triggs, and B. N. Fildes. Driving simulator validation for speed results. *Accident Analysis and Prevention*, 34(4):589–600, 1997.

[113] R. A. Monty and J. W. Senders. *Eye Movements and Psychological Processes*. National Research Council, Washington, D.C., 1976.

[114] P. Hallett. Eye movements and human visual perception. In K. R. Boff, L. Kaufman, and J. P. Thomas, editors, *Handbook of Perception and Human Performance*. Wiley, New York, 1986.

[115] N. Moray. Monitoring behavior and supervisory control. In K R Boff, L Kaufman, and J P Thomas, editors, *Handbook of Perception and Human Performance*, volume 2, pages Ch.40, 1–51. Wiley, New York, 1986.

[116] A. F. Sanders. Some aspects of the selective process in the functional field of view. *Ergonomics*, 13:101–117, 1970.

[117] K. T. Goode, K. K. Ball, M. Sloane, D. L. Roenker, D. L. Roth, R. S. Myers, and C. Owsley. Useful field of view and other neurocognitive indicators of crash risk in older adults. *Journal of Clinical Psychology in Medical Settings*, 5(4):425–440, 1998.

[118] U. Neisser. Visual search. *Scientific American*, 1964.

[119] M. P Eckstein. Visual search: A retrospective. *Journal of Vision*, 11:1–36, 2011.

[120] E. Lee and J. MacGregor. Minimizing user search time in menu retrieval systems. *Human Factors*, 27:157–162, 1985.

[121] M. Yeh and C. D. Wickens. Attentional filtering in the design of electronic map displays: A comparison of color coding, intensity coding, and decluttering techniques. *Human Factors: The Journal of the Human Factors and Ergonomics Society*, 43(4):543–562, 2001.

[122] P. Stager and R. Angus. "Locating crash sites in simulated air-to-ground visual search.". *Human Factors: The Journal of the Human Factors and Ergonomics Society*, 20(4):453–466, 1978.

[123] C. G. Drury. Inspection of sheet materials-model and data. *Human Factors*, 17(3):257–265, 1975.

[124] A. M. Treisman. Features and objects in visual processing. *Scientific American*, 255(11):114B–125, 1986.

[125] S. Yantis. Stimulus-driven attentional capture and attentional control settings. *Journal of Experimental Psychology: Human Perception and Performance*, 19(3):676–681, 1993.

[126] D. L. Fisher and K. C. Tan. Visual displays: The highlighting paradox. *Human Factors: The Journal of the Human Factors and Ergonomics Society*, 31(1):17–30, 1989.

[127] R. Parasuraman. Vigilance, monitoring, and search. In K R Boff, L Kaufman, and J P Thomas, editors, *Handbook of Perception and Human Performance*, volume 2, pages 1–39. Wiley, New York, 1986.

[128] A. K. Pradhan, A. P. Pollatsek, M. A. Knodler, and D. L. Fisher. Can younger drivers be trained to scan for information that will reduce their risk in traffic scenarios that are hard to identify as hazardous. *Ergonomics*, 52(6):657–673, 2009.

[129] R. A. Tyrrell, J. M. Wood, A. Chaparro, T. P. Carberry, B. S. Chu, and R. P. Marszalek. Seeing pedestrians at night: Visual clutter does not mask biological motion. *Accident Analysis & Prevention*, 41(3):506–512, 2009.

[130] I. Kwan and J. Mapstone. Visibility aids for pedestrians and cyclists: A systematic review of randomised controlled trials. *Accident Analysis and Prevention*, 36(3):305–312, 2004.

[131] M. Yeh and C. D. Wickens. Display signaling in augmented reality: Effects of cue reliability and image realism on attention allocation and trust calibration. *Human Factors*, 43:355–365, 2001.

[132] M. C. Schall, Jr., M. L. Rusch, J. D. Lee, J. D. Dawson, G. Thomas, N. S. Aksan, and M. Rizzo. Augmented reality cues and elderly driver hazard perception. *Human Factors*, 2012.

[133] J. Theeuwes. Endogenous and exogenous control of visual selection. *Perception*, 23(4):429–440, 1994.

[134] J. Theeuwes. Endogenous and exogenous control of visual selection. *Perception*, 23:429–440, 1994.

[135] W. P. Tanner Jr. and J. A. Swets. A decision-making theory of visual detection. *Psychological Review*, 61(6):401–409, 1954.

[136] N A Macmillan and C D Creelman. *Detection Theory: A user's guide*. Lawrence Erlbaum Associates, Mahwah, NJ, 2nd edition, 2005.

[137] T. D. Wickens. *Elementary Signal Detection Theory*. Oxford University Press, New York, 2002.

[138] A. Bisseret. Application of signal detection theory to decision making in supervisory control The effect of the operator's experience. *Ergonomics*, 24(2):81–94, feb 1981.

[139] D. B. Beringer. False alerts in the ATC conflict alert system: Is there a "cry wolf" effect? *Proceedings of the Human and Ergonomics Society Annual Meeting*, pages 91–95, 2009.

[140] T. Fawcett. An introduction to ROC analysis. *Pattern Recognition Letters*, 27(8):861–874, jun 2006.

[141] R. G. Swensson, S. J. Hessel, and P. G. Herman. Omissions in radiology: Faulty search or stringent reporting criteria? *Radiology*, 123(3):563–567, 1977.

[142] M. L. Kelly. A study of industrial inspection by method of paired comparisons. *Psychological Monographs: General and Applied*, 69(9):1–16, 1955.

[143] C. G. Drury. Human Factors and Automation in Test and Inspection. In G. Salvendy, editor, *Handbook of Industrial Engineering*, pages 1887–1920. John Wiley, 2001.

[144] C. E. Billings. Human-centered aviation automation: Principles and guidelines. Technical report, NASA, Moffet Field, CA, 1996.

[145] B. L. Lambert, K. Y. Chang, and S. J. Lin. Effect of orthographic and phonological similarity on false recognition of drug names. *Social Science & Medicine*, 52(12):1843–1857, 2001.

[146] R. Filik, K. Purdy, A. Gale, and D. Gerrett. Labeling of medicines and patient safety: evaluating methods of reducing drug name confusion. *Human factors*, 48(1):39–47, 2006.

[147] R. Filik, K. Purdy, A. Gale, and D. Gerrett. Drug name confusion: Evaluating the effectiveness of capital ("Tall Man") letters using eye movement data. *Social Science and Medicine*, 59(12):2597–2601, 2004.

[148] G A Miller. The magical number seven plus or minus two: Some limits on our capacity for processing information. *Psychological Review*, 63:81–97, 1956.

[149] G. C. Brainard, J. P. Hanifin, J. M. Greeson, B. Byrne, G. Glickman, E. Gerner, M. D. Rollag, N. C. Aggelopoulos, H. Meissl, M. Ahmad, A. R. Cashmore, J. Arendt, E. A. Boettner, J. R. Wolter, B. A. Richardson, T. S. King, S.A. Matthews, R. J. Reiter, A. J. Lewy, M. Menaker, L. S. Miller, R. H. Fredrickson, R. G. Weleber, V. Cassone, D. Hudson, J. Greeson, J. Hanifin, M. Rollag, G. van den Beld, B. Sanford, D. M. Bronstein, G. H. Jacobs, K. A. Haak, J. Neitz, L. D. Lytle, T. P. Coohill, C. A. Czeisler, T. L. Shanahan, E. B. Klerman, H. Martens, D. J. Brotman, J. S. Emens, T. Klein, J. F. Rizzo, S. E. Davis, P. J. Munson, M. L. Jaffe, D. Rodbard, O. Dkhissi-Benyahya, B. Sicard, H. M. Cooper, R. G. Foster, I. Provencio, S. Fiske, W. DeGrip, M. S. Freedman, RJ. Lucas, B. Soni, M. von Schantz, M. Muñoz, Z. David-Gray, J. R. Gaddy, E. A. Griffin, D. Staknis, C. J. Weitz, D. C. Klein, RY. Moore, S. M. Reppert, R. W. Lam, S. Lerman, A. J. Lewy, T. A. Wehr, F. K. Goodwin, D. A. Newsome, S. P. Markey, E. D. Lipson, R. G. Foster, E. F. MacNichol, J. S. Levine, R. J. W. Mansfield, L. E. Lipetz, BA. Collins, IM. McIntyre, TR. Norman, GD. Burrows, SM. Armstrong, Y. Miyamoto, A. Sancar, LP. Morin, DE. Nelson, JS. Takahashi, DA. Oren, JC. Partridge, WJ. De Grip, P. Pevet, G. Heth, A. Hiam, E. Nevo, PC. Podolin, MD. Rollag, G. Jiang, WP. Hayes, IR. Rodriguez, EF. Moreira, RJ. Reiter, RW. Rodieck, GD. Niswender, FL. Ruberg, DJ. Skene, JP. Hanifin, J. English, KC. Smith, BG. Soni, R. Stanewsky, M. Kaneko, P. Emery, B. Beretta, K. Wager-Smith, SA. Kay, M. Rosbash, JC. Hall, A. Stockman, LT. Sharpe, H. Sun, DJ. Gilbert, NG. Copeland, NA. Jenkins, J. Nathans, PJ. DeCoursey, L. Bauman, RJ. Thresher, MH. Vitaterna, A. Kazantsev, DS. Hsu, C. Petit, CP. Selby, L. Dawut, O. Smithies, GTJ. van der Horst, M. Muijtjens, K. Kobayashi, R. Takano, S. Kanno, M. Takao, J. de Wit, A. Verkerk, APM. Eker, D. van Leenen, R. Buijs, D. Bootsma, JHJ. Hoeijmakers, A. Yasui, F. Waldhauser, M. Dietzel, JY. Wang, SM. Webb, TH. Champney, AK. Lewinski, L. Wetterberg, T. Yoshimura, S. Ebihara, JM. Zeitzer, RE. Kronauer, D-J. Dijk, and EN. Brown. Action spectrum for melatonin regulation in humans: Evidence for a novel circadian photoreceptor. *Journal of Neuroscience*, 21(16):211–222, 2001.

[150] A. Chang, D. Aeschbach, J. F. Duffy, and C. A. Czeisler. Evening use of light-emitting eReaders negatively affects sleep, circadian timing, and next-morning alertness. *Proceedings of the National Academy of Sciences*, 112(4):1232–1237, 2015.

[151] A. C. Moller, A. J. Elliot, and M. A. Maier. Basic hue-meaning associations. *Emotion*, 9(6):898–902, 2009.

[152] A. J. Elliot and M. A. Maier. Color and psychological functioning. *Psychological Science*, 16(5):250–254, 2007.

[153] P. O'Donovan, A. Agarwala, and A. Hertzmann. Color compatibility from large datasets. *ACM Transactions on Graphics*, 30(4):1, 2011.

[154] M. U. Shankar, C. A. Levitan, and C. Spence. Grape expectations: The role of cognitive influences in color-flavor interactions. *Consciousness and Cognition*, 19(1):380–390, 2010.

[155] C. N. DuBose, A. V. Cardello, and O. Maller. Effects of colorants and flavorants on identification perceived flavor intensity, and hedonic quality of fruit-flavored beverages and cake, 1980.

[156] A. J Elliot and M. A. Maier. Color psychology: effects of perceiving color on psychological functioning in humans. *Annual Review of Psychology*, 65:95–120, 2014.

[157] T. Caelli and D Porter. On difficulties in localizing ambulance sirens. *Human Factors*, 22(6):719–24, dec 1980.

[158] OSHA. Guidelines for Noise Enforcement (Instruction CPL 2âĂŞ2.35). Technical report, Occupational Safety and Health Administration, Washington, D.C, 1983.

[159] S. Levey, T. Levey, and B. J. Fligor. Noise exposure estimates of urban MP3 player users. *Journal of Speech, Language, and Hearing Research*, 54:263–277, 2011.

[160] M. J. Crocker. Rating measures, descriptors, criteria, and procedures for determining human response to noise. In M. J. Crocker, editor, *Handbook of Noise and Vibration Control*, pages 394–413. Wiley, New York, 2007.

[161] M. Y. Park and J. G. Casali. A controlled investigation of in-field attenuation performance of selected insert, earmuff, and canal cap hearing protectors. *Human factors*, 33(6):693–714, dec 1991.

[162] K. D. Kryter. Effects of ear protective devices on the intelligibility of speech in noise. *The Journal of the Acoustical Society of America*, 18(2):413, 1946.

[163] K. D. Kryter. *The Effects of Noise on Man*. Academic Press, London, 1985.

[164] W. V. Summers, D. B. Pisoni, R. H. Bernacki, R. I. Pedlow, and M. A. Stokes. Effects of noise on speech production: Acoustic and perceptual analyses. *The Journal of the Acoustical Society of America*, 84(3):917, 1988.

[165] H. Brumm and S. A. Zollinger. The evolution of the Lombard effect: 100 years of psychoacoustic research. *Behaviour*, 148(11-13):1173–1198, 2011.

[166] M. J. Crocker, editor. *Handbook of noise and vibration control*. Wiley, Hobocken, NJ, 2007.

[167] F. H. Hawkins and H. W. Orlady. *Human Factors in Flight*. Ashgate, Burlington, VT, second edition, 1993.

[168] G. A. Miller and P. E. Nicely. An analysis of perceptual confusions among some English consonants. *The Journal of the Acoustical Society of America*, 27(2):338, 1955.

[169] W. A. Yost. *Fundamentals of Hearing: An Introduction*. Academic Press, New York, third edition, 2007.

[170] A. Quaranta, P. Portalatini, and D. Henderson. Temporary and permanent threshold shift: An overview. *Scandinavian Audiology. Supplementum*, 48:75–86, 1998.

[171] R. B. King and S. R. Oldfield. The impact of signal bandwidth on auditory localization: Implications for the design of three-dimensional audio displays. *Human Factors*, 39(2 RECAS):287–295, 1997.

[172] D. R. Begault and M. T. Pittman. Three-dimensional audio versus head-down traffic alert and collision avoidance system displays. *The International Journal of Aviation Psychology*, 6(1):79–93, jan 1996.

[173] R. S. Bolia, W. R. D'Angelo, and R. L. McKinley. Aurally aided visual search in three-dimensional space. *Human factors*, 41(4):664–669, 1999.

[174] S. P. Banbury, W. J. Macken, S. Tremblay, and D. M. Jones. Auditory distraction and short-term memory: Phenomena and practical implications, 2001.

[175] M. S. Wogalter, M. J. Kalsher, and B. M. Racicot. Behavioral compliance with warnings: Effects of voice, context, and location. *Safety Science*, 16(5):637–654, 1993.

[176] R. D. Patterson. Auditory warning sounds in the work-environment. *Philosophical Transactions of the Royal Society of London Series B-Biological Sciences*, 327(1241):485–492, 1990.

[177] J. Rivera, A. B. Talone, C. T. Boesser, F. Jentsch, and M. Yeh. Startle and surprise on the flight deck: Similarities, differences, and prevalence. *Proceedings of the Human Factors and Ergonomics Society Annual Meeting*, 58(1):1047–1051, 2014.

[178] F. J. Seagull and P. M. Sanderson. Anesthesia alarms in context: an observational study. *Human Factors*, 43:66–78, 2001.

[179] C. A. Simpson and D. H. Williams. Response time effects of alerting tone and semantic context for synthesized voice cockpit warnings. *Human factors*, 22(3):319–30, jun 1980.

[180] W. W. Gaver, R. B. Smith, and T. O'Shea. Effective sounds in complex systems: The Arkola simulation. In *Proceedings of CHI 1991*, pages 85–90. ACM, Reading, MA, 1991.

[181] S. Garzonis, S. Jones, T. Jay, and E. O'Neill. Auditory icon and earcon mobile service notifications: Intuitiveness, learnability, memorability and preference. *CHI -ACM*, pages 1513–1522, 2009.

[182] S. M. Belz, G S Robinson, and J. G. Casali. A new class of auditory warning signals for complex systems: Auditory icons. *Human Factors*, 41(4):608–618, 1999.

[183] J. Meyer and J. D. Lee. Trust, reliance, and compliance. In A Kirlik and J D Lee, editors, *The Oxford Handbook of Cognitive Engineering*, pages 109–124. Oxford University Press, New York, 2013.

[184] J. P. Bliss and R. D. Gilson. Emergency signal failure: implications and recommendations. *Ergonomics*, 41(1):57–72, 1998.

[185] C. D. Wickens, S. Rice, D. Keller, S. Hutchins, J. Hughes, and K. Clayton. False alerts in air traffic control conflict alerting system: is there a "cry wolf" effect? *Human factors*, 51(4):446–462, 2009.

[186] R. Parasuraman, P. A. Hancock, and O. Olofinboba. Alarm effectiveness in driver-centred collision-warning systems. *Ergonomics*, 40(3):390–399, 1997.

[187] R. Parasuraman and V. A. Riley. Humans and automation: Use, misuse, disuse, abuse. *Human Factors*, 39(2):230–253, 1997.

[188] G. Abe and J. Richardson. Alarm timing, trust and driver expectation for forward collision warning systems. *Applied Rrgonomics*, 37(5):577–86, sep 2006.

[189] J. D. Lee and K. A. See. Trust in automation: Designing for appropriate reliance. *Human Factors*, 46(1):50–80, 2004.

[190] K. A. Hoff and M. Bashir. Trust in automation: Integrating empirical evidence on factors that influence trust. *Human Factors*, 57'(3):407–434, 2015.

[191] J. Edworthy, E. Hellier, and R. Hards. The semantic associations of acoustic parameters commonly used in the design of auditory information and warning signals. *Ergonomics*, 38(11):2341–2361, 1995.

[192] D. C. Marshall, J. D. Lee, and R. A. Austria. Alerts for in-vehicle information systems: annoyance, urgency, and appropriateness. *Human Factors*, 49(1):145–57, 2007.

[193] R. D. Sorkin and D. D. Woods. Systems with human monitors: A signal detection analysis. *Human-Computer Interaction*, 1:49–75, 1985.

[194] D. D. Woods. The alarm problem and directed attention in dynamic fault management. *Ergonomics*, 38(11):2371–2393, 1995.

[195] J. M. Festen and R. Plomp. Effects of fluctuating noise and interfering speech on the speech-reception threshold for impaired and normal hearing. *The Journal of the Acoustical Society of America*, 88(4):1725, 1990.

[196] C. E. Shannon. Prediction and entropy of printed english, 1951.

[197] D. B. Pisoni. Perception of speech: The human listener as a cognitive interface. *Speech Technology*, 1(2):10–23, 1982.

[198] K. D. Kryter. Non-auditory effects of environmental noise. *Environmental Noise*, pages 389–398, 1972.

[199] W. H. Sumby and I. Pollack. Visual contribution to speech intelligibility in noise. *The Journal of the Acoustical Society of America*, 26(2):212, 1954.

[200] D. W. Massaro and M. M. Cohen. Perceiving talking faces. *Current Directions in Psychological Science*, 4(4):104–109, 1995.

[201] W. C. Meecham and N. Shaw. Effects of jet noise on mortality rates. *British Journal of Audiology*, 13(3):77–80, jan 1979.

[202] L. S. Finegold, C. S. Harris, and H. E. von Gierke. Community annoyance and sleep disturbance: Updated criteria for assessing the impacts of general transportation noise on people. *Noise Control Engineering Journal*, 42(1), 1994.

[203] H. M. E. Miedema and C. G. M. Oudshoorn. Annoyance from DNL and DENL and transportation their noise: Confidence relationships intervals with exposure metrics. *Environmental Health*, 109(4):409–416, 2010.

[204] L. S. Finegold and M. S. Finegold. Development of exposure-response relationships between transportation noise and community annoyance. *Annoyance Stress and Health Effects of Environmental Noise*, page 17, 2002.

[205] M. Campbell. Patent filings detail Retina MacBook Pro's quiet asymmetric fans, 2012.

[206] D. E. Broadbent. Individual differences in annoyance by noise. *British Journal of Audiology*, 6(3):56–61, 1972.

[207] W. S. Helton, G. Matthews, and J. S. Warm. Stress state mediation between environmental variables and performance: The case of noise and vigilance. *Acta Psychologica*, 130(3):204–213, 2009.

[208] L. Yu and J. Kang. Factors influencing the sound preference in urban open spaces. *Applied Acoustics*, 71(7):622–633, 2010.

[209] J. M. Loomis and S. J. Lederman. Tactual perception. In K. R. Boff, L. Kaufman, and J. P. Thomas, editors, *Handbook of Perception and Human Performance*, volume 2, pages 31–1 – 31–41. Wiley, New York, 1986.

[210] I. Dianat, C. M. Haslegrave, and A. W. Stedmon. Methodology for evaluating gloves in relation to the effects on hand performance capabilities: A literature review. *Ergonomics*, 55(11):1429–1451, 2012.

[211] K. S. Hale and K. M. Stanney, editors. *Handbook of Virtual Environments: Design, implementation, and Applications.* CRC Press, Boca Raton, FL, 2014.

[212] P. Bach-y Rita and S. W. Kercel. Sensory substitution and the human-machine interface. *Trends in Cognitive Sciences*, 7(12):541–546, 2003.

[213] S. A. Lu, C. D. Wickens, J. C. Prinet, S. D. Hutchins, N. B. Sarter, and A. Sebok. Supporting interruption management and multimodal interface design: Three meta-analyses of task performance as a Function of Interrupting Task Modality. *Human Factors*, pages 0018720813476298–, feb 2013.

[214] C. M. Oman. Motion sickness: A sythesis and evaluation of the sensory conflict theory. *Canadian Journal of Physiology and Pharmacology*, 68:294–303, 1990.

[215] J. T. Reason. Motion sickness adaptation: a neural mismatch model. *Journal of the Royal Society of Medicine*, 71(11):819–29, nov 1978.

[216] S. D Young, B. D Adelstein, and S. R Ellis. Demand characteristics in assessing motion sickness in a virtual environment: Or does taking a motion sickness questionnaire make you sick? *Ieee Transactions On Visualization And Computer Graphics*, 13(3):422–428, 2007.

[217] C. Diels and J. E. Bos. Self-driving carsickness. *Applied Ergonomics*, 53:374–382, 2016.

[218] H . J. Bullinger and M. Dangelmaier. Virtual prototyping and testing of in-vehicle interfaces. *Ergonomics*, 46(1-3):41–51, 2003.

[219] J. E. Bos and W. Bles. Mismatch detailed for vertical motions. *Brain Research Bulletin*, 47(5):537–542, 1999.

[220] S. Casey. *Set Phasers on Stun and other True Tales of Design, Technology, and Human Error.* Aegean Publishing, Santa Barbara, CA, 1998.

[221] N. J. Cooke and F. T. Durso. *Stories of Modern Technology Failures and Cognitive Engineering Successes.* CRC Press, 2007.

[222] E. L. Wiener. Controlled flight into terrain accidents: System-induced errors. *Human Factors: The Journal of the Human Factors and Ergonomics Society*, 19(2):171–181, 1977.

[223] M. A. Regan, J. D. Lee, and K. L. Young. *Driver Distraction: Theory, Effects and Mitigation.* CRC Press, Boca Raton, Florida, 2008.

[224] C. D. Wickens. Effort in human factors performance and decision making. *Human Factors: The Journal of the Human Factors and Ergonomics Society*, 56(8):1329–1336, 2014.

[225] C. D. Wickens, J Goh, J Helleberg, W. J. Horrey, and D A Talleur. Attentional Modeals of Multitask Pilot Performance Using Advanced Display Technology. *Human Factors*, 45(3):360–380, 2003.

[226] J. Driver and C. Spence. Multisensory perception: beyond modularity and convergence. *Current Biology*, 10(20):R731–5, oct 2000.

[227] N. B. Sarter. Multimodal support for interruption management: Models, empirical findings, and design recommendations. *Proceedings of the IEEE*, 101(9):2105–2112, 2013.

[228] R. A. Rensink. Change detection. *Annual Review of Psychology*, 53:245–277, 2002.

[229] C. D. Wickens, B. L. Hooey, B. F. Gore, A. Sebok, and C. S Koenicke. Identifying Black Swans in NextGen: Predicting Human Performance in Off-Nominal Conditions. *Human Factors: The Journal of the Human Factors and Ergonomics Society*, 51(5):638–651, 2009.

[230] D. J. Simons and D. Levin. Failure to detect changes to people during a real-world interaction. *Psychonomic Bulletin & Review*, 5(4):644–649, 1998.

[231] A. H. Bellenkes, C. D. Wickens, and A. F. Kramer. Visual scanning and pilot expertise: the role of attentional flexibility and mental model development. *Aviation, Space, and Environmental Medicine*, 1997.

[232] C. D. Wickens, J. Goh, J. Helleberg, W. J. Horrey, and D. A. Talleur. Attentional models of multitask pilot performance using advanced display technology. *Human Factors*, 45(3):360–380, 2003.

[233] D. G. Jones and M. R. Endsley. Sources of situation awareness errors in aviation. *Aviation, Space, and Environmental medicine*, 67(6):507–512, 1996.

[234] N. Moray. Attention in dichotic listening: affective cues and the influence of instructions. *Quarterly Journal of Experimental Psychology*, 11:56–60, 1959.

[235] A. R. Conway, N. Cowan, and M. F. Bunting. The cocktail party phenomenon revisited: the importance of working memory capacity. *Psychonomic bulletin & review*, 8(2):331–335, 2001.

[236] D. J. Simons and M. S. Ambinder. Change blindness:Theory and consequences. *Current directions in psychological science*, 14(1):44–48, 2005.

[237] W. J. Horrey, M F Lesch, and A Garabet. Assessing the awareness of performance decrements in distracted drivers. *Accident Analysis & Prevention*, 40(2):675–682, 2008.

[238] R. L. Goldstone. Perceptual learning. *Annual Review of Psychology*, 49:585–612, 1998.

[239] C. Simpson. Effects of linguistic redundancy on pilot's comprehension of synthesized speeds. *Proceedings of the 12th Annual Conference on Manual Control*, (NASA TM-X:170), 1976.

[240] J. Campbell, D. Hoffmeister, R. J. Kiefer, D. Selke, P. A. Green, and J. B. Richman. Comprehension testing of active safety symbols. *SAE International*, 2004.

[241] J. S. Wolff and M. S. Wogalter. Comprehension of Pictorial Symbols: Effects of Context and Test Method. *Human Factors: The Journal of the Human Factors and Ergonomics Society*, 40(2):173–186, 1998.

[242] H. H. Clark and W. G. Chase. On the process of comparing sentences against pictures. *Cognitive Psychology*, 3(3):472–517, 1972.

[243] D. Schacter. *The Seven Sins of Memory*. Houghton Mifflin, New York, 2001.

[244] L. T. C. Rolt. *Red For Danger*. Pan Books, London, 1955.

[245] N. Cowan. The magical number 4 in short-term memory: A reconsideration of mental storage capacity. *Behavioral and Brain Sciences*, 24:87–185, 2001.

[246] A. Baddeley. Working memory. *Science*, 255:556–559, 1992.

[247] A. Baddeley. Working memory: Theories, models, and controversies. *Annual Review of Psychology*, 63(1):1–29, 2012.

[248] R. H. Logie. The functional organization and capacity limits of working memory. *Current Directions in Psychological Science*, 20(4):240–245, 2005.

[249] Z. Shipstead, T. L. Harrison, and R. W. Engle. Working memory capacity and fluid intelligence: Maintenance and disengagement. *Perspectives on Psychological Science*, 11(6):771–799, 2016.

[250] M. Jipp. Expertise development with different types of automation: A function of different cognitive abilities. *Human Factors*, 58(1):92–106, 2016.

[251] F. I. M. Craik and R. S. Lockhart. Levels of processing: A framework for memory research. *Journal of Verbal Learning and Verbal Behavior*, 684:671–684, 1972.

[252] S. K. Card, T. P. Moran, and A. Newell. *The Psychology of Human-Computer Interaction*. Laurence Erlbaum Associates, Hillsdale, NJ, 1983.

[253] C. D. Wickens, A. F. Kramer, L. Vanasse, and E. Donchin. A psychophysiological analysis of the reciprocity of information-processing resources. *Science*, 221(4615):1080–1082, 1983.

[254] C. D. Wickens. Multiple resources and performance prediction. *Theoretical Issues in Ergonomics Science*, 3(2):159–177, apr 2002.

[255] G. R. Loftus, V. J. Dark, and D. Williams. Short-term memory factors in ground controller/pilot communication. *Human Factors*, 21(2):169–181, 1979.

[256] B. Peacock and G. Peacock-Goebel. Wrong number: They didn't listen to Miller. *Ergonomics in Design: The Quarterly of Human Factors Applications*, pages 21–22, 2016.

[257] W. D. Gray. The nature and processing of errors in interactive behavior. *Cognitive Science*, 24(2):205–248, 2000.

[258] F. Mathy and J. Feldman. What's magic about magic numbers? Chunking and data compression in short-term memory. *Cognition*, 122(3):346–362, 2012.

[259] S. C. Preczewski and D. L. Fisher. Selection of alpbanumeric code sequences. *Proceedings of the Human Factors Society 34th Annual Meeting*, pages 224–228, 1990.

[260] S. M. Hess, M. C. Detweiler, and R. D. Ellis. The utility of display space in keeping track of rapidly changing information. *Human Factors: The Journal of the Human Factors and Ergonomics Society*, 41(2):257–281, 1999.

[261] W. Kintsch and T. A. van Dijk. Toward a model of text comprehension and production. *Psychological Review*, 85(5):363–394, 1981.

[262] R. Carlson, M. Sullivan, and W. Schneider. Practice and working memory effects in building procedural skill. *Journal of Experimental Psychology: Learning, Memory, and Cognition*, 15(3):517–536, 1989.

[263] E Tulving. How many memory systems are there? *American Psychologist*, 40:385–398, 1985.

[264] F. C. Bartlett. *Remembering: An experimental and social study*. Cambridge University Press, Cambridge, 1932.

[265] E. F. Loftus. Make-believe memories. *The American Psychologist*, 58(11):867–73, nov 2003.

[266] J. T. Wixted. Dual-process theory and signal-detection theory of recognition memory. *Psychological Review*, 114(1):152–176, 2007.

[267] N. Steblay, J. Dysart, S. Fulero, and R. C. L. Lindsay. Eyewitness accuracy rates in sequential and simultaneous lineup presentations: A meta-analytic comparison. *Law and Human Behavior*, 25(5):459–473, 2001.

[268] D. B. Wright and G. M. Davies. Eyewitness testimony. In F. T. Durso, editor, *Handbook of Applied Cognition*, pages 789–818. John Wiley, New York, 1999.

[269] D. L. Schacter, J. Y. Chiao, and J. P. MItchell. The seven sins of memory. *Annals Of The New York Academy Of Sciences*, 7(1):226–239, 2003.

[270] D. B. Wright and A. T McDaid. Comparing system and estimator variables using data from real line-ups. *Applied Cognitive Psychology*, 10:75–84, 1996.

[271] L. Hope and D. Wright. Beyond the unusual? Examining the role of attention in weapon focus effect. *Applied Cognitive Psychology*, 21:951–961, 2007.

[272] R. P. Fisher and R. E. Geiselman. *Memory enhancing techniques for investigative interviewing: The cognitive interview*. Charles C Thomas Publisher, Springfield, IL, 1992.

[273] A. Memon, C. A. Meissner, and J. Fraser. The cognitive interview: A meta-analytic review and study space analysis of the past 25 years. *Psychology, Public Policy, and Law*, 16(4):340–372, 2010.

[274] A. F. Healy and L. E. Bourne Jr. Empirically valid principles for training in the real world. *The American Journal of Psychology*, 126(4):389–399, 2013.

[275] J. D. Karpicke and H. L. Roediger. The critical importance of retrieval for learning. *Science*, 319(5865):966–968, 2008.

[276] J. D. Karpicke and J. R. Blunt. Retrieval practice produces more learning than elaborative studying with concept mapping. *Science*, 331(6018):772–775, 2011.

[277] K.-P. L. Vu, B.-L. Tai, A. Bhargav, E. E. Schultz, and R. W. Proctor. Promoting memorability and security of passwords through sentence generation. *Proceedings of the Human Factors and Ergonomics Society Annual Meeting*, 48(13):1478–1482, 2004.

[278] D T Neal, W. Wood, and J M Quinn. Habits: A repeat performance. *Current Directions in Psychological Science*, 15(4):198–202, 2006.

[279] W. Wood and D. T. Neal. The habitual consumer. *Journal of Consumer Psychology*, 19(4):579–592, 2009.

[280] C. Duhigg. *The Power of Habit: Why we do what we do in life and business*. Random House, New York, 2012.

[281] P. N. Johnson-Laird. Mental Models in Cognitive Science. *Cognitive Science*, 4(1):71–115, jan 1980.

[282] K. R. Paap and R. J. Roske-Hofstrand. Design of menus. *In M. Helander (Ed.)*, Handbook o(pp. 205-235). Amsterdam):The Netherlands Elsevier Science Publishing Compan, 1988.

[283] K. S. Seidler and C. D. Wickens. Distance and organization in multifunction displays. *Human Factors*, 34(5)):555–570, 1992.

[284] J. C. Nesbit and O. O. Adesope. Learning with concept and knowledge maps: A meta-analysis. *Review of Educational Research*, 76:413–448, 2006.

[285] R. C. Schank and R. P. Abelson. *Scripts, plans, goals, and understanding*. Erlbaum, Hillsdale, NJ, 1977.

[286] D. Gentner and A. L. Stevens. *Mental Models*. Lawrence Erlbaum Associates, Hillsdale, NJ, 1983.

[287] W. B. Rouse and N. M. Morris. On looking into the black box: Prospects and limits in the search for mental models. *Psychological Bulletin and Review*, 100(3):349–363, 1986.

[288] J. R. Wilson and A. Rutherford. Mental Models: Theory and application in Human Factors. *Human Factors: The Journal of the Human Factors and Ergonomics Society*, 31(6):617–634, 1989.

[289] S. L. Smith. Exploring compatibility with words and pictures. *Human Factors*, 23(3):305–315, 1981.

[290] W. Chase. Cognitive skill: Implications for spatial skill in large-scale environments. *CHI Conference on Human Factors in Computing Systems*, 1979.

[291] M. J. Sholl. Cognitive maps as orienting schemata. *Journal of Experimental Psychology: Learning, memory, and cognition*, 13(4):615–28, 1987.

[292] N. Franklin and B. Tversky. Searching imagined environments. *Journal of Experimental Psychology: General*, 119(1):63–76, 1990.

[293] R. K. Dismukes. Prospective memory in workplace and everyday situations. *Current Directions in Psychological Science*, 21(4):215–220, 2012.

[294] J. E. Harris and A.J. Wilkins. Remembering to do things: A theoretical framework and an illustrative experiment. *Human Learning*, 1:123–136, 1982.

[295] D. Herrmann, B. Brubaker, C. Yoder, V. Sheets, and A. Tio. Devices that remind. In *Handbook of applied cognition*, pages 377–407. John Wiley, New York, 1999.

[296] A Degani and E. L. Wiener. Procedures in complex systems: The airline cockpit. *IEEE Transactions on Systems, Man, and Cybernetics*, SMC-this i(3):302–12, may 1997.

[297] J. R. Keebler, E. H. Lazzara, B. S. Patzer, E. M. Palmer, J. P. Plummer, D. C. Smith, V. Lew, S. Fouquet, Y. R. Chan, and R. Riss. Meta-Analyses of the effects of standardized handoff protocols on patient, provider, and organizational outcomes. *Human Factors: The Journal of the Human Factors and Ergonomics Society*, 58(8):1187–1205, 2016.

[298] A. Gawande and J. B. Lloyd. *The checklist manifesto: How to get things right*. Metropolitan Books, New York, 2010.

[299] A. B. Haynes, T. G. Weiser, W. R. Berry, S. R. Lipsitz, A. H. Breizat, E. P. Dellinger, T. Herbosa, S. Joseph, P. Kibatala, M. Lapitan, A. Merry, K. Moorthy, R. Reznick, B. Tayler, and A. Gawande. A surgical safety checklist to reduce morbidity and mortality in a global population. *New England Journal of Medicine*, 360(5):491–499, 2009.

[300] D. A. Norman. *Turn Signals Are the Facial Expressions of Automobiles*. Addison-Wesley, Reading, MA, 1992.

[301] V. A. Banks and N. A. Stanton. Keep the driver in control: Automating automobiles of the future. *Applied Ergonomics*, pages 1–7, 2015.

[302] T. H. Shaw, M. E. Funke, M. Dillard, G. J. Funke, J. S. Warm, and R. Parasuraman. Event-related cerebral hemodynamics reveal target-specific resource allocation for both "go" and "no-go" response-based vigilance tasks. *Brain and Cognition*, 82(3):265–273, 2013.

[303] P. M. Fitts and M. I. Posner. *Human Performance*. Brooks/Cole, Belmont, CA, 1969.

[304] R. M. Shiffrin and W. Schneider. Controlled and automatic human information processing: II. Perceptual learning, automatic attending and a general theory. *Psychological Review*, 84(2):127–190, 1977.

[305] G. S. Halford, W. H. Wilson, and S. Phillips. Processing capacity defined by relational complexity: implications for comparative, developmental, and cognitive psychology. *The Behavioral and Brain Sciences*, 21(6):803–864, 1998.

[306] D. J. Campbell. Task Complexity: A review and analysis. *Academy of Management Review*, 13(1):40–52, 1988.

[307] C. M. Carswell, D. Clarke, and W. B. Seales. Assessing mental workload during laparoscopic surgery. *Surgical Innovation*, 12(1):80–90, 2005.

[308] C. D. Wickens. Multiple resources and mental workload. *Human Factors*, 50(3):449–455, 2008.

[309] A. D. Milner and M. A. Goodale. Two visual systems re-viewed. *Neuropsychologia*, 46(3):774–785, 2008.

[310] D. Navon and D. Gopher. On the economy of the human processing system. *Psychological Review*, 86:254–285, 1979.

[311] T. Gillie and D. Broadbent. What makes interruptions disruptive? A study of length, similarity, and complexity. *Psychological Research*, 50(4):243–250, 1989.

[312] P. Lacherez, L. Donaldson, and J. S. Burt. Do learned alarm sounds interfere with working memory? *Human Factors*, page 0018720816662733, 2016.

[313] C. D. Wickens, R. S. Gutzwiller, and A. Santamaria. Discrete task switching in overload: A meta-analyses and a model. *International Journal of Human Computer Studies*, 79:79–84, 2015.

[314] J. G. Trafton and C. A. Monk. Task Interruptions. *Reviews of Human Factors and Ergonomics*, 3(1):111–126, 2007.

[315] C. D. Wickens, B. A Clegg, A. Z Vieane, and A. L Sebok. Complacency and automation bias in the use of imperfect automation. *Human Factors: The Journal of the Human Factors and Ergonomics Societyctors*, 57(5):728–39, 2015.

[316] P. C. Schutte and A. C. Trujillo. Flight Crew Task Management in Non-Normal Situations. *Proceedings of the Human Factors and Ergonomics Society Annual Meeting*, 40(4):244–248, 1996.

[317] A. Spink, M. Park, and S. Koshman. Factors affecting assigned information problem ordering during Web search: An exploratory study. *Information Processing and Management*, 42(5):1366–1378, 2006.

[318] R. Kurzban, A. Duckworth, J. W. Kable, and J. Myers. An opportunity cost model of subjective effort and task performance. *Behavioral and Brain Sciences*, 36(06):661–679, 2013.

[319] R. S. Gutzwiller, C. D. Wickens, and B. A. Clegg. The role of time on task in multi-task management. *Journal of Applied Research in Memory and Cognition*, 5(2):176–184, 2016.

[320] B. Zeigarnik. On finished and unfinished tasks. *A Source Book of Gestalt Psychology*, 1:1–15, 1938.

[321] L.D. Loukopoulos, R. K. Dismukes, and I. Barshi. *The Multitasking Myth: Handling complexity in real-world operations*. Routeledge, New York, 2016.

[322] P. D. Adamczyk and B. P. Bailey. If not now when?: the effects of interruption at different moments within task execution. *Proceedings of the SIGCHI conference on Human factors in computing systems*, 6(1):271–278, 2004.

[323] A. J. Rivera-Rodriguez and B.-T. Karsh. Interruptions and distractions in healthcare: review and reappraisal. *Quality & Safety in Health Care*, 19(4):304–312, aug 2010.

[324] NTSB. Northwest Airlines Inc McDonnell Douglas DC-9-82, N312RC Detroit Metropolitan Wayne County Airport, Romulus, Michigan, AUgust 16, 1987. Technical report, National Transportation Safety Board, Washington D.C., 1988.

[325] C. A. Monk, J. G. Trafton, and D. A. Boehm-Davis. The effect of interruption duration and demand on resuming suspended goals. *Journal of Experimental Psychology: Applied*, 14(4):299–313, 2008.

[326] E. M. Altmann and J. G. Trafton. Memory for goals: An activation-based model. *Cognitive Science: A Multidisciplinary Journal*, 26(1):39–83, jan 2002.

[327] S. T. Iqbal and B. P. Bailey. Oasis: A framework ofr linking notification delivery to the perceptual structure of goal-directed tasks. *ACM Transactions on Computer-Human Interaction*, 17(4):1–28, 2010.

[328] H. Sohn, J. D. Lee, D. L. Bricker, and J. D. Hoffman. A dynamic programming model for scheduling in-vehicle message display. *IEEE Transactions on Intelligent Transportation Systems*, 9(2):226–234, 2008.

[329] C. Y. Ho and M. I. Nikolic. Not now! Supporting interruption management by indicating the modality and urgency of pending tasks. *Human Factors: The Journal of the Human Factors and Ergonomics Society*, 2004.

[330] M. A. McDaniel, G. O. Einstein, T. Graham, and E. Rall. Delaying execution of intentions: Overcoming the costs of interruptions. *Applied Cognitive Psychology*, 18(5):533–547, 2004.

[331] D. Gopher. Emphasis change as a training protocol for high-demand tasks. In A. Kramer, D. Wiegmann, and A. Kirlik, editors, *Attention: From Theory to Practice*, pages 209–224. Oxford University Press, New York, 2007.

[332] R. Y. I. Koh, T. Park, C. D. Wickens, L. T. Ong, and S. N. Chia. Differences in attentional strategies by novice and experienced operating theatre scrub nurses. *Journal of Experimental Psychology: Applied*, 17(3):233–246, 2011.

[333] G. Underwood, D. Crundall, and P. Chapman. Selective searching while driving: the role of experience in hazard detection and general surveillance. *Ergonomics*, 45(1):1–12, 2002.

[334] G. Fogarty and L. Stankov. Abilities involved in performance on competing tasks. *Personality and Individual Differences*, 9(1):35–49, 1988.

[335] J. Reissland and D. Manzey. Serial or overlapping processing in multitasking as individual preference: Effects of stimulus preview on task switching and concurrent dual-task performance. *Acta Psychologica*, 168:27–40, 2016.

[336] D. L. Damos, T. F. Smist, and Jr. Bittner, A. C. Individual differences in multiple-task performance as a function of response strategy. *Human Factors*, 25(2):215–226, 1983.

[337] G. A. Klein, K. G. Ross, B. M. Moon, D. E. Klein, R. R. Hoffman, and E. Hollnagel. Macrocognition. *IEEE Intelligent Systems*, 18(3):81–85, 2003.

[338] J. D. Lee and A. Kirlik. Introduction to the handbook. In A Kirlik and J D Lee, editors, *The Oxford Handbook of Cognitive Engineering*, pages 3–16. Oxford University Press, New York, 2013.

[339] G. A. Klein, J. Orasanu, R. Calderwood, and C. E. Zsambok. *Decision making in action: Models and methods*. Ablex, Norwood, NJ, 1993.

[340] R. Lipshitz, G. A. Klein, J. Orasanu, and E. Salas. Focus article: Taking stock of naturalistic decision making. *Journal of Behavioral Decision Making*, 14(5):331–352, 2001.

[341] J. Rasmussen. *Information Processing and Human-Machine Interaction. An Approach to Cognitive Engineering*. North Holland, New York, 1986.

[342] K R Hammond, R M Hamm, J Grassia, and T Pearson. Direct comparison of the efficacy of intuitive and analytical cognition in expert judgment. *IEEE Transactions on Systems, Man, and Cybernetics*, SMC-17(5):753–770, 1987.

[343] J. S. B. T. Evans. Dual-processing accounts of reasoning, judgment, and social cognition. *Annual Review of Psychology*, 59:255–278, 2008.

[344] J. S. B. T. Evans and K. E. Stanovich. Dual-process theories of higher cognition: Advancing the debate. *Perspectives on Psychological Science*, 8(3):223–241, 2013.

[345] D. Kahneman. *Thinking, fast and slow*. Macmillan, New York, 2011.

[346] A. R. Damasio. *Descartes' Error: Emotion, Reason, and the Human Brain*. A Grosset/Putnam Book, Published by G.P. Putnam's Sons, New York, 1994.

[347] D. Kahneman, P. Slovic, and A. Tversky. Judgment Under Uncertainty: Heuristics and Biases. *Science*, 185:1124–1131, 1982.

[348] A. Tversky and D. Kahneman. Judgment under uncertainty: Heuristics and biases. *Science*, 185(4157):1124–1131, 1974.

[349] G. A. Klein, R. Calderwood, and D. MacGregor. Critical decision method for eliciting knowledge. *IEEE Transactions on Systems, Man, and Cybernetics*, 19(3):462–472, 1989.

[350] G. A. Klein. Naturalistic Decision Making. *Human Factors*, 50(3):456–460, 2008.

[351] D. Kahneman and G. A. Klein. Conditions for intuitive expertise: A failure to disagree. *American Psychologist*, 64(6):515–526, sep 2009.

[352] J. Rasmussen. Skills, rules, and knowledge: Signals, signs, and symbols, and other distinctions in human performance models. *IEEE Transactions on Systems, Man, and Cybernetics*, SMC-13(3):257–266, 1983.

[353] J. Rasmussen. Deciding and doing: Decision making in natural contexts. In G A Klein, J Orasanu, R Calderwood, and C E Zsambok, editors, *Decision Making in Action: Models and Methods,* pages 158–171. Ablex, Norwood, New Jersey, 1993.

[354] J. Reason. *Human Error.* Cambridge University Press, Cambridge, England, 1991.

[355] J. B. Soll, K. L. Milkman, and J. W. Payne. A user's guide to debiassing. In *The Wiley Blackwell Handbook of Judgment and Decision Making,* pages 1–26. John Wiley & Sons, Ltd, Chichester, UK, dec 2015.

[356] C. I. Canfield, B. Fischhoff, and A. Davis. Quantifying phishing susceptibility for detection and behavior decisions. *Human Factors: The Journal of the Human Factors and Ergonomics Society,* 58(8):1158–1172, 2016.

[357] R. M. Hogarth and H. J. Einhorn. Behavioral decision theory. *Annual Review of Psychology,* 32:53–88, 1980.

[358] H. A. Simon. Rational decision making in business organizations. *American Economic Association,* 69(4):493–513, 1979.

[359] H. A. Simon. Bounded rationality in social science: Today and tomorrow. *Mind and Society,* 1:25–39, 2000.

[360] G. Gigerenzer and P. M. Todd. *Simple Heuristics That Make Us Smart.* Oxford University Press, New York, 1999.

[361] W. Edwards. Decision making. In G Salvendy, editor, *Handbook of Human Factors.* Wiley, New York, 1987.

[362] B. Kleinmuntz. Why We Still Use Our Heads Instead of Formulas - Toward an Integrative Approach. *Psychological Bulletin and Review,* 107(3):296–310, 1990.

[363] R/ Accorsi, E. Zio, and G. E. Apostolakis. Developing utility functions for environmental decision making. *Progress in Nuclear Energy,* 34(4):387–411, 1999.

[364] A. A. Aly and M. Subramaniam. Design of an FMS decision-support system. *International Journal of Production Research,* 31(10):2257–2273, 1993.

[365] A. Webb. *Data, a Love Story: How I cracked the online dating code to meet my match.* Penguin, New York, 2013.

[366] S. Frederick. Cognitive reflection and decision making. *The Journal of Economic Perspectives,* 19(4):25–42, dec 2005.

[367] H. A. Simon. *Models of Man.* Wiley, New York, 1957.

[368] G. A. Klein. Recognition-primed decisions. In W B Rouse, editor, *Advances in Man-Machine Systems Research,* volume 5, pages 47–92. JAI Press, Greenwich, CT, 1989.

[369] M. S. Pfaff, G. A. Klein, J. L. Drury, S. P. Moon, Y. Liu, and S. O. Entezari. Supporting complex decision making through option awareness. *Journal of Cognitive Engineering and Decision Making,* 7(2):155–178, 2013.

[370] J. Orasanu. Decision-making in the cockpit. In E L Weiner, B G Kanki, and R L Helmreich, editors, *Cockpit Resource Management,* pages 137–168. Academic Press, San Diego, CA, 1993.

[371] E. M. Roth. Analysis of decision making in nuclear power plant emergencies: An investigation of aided decision making. In C E Zsambok and G Klein, editors, *Naturalistic Decision Making,* pages 175–182. Erlbaum, Mahwah, NJ, 1997.

[372] G. Gigerenzer and W. Gaissmaier. Heuristic decision making. *Annual Review of Psychology,* 62:451–482, 2011.

[373] T. Gilovich, D. Griffin, and D. Kahneman, editors. *Heuristics and Biases: The Psychology of Intuitive Judgment.* Cambridge University Press, Cambridge, England, 2002.

[374] D. Arnott. Cognitive biases and decision support systems development: A design science approach. *Information Systems Journal,* 16:55–78, 2006.

[375] L. Adelman, T. Bresnick, P. K. Black, F. F. Marvin, and S. G. Sak. Research with patriot air defense officers: Examining information order effects. *Human Factors: The Journal of the Human Factors and Ergonomics Society,* 38(2):250–261, 1996.

[376] G. R. Bergus, I. P. Levin, and A. S. Elstein. Presenting risks and benefits to patients - The effect of information order on decision making. *Journal of General Internal Medicine,* 17(8):612–617, 2002.

[377] M. R. Endsley. Toward a theory of situation awareness in dynamic systems. *Human Factors*, 37(1):32–64, 1995.

[378] E. M. Johnson, R. C. Cavanagh, R. L. Spooner, and M. G. Samet. Utilization of reliability measurements in Bayesian inference: Models and human performance. *IEEE Transactions on Reliability*, R-22(3):176–183, aug 1973.

[379] D. A. Schum. The weighing of testimony in judicial proceedings from sources having reduced credibility. *Human Factors*, 17(2):172–182, 1975.

[380] A. Tversky and D. Kahneman. Availability : A heuristic for judging frequency. *Cognitive Psychology*, 5:207–232, 1973.

[381] M. Wanke, N. Schwarz, and H. Bless. The availability heuristic revisited: Experienced ease of retrieval in mundane frequency estimates. *Acta Psychologica*, 89:83–90, 1995.

[382] N. Schwarz and F. Strack. Beyond âĂIJwhatâĂİ comes to mind: Experiential and conversational determinants of information use. *Current Opinion in Psychology*, 12:89–93, 2016.

[383] J. N. Braga, M. B. Ferreira, and S. J. Sherman. The effects of construal level on heuristic reasoning: The case of representativeness and availability. *Decision*, 2(3):216–227, 2015.

[384] T. Mehle. Hypothesis generation in an automobile malfunction inference task. *Acta Psychologica*, 52:87–116, 1982.

[385] R. A. Bjork. Assessing our own competence: Heuristics and illusions. In A Koriat, editor, *Attention and performance XVII*, pages 435–459. Bradford Book., Cambridge, MA, 1999.

[386] D. J. Simons. Unskilled and optimistic: Overconfident predictions despite calibrated knowledge of relative skill. *Psychonomic Bulletin & Review*, 20:601–7, 2013.

[387] R. Buehler, D. Griffin, and M. Ross. Exploring the "planning fallacy": Why people underestimate their task completion times. *Journal of Personality and Social Psychology*, 67(3):366–381, 1994.

[388] R. Buehler, D. Griffin, and J. Peetz. *The Planning Fallacy. Cognitive, Motivational, and Social Origins*, volume 43. Elsevier Inc., 2010.

[389] R. I. Cook and D. D. Woods. Operating at the sharp end: The complexity of human error. In M S Bogner, editor, *Human Error in Medicine*. Lawrence Erlbaum, New Jersey, 1994.

[390] Y. Xiao and C. F. Mackenzie. Decision making in dynamic environments. *Proceedings of the Human and Ergonomics Society 39th Annual Meeting*, pages 469–473, 1995.

[391] T. Rubinstein and A. F. Mason. The accident that shouldn't have happened: an analysis of Three Mile Island. *IEEE Spectrum*, Nov.:33–57, 1979.

[392] S. S. Iyengar and M. R. Lepper. Rethinking the value of choice: a cultural perspective on intrinsic motivation. *Journal of personality and social psychology*, 76(3):349–366, 1999.

[393] S. S. Iyengar and E. Kamenica. Choice proliferation, simplicity seeking, and asset allocation. *Journal of Public Economics*, 94(7-8):530–539, 2010.

[394] H. J. Einhorn and R. M. Hogarth. Confidence in judgment: Persistence of the illusion of validity. *Psychological Review*, 85:395–416, 1978.

[395] R. S. Nickerson. Confirmation bias: A ubiquitous phenomenon in many guises. *Review of General Psychology*, 2(2):175–220, 1998.

[396] H. Arkes and R. R. Harkness. The effect of making a diagnosis on subsequent recognition of symptoms. *Journal of Experimental Psychology: Human Learning and Memory*, 6:568–575, 1980.

[397] I. L. Janis. Decision making under stress. In L Goldberger and S Breznitz, editors, *Handbook of Stress: Theoretical and Clinical Aspects*, pages 69–87. Free Press, New York, 1982.

[398] P. Wright. The harassed decision maker: Time pressures, distractions, and the use of evidence. *Journal of Applied Psychology*, 59:555–561, 1974.

[399] A Degani and E. L. Wiener. Cockpit checklists: Concepts, design, and use. *Human Factors*, 35(2):345–359, 1993.

[400] J. E. Driskell and E. Salas. Group decision making under stress. *Journal of Applied Psychology*, 76(3):473, 1991.

[401] D. D. Woods, L. J. Johannesen, R. I. Cook, and N. B. Sarter. *Behind human error: Cognitive systems, computers, and hindsight*. Crew Systems Ergonomics Information Analysis Center (SOAR/CERIAC), Wright-Patterson AFB, OH, 1994.

[402] N. J. Roese and K. D. Vohs. Hindsight bias. *Perspectives on Pychological Science*, 7(5):411–26, 2012.

[403] B. Fischhoff. Hindsight foresight: The effect of outcome knowledge on judgment under uncertainty. *Journal of Experimental Psychology: Human Perception and Performance*, 1:288–299, 1975.

[404] A. Tversky and D. Kahneman. The framing of decisions and the psychology of choice. *Science*, 211(4481):453–458, 1981.

[405] I. P. Levin, G. J. Gaeth, J. Schreiber, and M. Lauriola. A new look at framing effects: Distribution of effect sizes, individual differences, and independence of types of effects. *Organizational Behavior and Human Decision Processes*, 88(1):411–429, 2002.

[406] B. J. McNeil, S. G. Pauker, H. C. Jr. Cox, and A. Tversky. On the elicitation of preferences for alternative therapies. *New England Journal of Medicine*, 306:1259–1262, 1982.

[407] T. Garling, E. Kirchler, A. Lewis, and F. van Raaij. Psychology, financial decision making, and financial crises. *Psychological Science in the Public Interest*, 10(1):1–47, 2009.

[408] H. R. Arkes and L. Hutzel. The role of probability of success estimates in the sunk cost effect. *Journal of Behavioral Decision Making*, 13(3):295–306, 2000.

[409] D. C. Molden and C. M. Hui. Promoting de-escalation of commitment: A regulatory-focus perspective on sunk costs. *Psychological Science*, 22(1):8–12, 2011.

[410] E. J. Johnson and D. G. Goldstein. Do defaults save lives? *Science*, 302:1338–1339, 2003.

[411] E. Duflo and E. Saez. Participation and investment decisions in a retirement plan: The influence of colleagues' choices. *Journal of Public Economics*, 85(1):121–148, jul 2002.

[412] R. H. Thaler and S. Benartzi. Save More TomorrowâĎć: Using behavioral economics to increase employee saving. *Journal of Political Economy*, 112(February 2004):S164–S187, 2004.

[413] R. H. Thaler and C. R. Sunstein. *Nudge: Improving Decisions about Health, Wealth, and Happiness*. Penguin Books, New York, 2008.

[414] E. J. Johnson, S. B. Shu, B. G. C. Dellaert, C. Fox, D. G. Goldstein, G. Häubl, R. P. Larrick, J. W. Payne, E. Peters, D. Schkade, B. Wansink, and E. U. Weber. Beyond nudges: Tools of a choice architecture. *Marketing Letters*, 23(2):487–504, may 2012.

[415] E Peters, D Vastfjall, P. Slovic, C K Mertz, K Mazzocco, and S Dickert. Numeracy and decision making. *Psychological Science*, 17(5):407–413, 2006.

[416] R. H. Thaler, C. R. Sunstein, and J. P. Balz. Choice architecture. *Social Science Research Network*, pages 428–439, 2010.

[417] H. E Hershfield, D. G. Goldstein, W. F Sharpe, J. Fox, L. Yeykelis, L. L Carstensen, and J. N. Bailenson. Increasing saving behavior through age-progressed renderings of the future self. *JMR, Journal of marketing research*, 48:S23–S37, 2011.

[418] R. P. Larrick and J. B. Soll. The MPG illusion. *Science*, 320(5883):1593–1594, 2008.

[419] L. L. Lopes. Procedural debiasing. *Acta Psychologica*, 64(2):167–185, 1987.

[420] R. L. Keeney. A decision analysis with multiple objectives: The Mexico City airport. *The Bell Journal of Economics and Management Science*, 4(1):101, 1973.

[421] L. A. Greening and S. Bernow. Design of coordinated energy and environmental policies: Use of multi-criteria decision-making. *Energy Policy*, 32(6):721–735, 2004.

[422] D. N. Ricchiute. Evidence, memory, and causal order in a complex audit decision task. *Journal of Experimental Psychology: Applied*, 4(1):3–15, 1998.

[423] P. Humphreys and W. McFadden. Experiences with MAUD: Aiding decision structuring versus bootstrapping the decision maker. *Acta Psychologica*, 45(1):51–69, 1980.

[424] E. F. Cabrera and N. S. Raju. Utility analysis: Current trends and future directions. *International Journal of Selection and Assessment*, 9(1-2):92–102, 2001.

[425] K. V. Katsikopoulos and G. Gigerenzer. Modeling decision heuristics. In J. D. Lee and A. Kirlik, editors, *The Oxford Handbook of Cognitive Engineering*, pages 1–13. Oxford University Press, 2013.

[426] N. Phillips. FFTrees: Generate, visualise, and compare fast and frugal decision trees, 2016.

[427] B. Fischhoff. Debiasing. In D. Kahneman, P. Slovic, and A. Tversky, editors, *Judgment Under Uncertainty: Heuristics and Biases*, volume P. Slovic, page & A. Cambridge University Press., Cambridge, England, 1982.

[428] W. B. Rouse and R. M. Hunt. Human problem solving performance in fault diagnosis tasks. In W B Rouse, editor, *Advances in Man-Machine Systems Research*, volume 1. JAI Press, Greenwich, CT, 1983.

[429] T. Mussweiler and T. Pfeif. Over coming the Inevitable anchoring effect: Considering the opposite compensates for s lective accessibility. *Personality and Social Psychology Bulletin*, 26(9):1142–1150, 2000.

[430] M. S. Cohen, J. T. Freeman, and B. B. Thompson. Training the naturalistic decision maker. In C E Zsambok and G Klein, editors, *Naturalistic Decision Making*, pages 257–268. Erlbaum, Mahwah, NJ, 1997.

[431] A. H. Murphy and R. L. Winkler. A general framework for forecast verification. *Monthly Weather Review*, 115(7):1330–1338, 1987.

[432] G. A. Klein. Performing a project premortem. *Harvard Business Review*, 85(9):18–19, 2007.

[433] E. Dayan and M. Bar-Hillel. Nudge to nobesity II: Menu positions influence food orders. *Judgment*, 6(4):333–342, 2011.

[434] E. R. Stone, J. F. Yates, and A. M. Parker. Effects of numerical and graphical displays on professed risk-taking behavior. *Journal of Experimental Psychology: Applied*, 3(4):243–256, 1997.

[435] D. A. Schkade and D. N. Kleinmuntz. Information displays and choice processes: Differential effects of organization, form, and sequence. *Organizational Behavior and Human Decision Processes*, 57:319–337, 1994.

[436] M. B. Cook and H. S. Smallman. Human Factors of the confirmation bias in intelligence analysis: Decision support from graphical evidence landscapes. *Human Factors: The Journal of the Human Factors and Ergonomics Society*, 50(5):745–754, 2008.

[437] B. J. Barnett and C. D. Wickens. Display proximity in multicue information integration: The benefit of boxes. *Human Factors*, 30:15–24, 1988.

[438] K. B. Bennett, J. M. Flach, T. R. McEwen, and O. Fox. Enhancing creative problem solving through visual display design. In D. A. Boehm-Davis, F. T. Durso, and J. D. Lee, editors, *APA Handbook of Human Systems Integration*. APA Press, Washington, DC, 2015.

[439] C. M. Burns, G. Skraaning, G. A Jamieson, N. Lau, J. Kwok, R. Welch, and G. Andresen. Evaluation of ecological interface design for nuclear process control: Situation awareness effects. *Human Factors*, 50(4):663–679, 2008.

[440] A. Garg, N. Adhikari, H. McDonald, M. P. Rosas-Arellano, P. J. Devereaux, J. Beyene, J. Sam, and R. B. Haynes. Effects of computerized clinical decision support systems on practitioner performance and patient outcomes. *American Medical Association*, 293(10):1223–1238, 2005.

[441] D G Morrow, R. North, and C. D. Wickens. Reducing and mitigating human error in medicine. *Reviews of Human Factors and Ergonomics*, 1(1):254–296, 2005.

[442] F. T. Durso and S. D. Gronlund. Situation awareness. *Handbook of Applied Cognition*, 1999.

[443] M. J. Adams, Y. J. Tenney, and R. W. Pew. State of the Art Report. Strategic workload and the cognitive management of advanced multi-task systems. Technical report, Airforce Research Laboratory, Wright-Patterson AFB, OH, 1991.

[444] M. R. Endsley. Situation Awareness. In J. D. Lee and A. Kirlik, editors, *The Oxford Handbook of Cognitive Engineering*, chapter 5. Oxford University Press, 2013.

[445] B. Strauch. Decision errors and accidents: Applying naturalistic decision making to accident investigations. *Journal of Cognitive Engineering and Decision Making*, 10(3):281–290, 2016.

[446] M. R. Endsley. Toward a Theory of Situation Awareness in Dynamic Systems. *Human Factors: The Journal of the Human Factors and Ergonomics Society*, 37(1):32–64, 1995.

[447] M. R. Endsley and D. J. Garland. Theoretical underpinnings of situation awareness: A critical review. In M. R. Endsley, editor, *Situation Awareness Analysis and Measurement*, pages 3–32. Lawrence Earlbaum Associates, Mahwah, N.J., 2000.

[448] M. R. Endsley. Measurement of situation awareness in dynamic systems. *Human Factors*, 37(1):65–84, 1995.

[449] L. J. Gugerty. Situation awareness during driving: Explicit and implicit knowledge in dynamic spatial memory. *Journal of Experimental Psychology: Applied*, 3(1):42–66, 1997.

[450] E. Fioratou, R. Flin, R. Glavin, and R. Patey. Beyond monitoring: Distributed situation awareness in anaesthesia. *British Journal of Anaesthesia*, 105(1):83–90, 2010.

[451] C. D. Wickens. Situation awareness and workload in aviation. *Current Directions in Psychological Science*, 11(4):128–133, 2002.

[452] F. T. Durso and A. R. Dattel. SPAM: The real-time assessment of SA. In S. Banbury and S. Tremblay, editors, *A Cognitive Approach to Situation Awareness: Theory and Application*, pages 137–154. Ashgate, Aldershot, UK, 2004.

[453] S. J. Selcn, R. M. Taylor, and E. Koritas. Workload or situation awareness? TLX vs SART aerospace systems design evaluation. *Proceedings of the Human and Ergonomics Society 35th Annual Meeting*, pages 62–66, 1991.

[454] C. D. Wickens. The trade-off of design for routine and unexpected performance: Implications of situation awareness. In M. R. Endsley and D. J. Garland, editors, *Situation awareness analysis and measurement*, pages 211–225. Lawrence Erlbaum Associates, Mahwah, N.J., 2000.

[455] S. M. Casner. Understanding the determinants of problem-solving behavior in a complex environment. *Human Factors*, 36(4):580–596, 1994.

[456] R. C. Teague and J. A. Allen. The reduction of uncertainty and troubleshooting performance. *Human Factors: The Journal of the Human Factors and Ergonomics Society*, 39(2):254–267, 1997.

[457] J. Rasmussen. Models of mental strategies in process plant diagnosis. In J. Rasmussen and W B Rouse, editors, *Human Detection and Diagnosis of System Failures*, pages 241–258. Springer, 1981.

[458] D. D. Woods and R. I. Cook. Perspectives on human error: Hindsight bias and local rationality. *Handbook of Applied Cognition*, pages 141–172, 1999.

[459] J. G. Wohl. Maintainability prediction revisited: diagnostic behavior, system complexity, and repair time. *IEEE Transaction on System, Man, and Cybernetics*, 12(3):241–250, 1982.

[460] P. M. Sanderson and J. M. Murtagh. Predicting fault diagnosis performance: Why are some bugs hard to find? *IEEE Transactions On Systems Man and Cybernetics*, 20(1):121–159, 1990.

[461] R. Flin, G. Slaven, and K. Stewart. Emergency decision making in the offshore oil and gas industry. *Human Factors*, 38:262–277, 1996.

[462] M. K. Tulga and T. B. Sheridan. Dynamic decisions and work load in multitask supervisory control. *IEEE Transactions on Systems, Man, and Cybernetics*, SMC-10(5):217–232, 1980.

[463] D. A. Wiegmann, J. Goh, and D. O'Hare. The role of situation assessment and flight experience in pilots' decisions to continue visual flight rules flight into adverse weather. *Hum Factors*, 44(2):189–197, 2002.

[464] E. K. Muthard and C. D. Wickens. Change detection after preliminary flight decisions: Linking planning errors to biases in plan monitoring. *Proceedings of the Human Factors and Ergonomics Society Annual Meeting*, 46(1):91–95, 2002.

[465] P. M. Moertl, J. M. Canning, S. D. Gronlund, M. R. P. Dougherty, J. Johansson, and S. H. Mills. Aiding planning in air traffic control: An experimental investigation of the effects of perceptual information integration. *Human Factors*, 44(3):404–412, 2002.

[466] C. D. Wickens, K. Gempler, and M. E. Morphew. Workload and reliability of predictor displays in aircraft traffic avoidance. *Transportation Human Factors*, 2(2):99–126, 2000.

[467] C. Layton, P. J. Smith, and C. E. McCoy. Design of a cooperative problem-solving system for enroute flight planning: An empirical evaluation. *Human Factors*, 36(1):94–119, 1994.

[468] T. B. Sheridan. *Humans and Automation*. John Wiley, New York, 2002.

[469] S. D. Gronlund, M. R. P. Dougherty, F. T. Durso, J. M. Canning, and S. H. Mills. Planning in air traffic control: Impact of problem type. *International Journal of Aviation Psychology*, 15(3):269–293, 2005.

[470] C. D. Wickens, N. Herdener, B. A Clegg, and C. A. P. Smith. Purchasing information to reduce uncertainty in trajectory prediction. *Proceedings of the Human Factors & Ergonomics Society Annual Meeting*, pages 323–327, 2016.

[471] W. Fu and W. D. Gray. Modeling cognitive versus perceptual-motor tradeoffs using ACT-R/PM. *In Proceedings of the Fourth International Conference on Cognitive Modeling*, pages 247–248, 2001.

[472] J. M. Hammer. Human factors of functionality and intelligent avionics. In J. WIse, V D Hopkin, and D J Garland, editors, *Handbook of Human Factors in Aviation*, pages 549–565. CRC Press, Boca Raton, FL, 1999.

[473] S. Danziger, J. Levav, and L. Avnaim-Pesso. Extraneous factors in judicial decisions. *Proceedings of the National Academy of Sciences of the United States of America*, 108(17):6889–92, 2011.

[474] J. A. Linder. Time of day and the decision to prescribe antibiotics. *JAMA Internal Medicine*, 174(12):2029–2031, 2014.

[475] J. T. Reason. *Human Error and Managing the Risks of Organizational Accidents*. Ashgate, Burlington, VT, 1997.

[476] N. Meshkati. Human factors in large-scale technological systems' accidents: Three Mile Island, Bhopal, Chernobyl. *Organization & Environment*, 5(2):133–154, 1991.

[477] C. D. Wickens and J. S. McCarley. *Applied Attention Theory*. CRC Press, Boca Raton, FL, 2008.

[478] D. C. Marshall, J. D. Lee, and P. A. Austria. Alerts for in-vehicle information systems: Annoyance, urgency, and appropriateness. *Human Factors*, 49(1):145–157, 2007.

[479] E. C. Haas and J. Edworthy. Designing urgency into auditory warnings using pitch, speed and loudness. *Computing & Control Engineering Journal*, 7(4):193–198, 1996.

[480] P. Kroft and C. D. Wickens. Paul Kroft and Christopher D . Wickens Displaying multi-domain graphical database information. *Information Design Journal*, 11:44–52, 2003.

[481] C. D. Wickens and C. M. Carswell. The proximity compatibility principle: Its psychological foundation and relevance to display design. *Human Factors*, 37(3):473–494, 1995.

[482] S. N. Roscoe. Airborne Displays for Flight and Navigation. *Human Factors: The Journal of the Human Factors and Ergonomics Society*, 10(4):321–332, 1968.

[483] R. D. Sorkin, B. H. Kantowitz, and S. C. Kantowitz. Likelihood alarm displays. *Human Factors*, 30:445–459, 1988.

[484] D. A. Norman. Categorization of action slips. *Psychological Review*, 88(1):1–15, 1981.

[485] M. G. Helander. Design of visual displays. *In G. Salvendy (Ed.)*, Handbook o:(pp, 1987.

[486] K. R. Boff and J. R. Lincoln. *Engineering Data Compendium Human Perception and Performance*. Harry G. Armstrong Aerospace Medical Research Laboratory, Wright Patterson Air Force Base, OH, 1988.

[487] A. H. S. Chan and W. H. Chan. Movement compatibility for circular display and rotary controls positioned at peculiar positions. *International Journal of Industrial Ergonomics*, 36(8):737–745, 2006.

[488] N. Herdener, C. D. Wickens, B. A. Clegg, and C. A. P. Smith. Overconfidence in Projecting Uncertain Spatial Trajectories. *Human Factors: The Journal of the Human Factors and Ergonomics Society*, 58(6):899–914, 2016.

[489] C. D. Wickens, M. A. Vincow, A. W. Schopper, and J. E. Lincoln. Computational models of human performance in the design and layout of controls and displays (CSERIAC SOAR Report 97-22). Technical report, Wright Patterson Air Force Base, OH: Crew System Ergonomics Information Analysis Center, Dayton, OH, 1997.

[490] J. M. Flach and K. B. Bennett. *Display and Interface Design: Subtle Science, Exact Art.* CRC Press, Boca Raton, FL, 2011.

[491] A. D. Andre and C. D. Wickens. A Computational Approach to Display Layout Analysis. Technical report, NASA Ames Research Center, Moffet Field, CA, 1991.

[492] D. J. Weintraub and M. J. Ensing. The book of HUD: A headup display state of the art report (CSERIAC state of the art report). Technical report, " Wright-Patterson Air Force Base, OH: Armstrong Aerospace Medical Research Lab, 1992.

[493] R. L Newman. Helmet-Mounted Display Symbology and Stabilization Concepts Helmet-Mounted Display Symbology and Stabilization Concepts. Technical report, NASA/USAATCOM, Moffet Field, CA, 1995.

[494] S. Fadden, P. M. Ververs, and C. D. Wickens. Pathway HUDs: are they viable? *Human Factors*, 43(2):173–193, 2001.

[495] B. L. Harrison, I. Hiroshi, K. J. Vicente, and W. A. S. Buxton. Transparent layered user interfaces: an evaluation of a display design to enhance focused and divided attention. *Proceeding CHI '95 Proceedings of the SIGCHI Conference on Human Factors in Computing Systems*, pages 317–324, 1995.

[496] C. D. Wickens and J. Long. Object versus space-based models visual of attention: Implications for design of head-up displays. *Journal of Experimental Psychology: Applied*, 1(3):179–193, 1995.

[497] R. J. Kiefer. Effect of Head-Up versus Head-Down digital speedometer on visual sampling behavior and speed control performance during daytime automobile driving. *SAE Technical paper 910111, Warrendale, PA: Society of Automotive Engineers*, 1991.

[498] R. J. Sojourner and J. F. Antin. The effects of a simulated head-up display speedometer on perceptual task performance. *Human Factors*, 32(3)):329–339, 1990.

[499] W. J. Horrey and C. D. Wickens. Driving and Side Task Performance: The Effects of Display Clutter, Separation, and Modality. *Human Factors: The Journal of the Human Factors and Ergonomics Society*, 46(4):611–624, 2004.

[500] R.. F. Haines, E. Fischer, and T. A. Price. Head-up transition behavior of pilots with and without head-up display in simulated low-visibility approaches. Technical Report 10, NASA, Moffet Field, CA, 1980.

[501] A. Woodham, M. Billinghurst, and W. S Helton. Climbing with a head-mounted display: Dual-task costs. *Human factors*, 58(3):452–61, 2016.

[502] M. Yeh, C. D. Wickens, and F. J. Seagull. Target cuing in visual search: The effects of conformality and display location on the allocation of visual attention. *Human Factors*, 41(4):524–542, 1999.

[503] M. Yeh, J. L Merlo, C. D. Wickens, and D. L Brandenburg. Head up versus head down: The costs of imprecision, unreliability, and visual clutter on cue effectiveness for display signaling. *Human Factors*, 45(3):390–407, 2003.

[504] R. Blake. A primer on binocular rivalry, including current controversies. *Brain and Mind*, 2:5–38, 2001.

[505] N. I. Durlach and A. S. Mavor, editors. *Virtual Reality: Scientific and technological challenges.* National Academies Press, Washington D.C., 1994.

[506] F. J. Seagull and D. Gopher. Training head movement in visual scanning: An embedded approach to the development of piloting skills with helmet-mounted displays. *Journal of Experimental Psychology: Applied*, 3(3):163–180, 1997.

[507] W. G. Cole. Cognitive strain, cognitive graphics, and medical cognitive science. In *Proceedings of the 1986 Congress of the American Association of Medical Systems and Informatics (AAMSI)*, pages 288–292, 1986.

[508] J. R. Pomerantz and E. A. Pristach. Emergent features, attention, and perceptual glue in visual form perception. *Journal of Experimental Psychology. Human perception and performance*, 15(4):635–649, 1989.

[509] J. Holt, K. B. Bennett, and J. M. Flach. Emergent features and perceptual objects: re-examining fundamental principles in analogical display design. *Ergonomics*, 58(12):1960–1973, 2015.

[510] D. D. Woods, J. A. Wise, and L. F. Hanes. An evaluation of nuclear power plan safety parameter display systems. *Proceedings of the Human Factors 25th Annual Meeting*, pages 110–114, 1981.

[511] K. B. Bennett and J. M. Flach. Graphical displays: Implications for divided attention, focused attention, and problem solving. *Human Factors*, 34(5):513–533, 1992.

[512] J. M. Flach, K. B. Bennett, R. J. Jagacinski, M. Mulder, M. M. van Paassen, and R. van Paassen. The closed-loop dynamics of cognitive work. In J. D. Lee and Alex Kirlik, editors, *The Oxford Handbook of Cognitive Engineering*, chapter 1. Oxford University Press, Oxford, England, 2013.

[513] K. J. Vicente. Ecological interface design: Progress and challenges. *Human Factors*, 44(1):62–78, 2002.

[514] K. J. Vicente and J. Rasmussen. Ecological interface design: Theoretical foundations. *IEEE Transactions on Systems, Man, and Cybernetics*, SCM-22(4):589–606, 1992.

[515] C. Borst, J. M. Flach, and J. Ellerbroek. Beyond ecological interface design: Lessons from concerns and misconceptions. *IEEE Transactions on Human-Machine Systems*, 45(2):164–175, 2015.

[516] C. M. Burns. Putting it all together: Improving display integration in ecological displays. *Human Factors*, 42(2):226–241, jan 2000.

[517] D. J. Garland and M. R. Endsley. Proceedings of an international conference on experimental analysis and measurement of situation awareness., 1995.

[518] L. A. Streeter, D. Vitello, and S. A. Wonsiewicz. How to tell people where to go: Comparing navigational aids, *International Journal of Man-Machine Studies*, 22:549–562, 1985.

[519] S. Silva, B. Sousa Santos, and J. Madeira. Using color in visualization: A survey. *Computers and Graphics*, 35(2):320–333, 2011.

[520] L. Reynolds. Colour for air traffic control displays. *Displays*, 15(4):215–225, 1994.

[521] N. Moacdieh and N. B. Sarter. Clutter in Electronic Medical Records: Examining Its Performance and Attentional Costs Using Eye Tracking. *Human factors*, 57(4):591–606, 2015.

[522] R. Rosenholtz, A. Dorai, and R. Freeman. Do predictions of visual perception aid design? *ACM Transactions on Applied Perception*, 8(2):1–20, 2011.

[523] S. H. Yoon, J. Lim, and Y. G. Ji. Assessment model for perceived visual complexity of automotive instrument cluster. *Applied Ergonomics*, 46:76–83, 2015.

[524] M. Mykityshyn, J. K. Kuchar, and R. J. Hansman. Experimental study of electronically based instrument approach plates. *The International Journal of Aviation Psychology*, 4(2):141–166, 1994.

[525] M. St John, H. S. Smallman, D. I. Manes, B. A. Feher, and J. G. Morrison. Heuristic automation for decluttering tactical displays. *Human factors*, 47(3):509–25, jan 2005.

[526] M. Levine. You-are-here maps psychological considerations. *Environment and Behavior*, 14(2):221–237, 1982.

[527] C. D. Wickens, M. A. Vincow, and M. Yeh. Design applications of visual spatial thinking. In A. Miyaki and P. Shah, editors, *Handbook of Visual Spatial Thinking*. Cambridge University Press, 2005.

[528] A. J. Aretz. The design of electronic map displays. *Human Factors*, 33(1)):85–101, 1991.

[529] M. Levine, I. Marchon, and G. Hanley. The placement and misplacement of you-are-here maps. *Environment and Behavior*, 16(2)):139–157, 1984.

[530] D. D. Woods. Visual momentum: A concept to improve the cognitive coupling of person and computer. *International Journal of Man-Machine Studies*, 21:229–244, 1984.

[531] C. D. Wickens, C. C. Liang, T. T. Prevett, and O. Olmos. Egocentric and exocentric displays for terminal area navigation. *International Journal of Aviation Psychology*, 6(3):241–271, 1996.

[532] C. D. Wickens, L. C. Thomas, and R. Young. Frames of reference for the display of battlefield information: Judgment-display dependencies. *Human Factors*, 42(4):660–675, 2000.

[533] M. St John, M. B. Cowen, H. S. Smallman, and H. M. Oonk. The use of 2D and 3D displays for shape-understanding versus relative-position tasks. *Human factors*, 43(1):79–98, 2001.

[534] C. D. Wickens, A. Aretz, and K. Harwood. Frame of reference for electronic maps: The relevance of spatial cognition, mental rotation, and componential task analysis. *International Symposium on Aviation Psychology*, pages 245–250, 1989.

[535] R. L. Sollenberger and P. Milgram. Effects of stereoscopic and rotational displays in a three-dimensional path-tracing task. *Human Factors*, 35(3):483–99, 1993.

[536] D. J. Gillan, C. D. Wickens, J. G. Hollands, and C. M. Carswell. Guidelines for presenting quantitative data in HFES publications. *Human Factors*, 40(1):28–41, 1998.

[537] E. R. Tufte. *Envisioning Information*. Graphics Press, Cheshire, CT, 1990.

[538] E. R. Tufte. *Visual Explanations: Images and Quantities, Evidence and Narrative*. Graphics Press, Cheshire, CT, 1997.

[539] E. R. Tufte. *The Visual Display of Quantitative Information*. Graphics Press, Cheshire, CT, 1983.

[540] S. M. Kosslyn. *Elements of Graph Design*. W. H. Freeman and Company, New York, 1994.

[541] T. Munzner. *Visualization Analysis and Design*. CRC Press, Boca Raton, FL, 2014.

[542] S. Few. *Now You See It: Simple Visualization Techniques for Quantitative Analysis*. Analytics Press, apr 2009.

[543] W. S. Cleveland and R. McGill. Graphical perception and graphical methods for analyzing scientific data. *Science*, 229(4716):828–833, 1985.

[544] G. L. Lohse. A cognitive model for understanding graphical perception. *Human-Computer Interaction,*, 8(4):353–388, 1993.

[545] FHWA. 2009 National Household Travel Survey User's Guide. Technical Report February, Federal Highway Administration, Washington D.C., 2011.

[546] C. North. Toward measuring visualization insight. *IEEE Computer Graphics and Applications*, 26(3):6–9, 2006.

[547] C. North. Information Visualization. In G. Salvendy, editor, *Hanbook of Human Factors and Ergonomics*, pages 1209–1236. Wiley & Sons, Hobocken, NJ, fourth edition, 2012.

[548] J. H. Ely, R. M. Thomson, and J. Orlansky. Design of Controls. Technical report, Wright Air Development Center Air Research adn Development Command, Springfield, OH, 1956.

[549] P. M. Fitts and C M Seeger. S-R compatibility: Spatial characteristics of response codes. *Journal of Experimental Psychology*, 46:199–210, 1953.

[550] S. N. H. Tsang, J. K. L. Ho, and A. H. S. Chan. Interface design and display-control compatibility. *Measurement & Control*, 48(3):81–86, 2015.

[551] C. D. Wickens and C. M. Carswell. Information processing. In G. Salvendy, editor, *Handbook of Human Factors and Ergonomics*, pages 117–161. John Wiley & Sons, Hobocken, NJ, fourth edition, 2012.

[552] A. H. S. Chan and R. R. Hoffmann. Movement compatibility for configurations of displays located in three cardinal orientations and ipsilateral, contralateral and overhead controls. *Applied Ergonomics*, 43(1):128–140, 2012.

[553] C. J. Worringham and D. B. Beringer. Directional stimulus-response compatibility: A test of three alternative principles. *Ergonomics*, 41(6):864–880, 1998.

[554] C. D. Wickens, J. W. Keller, and R. L. Small. Left. No, Right! Development of the frame of reference transformation tool (FORT). *Proceedings of the Human Factors and Ergonomics Society Annual Meeting*, 54(13):1022–1026, 2010.

[555] W. E. Hick. On the rate gain of information. *Quarterly Journal of Experimental Psychology*, 4:11–26, 1952.

[556] R. Hyman. Stimulus information as a determinant of reaction time. *Journal of Experimental Psychology*, 45(3):188–196, 1953.

[557] P. M. Fitts. The information capacity of the human motor system in controlling the amplitude of movement. *Journal of Experimental Biology*, 47(6):381–391, 1954.

[558] R. J. Jagacinski and J. M. Flach. *Control Theory for Humans: Quantitative Approaches to Modeling Performance*. Lawrence Erlbaum Associates Publishers, Mahwah, New Jersey, 2003.

[559] S. K. Card, W. K. English, and B. Burr. Evaluation of Mouse, rate-controlled isometric joystick, step keys, and text keys for text selection on a CRT. *Ergonomics*, 21(8):601–613, 1978.

[560] C. G. Drury. Application of Fitts' law to foot pedal design. *Human Factors*, 17(4):368–373, 1975.

[561] G. D. Langolf, D. B. Chaffin, and J. A. Foulke. An investigation of Fitts' law using a wide range of movement amplitudes. *Journal of Motor Behavior*, 8(2):113–128, 1976.

[562] T. Kaaresoja, S. Brewster, and V. Lantz. Towards the temporally perfect virtual button: Touch-feedback simultaneity and perceived quality in mobile touchscreen press interactions. *ACM Transactions on Applied Perception*, 11(2):1–25, 2014.

[563] C. D. Wickens, D L Sandry, and M. A. Vidulich. Compatibility and resource competition between modalities of input, central processing, and output: Testing the model of complex task performance. *Human Factors*, 25(2):227–248, 1983.

[564] C. Baber. *Beyond the Desktop: Designing and using interaction devices*. Academic Press, San Diego, 1997.

[565] G. C. Vanderheiden and J. B. Jordan. Design for people with functional limitations. In G. Salvendy, editor, *Handbook of Human Factors and Ergonomics*, pages 1407–1441. Wiley & Sons, New York, fourth edi edition, 2012.

[566] R. Seibel. Data entry through chord, parallel entry devices. *Human Factors*, 6(2):189–192, 1964.

[567] D. Gopher and D. Raij. Typing with a two-hand chord keyboard: Will the QWERTY become obsolete? *IEEE Transactions on Systems, Man and Cybernetics*, 18(4):601–609, 1988.

[568] R. Conrad and D. J. A. Longman. Standard typewriter versus chord keyboard: An experimental comparison. *Ergonomics*, 8(1):77–88, jan 1965.

[569] J. Noyes. Chord keyboards. *Applied Ergonomics*, 14(1):55–59, 1983.

[570] Srinath Sridhar, Anna Maria Feit, Christian Theobalt, and Antti Oulasvirta. Investigating the dexterity of multi-finger Input for mid-air text entry. *Proceedings of the ACM CHI'15 Conference on Human Factors in Computing Systems*, 1:3643–3652, 2015.

[571] J. Shutko and L. Tijerina. Ergonomics in Design: The Quarterly of Human Factors Applications. *Ford's approach to managing driver attention: SYNC and MyFord Touch*, 19(4):13–16, 2011.

[572] S. Oviatt. Multimodal Interfaces. In J. A. Jacko, editor, *The Human-Computer Interaction Handbook: Fundamentals, Evolving Technologies and Emerging Applications*. CRC Press, Boca Raton, FL, 2012.

[573] G. R. McMillan, R. G. Eggleston, and T. R. Anderson. Nonconventional controls. In G. Salvendy, editor, *Handbook of Human Factors*. Wiley & Sons, New York, 1997.

[574] C. L. Giddens, K. W. Barron, J. Byrd-Craven, K. F. Clark, and A. S. Winter. Vocal indices of stress: A review. *Journal of Voice*, 27(3), 2013.

[575] C. D. Wickens. Processing resources and attention. In R Parasuraman and R Davies, editors, *Varieties of Attention*, pages 63–102. Academic Press, New York, 1984.

[576] H. Mitchard and J. Winkles. Experimental comparisons of data entry by automated speech recognition, keyboard, and mouse. *Human Factors*, 44(2):198–209, 2002.

[577] C. D. Wickens. The effects of control dynamics on performance. In K. R. Boff, L. Kaufman, and J. Homas, editors, *Handbook of Perception and Human Performance, Vol. 2: Cognitive processes and performance*, pages 1–60. John Wiley & Sons, Oxford, England, 1986.

[578] T. B. Sheridan. Space teleoperation through time-delay: Review and prognosis. *IEEE Transactions on Robotics and Automation*, 9(5):592–606, 1993.

[579] D. T. McRuer and H. R. Jex. A review of quasi-linear pilot models. *IEEE Transactions on Human Factors in Electronics*, HFE-8(3):231–249, 1967.

[580] R. A. Hess. Feedback control models: Manual control and tracking. In G Salvendy, editor, *Handbook of Human Factors and Ergonomics*, pages 1250–1292. New York, John Wiley, 1997.

[581] R. Pausch. Virtual reality on five dollars a day. *Proceedings of the SIGCHI Conference on Human Factors in Computing Systems, ACM.*, pages 265–270, 1991.

[582] D. W. F van Krevelen and R. Poelman. A survey of augmented reality technologies, applications and limitations. *The International Journal of Virtual Reality*, 9(2):1–20, 2010.

[583] R. J. Jagacinski and R. A Miller. Describing the Human Operator's Internal Model of a Dynamic System. *Human Factors: The Journal of the Human Factors and Ergonomics Society*, 20(4):425–433, 1978.

[584] R. W. Proctor and K. L. Vu. Selection and control of action. In *Hanbook of Human Factors and Ergonomics*, pages 95–112. Wiley, New York, 2013.

[585] W. S. Marras. Basic biomechanics of workstation design. In G. Salvendy, editor, *Handbook of Human Factors and Ergonomics*, pages 347–381. Wiley, Hobocken, NJ, fourth edition, 2012.

[586] S. N. Roscoe, L. Corl, and R. S Jensen. Flight Display Dynamics Revisited. *Human Factors*, 23(3):341–353, 1981.

[587] J. Y. C. Chen, E. C. Haas, and M. J. Barnes. Human performance issues and user interface design for teleoperated robots. *Systems, Man and Cybernetics, Part C (Applications and Reviews)*, 37(6):1231–1245, 2007.

[588] T. B. Sheridan. Human-robot interaction: Status and challenges. *Human factors*, 58(4):525–32, 2016.

[589] R. Pausch, D. Proffitt, and G. Williams. Quantifying immersion in virtual reality. *Proceedings of the 24th annual conference on Computer graphics and interactive techniques - SIGGRAPH '97*, pages 13–18, 1997.

[590] K. M. Stanney and J. V. Cohn. Virtual environments. In J. A. Jacko, editor, *The Human-Computer Interaction Handbook: Fundamentals, Evolving Technologies and Emerging Applications*. CRC Press, Boca Raton, 2012.

[591] H. G. Stassen and G. J. F. Smets. Telemanipulation and telepresence. *Control Engineering Practice*, 5(3):363–374, 1997.

[592] R.E. Ellis, O.M. Ismaeil, and M.G. Lipsett. Design and evaluation of a high-performance haptic interface. *Robotica*, 14(03):321, 1996.

[593] J. F. T. Bos, H. G. Stassen, and A. van Lunteren. Aiding the operator in the manual control of a space manipulator. *Control Engineering Practice*, 3(2):223–230, 1995.

[594] T. B. Sheridan. Human supervisory control of robot systems. In *Proceedings of the IEEE International Conference on Robotics and Automation*, pages 808–812, 1986.

[595] J. A. Jacko, editor. *The Human-Computer Interaction Handbook*. CRC Press, Boca Raton, FL, 2012.

[596] D. A. Norman. *Things That Make Us Smart*. Addison-Wesley, Reading, MA, 1993.

[597] Jonathan Grudin. Introduction: A moving target - the evolution of human-computer interaction. *Human-computer interaction handbook: Fundamentals, evolving technologies and emerging applications*, pages xxvii – lxi, 2012.

[598] Q. Gong and G. Salvendy. Design of skill-based adaptive interface: The effect of a gentle push. *Proceedings of the Human Factors and Ergonomics Society Annual Meeting*, 38(4):295–299, 1994.

[599] F. Lotte, M. Congedo, and L. Anatole. A review of classification algorithms for EEG-based brain âĂŞ computer interfaces : A review of classification algorithms for EEG-based brain-computer interfaces. *Journal of Neural Engineering*, 4(2), 2007.

[600] C-M. Karat, J. Lai, S. Osamuyimen, and N. Yankelovich. Speech and language interfaces, applications, and technologies. In J. A. Jacko, editor, *The Human-Computer Interaction Handbook: Fundamentals, Evolving Technologies and Emerging Applications*, pages 367–386. CRC Press, Boca Raton, FL, 2012.

[601] D. L. Fisher, E. J. Yungkurth, and S. M. Moss. Optimal menu hierarchy design: Syntax and semantics. *Human Factors: The Journal of the Human Factors and Ergonomics Society*, 32(6):665–683, 1990.

[602] E. L. Hutchins, J D Hollan, and D. A. Norman. Direct manipulation interfaces. *User centered system design*, 1:311–338, 1985.

[603] W. C. Westerman and J. G. Elias. Multi-touch system and method for emulating modifier keys via fingertip, 2003.

[604] A. Markussen, M. R. Jakobsen, and K. Hornbæk. Vulture: A mid-air word-gesture keyboard. *Proceedings of the 32nd annual ACM conference on Human factors in computing systems - CHI '14*, pages 1073–1082, 2014.

[605] B. Dumas, D. Lalanne, and S. Oviatt. Multimodal interfaces: A survey of principles, models and frameworks. In D. Hutchison and J. C Mitchell, editors, *Human Machine Interaction*, pages 3–26. Springer, Berlin, 2009.

[606] W. Yee. Potential limitations of multi-touch gesture vocabulary: Differentiation, adoption, fatigue. In *International Conference on Human-Computer Interaction*, pages 291–300. Springer Berlin Heidelberg, 2009.

[607] D. A. Bowman, E. Kruijff, J. J. LaViola, and I. Poupyrev. An Introduction to 3-D User Interface Design. *Presence: Teleoperators and Virtual Environments*, 10(1):96–108, 2001.

[608] C. R Harris and H. E. Pashler. Attention and the processing of emotional words and names: not so special after all. *Psychological Science*, 15(3):171–8, mar 2004.

[609] M. McTear, Z. Callejas, and D. Griol. *The Conversational Interface: Talking to Smart Devices.* Springer, 2016.

[610] K. S. Nagel and G. D. Abowd. Designing for Intimacy: Bridging the Interaction Challenges of Conversation. *Intimate Ubiquitous Computing Workshop (UbiComp)*, pages 34–37, 2003.

[611] D. A. Norman. Cognitive engineering. *User centered system design*, pages 31–61, 1986.

[612] D. R. Mayhew. *Principles and guidelines in software user interface design.* Prentice Hall, Englewood Cliffs, New Jersey, 1992.

[613] L. A. Wozny. The application of metaphor, analogy, and conceptual models in computer systems. *Interacting with Computers*, 1(3):273–283, 1989.

[614] R. W. Picard. *Affective Computing.* MIT Press, Cambridge, Mass., 1997.

[615] B. Reeves and C. Nass. *The Media Equation: How people treat computers, television, and new media like real people and places.* Cambridge University Press, New York, 1996.

[616] C. Nass, Y. Moon, B. J. Fogg, B. Reeves, and D. C. Dryer. Can computer personalities be human personalities? *International Journal of Human-Computer Studies*, 43:223–239, 1995.

[617] Y. Liu. Engineering aesthetics and aesthetic ergonomics: Theoretical foundations and a dual-process research methodology. *Ergonomics*, 46(13-14):1273–92, 2003.

[618] N. Tractinsky, A. S. Katz, and D. Ikar. What is beautiful is usable. *Interacting with Computers*, 13(2):127–145, 2000.

[619] D. A. Norman, A. Ortony, and D. M. Russell. Affect and machine design: Lessons for the development of autonomous machines. *IBM Systems Journal*, 42(1):38–44, 2003.

[620] S. Tseng and B. J. Fogg. Credibility and computing technology - Users want to trust, and generally do. But that trust is undermined, often forever, when the system delivers erroneous information. *Communications of the Acm*, 42(5):39–44, 1999.

[621] J. Kim and J. Y. Moon. Designing towards emotional usability in customer interfaces - trustworthiness of cyber-banking system interfaces. *Interacting with Computers*, 10(1):1–29, 1998.

[622] K. Karvonen and J. Parkkinen. Signs of trust: A semiotic study of trust formation in the web. In Michael J Smith, Gavriel Salvendy, Don Harris, and Richard J Koubek, editors, *1st International Conference on Universal Access in Human-Computer Interaction*, volume 1-Usabilit, pages 1076–1080, New Orleans, LA, 2001. Lawrence Erlbaum Associates.

[623] B. P. Bailey and J A Konstan. On the need for attention-aware systems: Measuring effects of interruption on task performance, error rate, and affective state. *Computers in Human Behavior*, 22(4):685–708, 2006.

[624] S. Y. W. Li, F. Magrabi, and E. Coiera. A systematic review of the psychological literature on interruption and its patient safety implications. *Journal of the American Medical Informatics Association : JAMIA*, 19(1):6–12, 2011.

[625] W. D. Gray and W. T. Fu. Soft constraints in interactive behavior: The case of ignoring perfect knowledge in-the-world for imperfect knowledge in-the-head. *Cognitive Science*, 28(3):359–382, 2004.

[626] D. A. Norman and B. Tognazzini. How apple Is giving design a bad name, 2015.

[627] G. Nielson. Visualization in scientific computing. *Computer*, 22(8):10–11, 1989.

[628] B. J. Fogg. *Persuasive Technology: Using computers to change what we do and think.* Morgan Kaaufmann, New York, 2003.

[629] B. Shneiderman. The eyes have it: a task by data type taxonomy for informatio nvisualizations. *Proceedings 1996 IEEE Symposium on Visual Languages*, pages 336–343, 1996.

[630] R. A. Becker, W. S. Cleveland, and A. R. Wilks. Dynamic graphics for data analysis. *Statistical Science*, 2:355–395, 1987.

[631] SR Klemmer, B Hartmann, and L Takayama. How bodies matter: five themes for interaction design. *Proceedings of the 6th …*, 2006.

[632] H. Ishii and B. Ullmer. Tangible bits: Towards seamless interfaces between people, bits and atoms. *Proceedings of the ACM SIGCHI Conference on Human factors in computing systems (ACM)*, pages 234–241, 1997.

[633] M. Koelle. Tangible user interfaces. In J. A. Jacko, editor, *The Human-Computer Interaction Handbook: Fundamentals, Evolving Technologies and Emerging Applications*, pages 2–4. CRC Press, Boca Raton, FL, 2012.

[634] C. Kang. Disney's $1 Billion Bet on a Magical Wristband | WIRED, 2015.

[635] N.G. Leveson. *Safeware: System Safety and Computers*. Addison-Wesley, New York, NY, 1995.

[636] D Hughes and M A Dornheim. Accidents direct focus on cockpit automation. *Aviation Week & Space Technology*, January 30:52–54, 1995.

[637] N. B. Sarter and D. D. Woods. Team play with a powerful and independent agent: Operational experiences and automation surprises on the Airbus A-320. *Human Factors*, 39(3):390–402, 1997.

[638] A. Degani. *Taming HAL: Designing interfaces beyond 2001*. Springer, New York, 2004.

[639] J. Prinet, Y. Wan, and N. Sarter. Tactile spatial guidance for collision avoidance in NextGen flight operations. *Proceedings of the 60th Annual Meeting of the Human Factors and Ergonomics Society (HFES)*, pages 303–307, 2016.

[640] M. Lewis. *Flash Boys: A Wall Street revolt*. Norton, New York, 2014.

[641] A. Sebok and C. D. Wickens. Implementing lumberjacks and black swans into model-based tools to support human-automation interaction. *Human Factors*, pages 1–15, 2016.

[642] R. Parasuraman. Human-computer monitoring. *Human Factors*, 29(6):695–706, 1987.

[643] L. Bainbridge. Ironies of automation. *Automatica*, 19(6):775–779, nov 1983.

[644] O'Neil. *Weapons of Math Destruction: How big data increases inequality and threatens democracy*. Crown, New York, 2016.

[645] R. Parasuraman, T. B. Sheridan, and C. D. Wickens. A model for types and levels of human interaction with automation. *IEEE Transactions on Systems Man and Cybernetics Part A Systems and Humans*, 30(3):286–297, 2000.

[646] R. Parasuraman and C. D. Wickens. Humans: still vital after all these years of automation. *Human Factors*, 50(3):511–520, 2008.

[647] L. Onnasch, C. D. Wickens, H. Li, and D. Manzey. Human performance consequences of stages and levels of automation: An integrated meta-analysis. *Human Factors: The Journal of the Human Factors and Ergonomics Society*, 56(3):476–488, aug 2013.

[648] M. T. Dzindolet, L. G. Pierce, H. P. Beck, and L. A. Dawe. The perceived utility of human and automated aids in a visual detection task. *Human Factors*, 44(1):79–94, jan 2002.

[649] A. J. Reiner, J. G. Hollands, and G. A. Jamieson. Target detection and identification performance using an automatic target detection system. *Human Factors: The Journal of the Human Factors and Ergonomics Society*, 2016.

[650] K. L. Mosier, L J Skitka, S Heers, and M Burdick. Automation bias: Decision making and performance in high-tech cockpits. *International Journal of Aviation Psychology*, 8(1):47–63, 1998.

[651] J. Y. C. Chen, M. J. Barnes, and M. Harper-Sciarini. Supervisory control of multiple robots: Human-performance issues and user-interface design. *IEEE Transactions on Systems, Man, and Cybernetics, Part C: Applications and Reviews*, 41(4):435–454, jul 2011.

[652] DSB. The Role of Autonomy in DoD Systems. Technical report, Department of Defense, Defence Science Board, 2012.

[653] C. E. Billings. *Aviation Automation: The Search for a Human-Centered Approach*. Erlbaum, Mahwah, NJ, 1997.

[654] A. E. Diehl. Human performance and systems safety considerations in aviation mishaps. *The International Journal of Aviation Psychology*, 1(2):97–106, 1991.

[655] L Lin, K. J. Vicente, and D J Doyle. Patient safety, potential adverse drug events, and medical device design: A human factors engineering approach. *Journal of Biomedical Informatics*, 34(4):274–284, aug 2001.

[656] N. B. Sarter and D. D. Woods. Team play with a powerful and independent agent: A full-mission simulation study. *Human Factors*, 42(3):390–402, 2000.

[657] N. B. Sarter, D. D. Woods, and C. E. Billings. Automation surprises. In G Salvendy, editor, *Handbook of Human Factors and Ergonomics*, pages 1926–1943. Wiley, New York, 2nd edition, 1997.

[658] M A Dornheim. Dramatic incidents highlight mode problems in cockpits. *Aviation Week & Space Technology*, January 30:57–59, 1995.

[659] C. D. Wickens and S R Dixon. The benefits of imperfect diagnostic automation: A synthesis of the literature. *Theoretical Issues in Ergonomics Science*, 8(3):201–212, 2007.

[660] B M Muir. Trust between humans and machines, and the design of decision aids. *International Journal of Man-Machine Studies*, 27:527–539, 1987.

[661] J Lee and N. Moray. Trust, control strategies and allocation of function in human-machine systems. *Ergonomics*, 35(10):1243–1270, 1992.

[662] Y. Seong and A. M. Bisantz. The impact of cognitive feedback on judgment performance and trust with decision aids. *International Journal of Industrial Ergonomics*, 38(7-8):608–625, jul 2008.

[663] B. H. Kantowitz, R. J. Hanowski, and S. C. Kantowitz. Driver acceptance of unreliable traffic information in familiar and unfamiliar settings. *Human Factors*, 39(2):164–176, 1997.

[664] S. Lewandowsky, M. Mundy, and G. P. A. Tan. The dynamics of trust: Comparing humans to automation. *Journal of Experimental Psychology: Applied*, 6(2):104–123, 2000.

[665] Yili L Liu, Robert Fuld, and C. D. Wickens. Monitoring behavior in manual and automated scheduling systems. *International Journal of Man-Machine Studies*, 39(6):1015–1029, 1993.

[666] S Rice. Examining single-and multiple-process theories of trust in automation. *The Journal of General Psychology*, 136(3):303–319, 2009.

[667] R. D. Sorkin. Why are people turning off our alarms? *Journal of the Acoustical Society of America*, 84(3):1107–1108, 1988.

[668] S. R. Dixon, C. D. Wickens, and J. S. McCarley. On the independence of compliance and reliance: are automation false alarms worse than misses? *Human Factors*, 49(4):564–72, 2007.

[669] R. Parasuraman, R. Molloy, and I. L. Singh. Performance consequences of automation-induced "complacency". *International Journal of Aviation Psychology*, 3(1):1–23, 1993.

[670] N. Moray. Monitoring, complacency, scepticism and eutactic behaviour. *International Journal of Industrial Ergonomics*, 31(3):175–178, 2003.

[671] P. M. Sanderson, Marcus O Watson, and W John Russell. Advanced patient monitoring displays: tools for continuous informing. *Anesthesia & Analgesia*, 101(1):161–168, jul 2005.

[672] R. Parasuraman, I. L. Singh, and R Molloy. Automation-related complacency: A source of vulnerability in contemporary organizations. *IFIP Transactions A-Computer Science and Technology*, 13:426–432, 1992.

[673] R. Parasuraman, M. Mouloua, and R. Molloy. Monitoring automation failures in human-machine systems. In M Mouloua and R Parasuraman, editors, *Human Performance in Automated Systems: Current Research and Trends*, pages 45–49. Lawrence Erlbaum Associates, Hillsdale, NJ, 1994.

[674] R. Parasuraman, M Mouloua, and R Molloy. Effects of adaptive task allocation on monitoring of automated systems. *Human Factors*, 38(4):665–679, 1996.

[675] N. J. Slamecka and P. Graf. The generation effect: Delineation of a phenomenon. *Journal of experimental psychology. Human learning and memory*, 4(6):592–604, 1978.

[676] M. R. Endsley and E. O. Kiris. The out-of-the-loop performance problem and level of control in automation. *Human Factors*, 37(2):381–394, jun 1995.

[677] M Wilson. Six views of embodied cognition. *Psychonomic Bulletin & Review*, 9(4):625–636, 2002.

[678] S. M. Casner, R. W. Geven, M. P. Recker, and J. W. Schooler. The Retention of Manual Flying Skills in the Automated Cockpit. *Human Factors: The Journal of the Human Factors and Ergonomics Society*, 56(8):1506–1516, 2014.

[679] E. L. Wiener and D C Nagel. *Human factors in aviation*. Academic Press, New York, 1988.

[680] J. D. Lee and N. Moray. Trust, self-confidence, and operators' adaptation to automation. *International Journal of Human-Computer Studies*, 40(1):153–184, 1994.

[681] C. Gold, M. Korber, D. Lechner, and K. Bengler. Taking over control from highly automated vehicles in complex traffic situations: The role of traffic density. *Human Factors: The Journal of the Human Factors and Ergonomics Society*, 58(4):642–652, 2016.

[682] P. A. Hancock, R. Parasuraman, and E A Byrne. Driver-centered issues in advanced automation for motor vehicles. In R Parasuraman and M Mouloua, editors, *Automation and Human Performance: Theory and Applications*, pages 337–364. Lawrence Erlbaum Associates, Mahwah, NJ, 1996.

[683] G H Walker, N A Stanton, and M. S. Young. Where is computing driving cars? *International Journal of Human-Computer Interaction*, 13(2):203–229, 2001.

[684] E. S. Patterson, E. M. Roth, D. D. Woods, R. Chow, and J. O. Gomes. Handoff strategies in settings with high consequences for failure: Lessons for health care operations. *International Journal for Quality in Health Care*, 16(2):125–32, apr 2004.

[685] A. W. Stedmon, S. Sharples, R. Littlewood, G. Cox, H. Patel, and J. R. Wilson. Datalink in air traffic management: Human factors issues in communications. *Applied Ergonomics*, 38(4):473–480, 2007.

[686] C. Navarro and S. Sikorski. Datalink communication in flight deck operations: A synthesis of recent studies. *The International Journal of Aviation Psychology*, 9(4):361–376, 1999.

[687] K Kerns. Data-link communication between controllers and pilots: A review and synthesis of the simulation literature. *The International Journal of Aviation Psychology*, 1(3):181–204, 1991.

[688] S. Zuboff. *In the Age of Smart Machines: The future of work, technology and power*. Basic Books, New York, 1988.

[689] C Perrow. *Normal accidents*. Basic Books, New York, 1984.

[690] J. D. Lee and T. F. Sanquist. Augmenting the operator function model with cognitive operations: Assessing the cognitive demands of technological innovation in ship navigation. *IEEE Transactions on Systems Man and Cybernetics -Part A: Systems and Humans*, 30(3):273–285, 2000.

[691] P. M. Fitts. Human engineering for an effective air-navigation and traffic-control system. Technical report, National Research Council, Division of Anthropology and Psychology, Washington D.C., 1951.

[692] R B Fuld. The fiction of function allocation, revisited. *International Journal of Human-Computer Studies*, 52(2):217–233, 2000.

[693] H E Price. The allocation of functions in systems. *Human Factors*, 27(1):33–45, 1985.

[694] J C F de Winter and D Dodou. Why the Fitts list has persisted throughout the history of function allocation. *Cognition, Technology {&} Work*, 16(1):1–11, 2014.

[695] A. R. Pritchett, S. Y. Kim, and K. M. Feigh. Modeling Human-Automation Function Allocation. *Journal of Cognitive Engineering and Decision Making*, 8(1):33–51, 2014.

[696] S. Verma, T. Kozon, D. Ballinger, and A. Farrahi. Functional allocation of roles between humans and automation for a pairing tool used for simultaneous approaches. *The International Journal of Aviation Psychology*, 23(4):335–367, 2013.

[697] N. B. Sarter and D. D. Woods. Decomposing automation: Autonomy, authority, observability and perceived animacy. In M Mouloua and R Parasuraman, editors, *Human Performance in Automated Systems: Current Research and Trends*, pages 22–27. Lawrence Erlbaum Associates, Hillsdale, NJ, 1994.

[698] D. D. Woods. Decomposing automation: Apparent simplicity, real complexity. *Automation and Human Performance: Theory and Applications*, pages 3–17, 1996.

[699] Asaf Degani, M. Shafto, and A. Kirlik. Modes in automated cockpits: Problems, data analysis, and a modeling framework. In *Proceedings of the 36th Israel Annual Conference on Aerospace Science*, Haifa, Israel, 1995.

[700] V. A. Riley. A new language for pilot interfaces. *Ergonomics in Design*, 9(2):21–27, 2001.

[701] R. Parasuraman and CA A Miller. Trust and etiquette in high-criticallity automated systems. *Communications of the ACM*, 47(4):51–55, 2004.

[702] E. L. Wiener and R E Curry. Flight deck automamtion: Promises and problems. *Ergonomics*, 23(10):995–1011, 1980.

[703] G. A. Klein, D. D. Woods, J. M. Bradshaw, R. R Hoffman, P. Feltovich, R. R. Hoffman, P. J Hayes, and K. M. Ford. Ten challenges for making automation a "Team Player" in joint human-agent activity. *IEEE Intelligent Systems*, 19(6):91–95, 2004.

[704] B. D. Seppelt and J. D. Lee. Making adaptive cruise control (ACC) limits visible. *International Journal of Human-Computer Studies*, 65(3):192–205, 2007.

[705] J. E. Mercado, M. A. Rupp, Y. J. C. Chen, M. J. Barnes, D. Barber, and K. Procci. Intelligent agent transparency in human-agent teaming for multi-UxV management. *Human factors*, 58(3):401–15, 2016.

[706] A. Sebok, C. D. Wickens, B. Clegg, and R. Sargent. Using Empirical Research and Computational Modeling to Predict Operator Response to Unexpected Events. *Proceedings of the Human Factors and Ergonomics Society Annual Meeting*, 58(1):834–838, 2014.

[707] R. Parasuraman, T. B. Sheridan, and C. D. Wickens. A model for types and levels of human interaction with automation. *IEEE Transactions on Systems Man and Cybernetics - Part A: Systems and Humans*, 30(3):286–297, 2000.

[708] N. B. Sarter and D. D. Woods. How in the world did we ever get into that mode? Mode error and awareness in supervisory control. *Human Factors*, 37(1):5–19, 1995.

[709] W. B. Rouse. Adaptive aiding for human/computer control. *Human Factors*, 30:431–443, 1988.

[710] M W Scerbo. Theoretical perspectives on adaptive automation. In R Parasuraman and M Mouloua, editors, *Automation and Human Performance: Theory and Applications*, pages 37–63. Lawrence Erlbaum Associates, Mahwah, NJ, 1996.

[711] M. Vagia, A. A. Transeth, and S. A. Fjerdingen. A literature review on the levels of automation during the years: What are the different taxonomies that have been proposed? *Applied Ergonomics*, 53:190–202, 2016.

[712] D. B. Kaber. Adaptive automation. In J. D. Lee and A. Kirlik, editors, *The Oxford Handbook of Cognitive Engineering*, pages 594–609. Oxford University Press, New York, 2013.

[713] J. Sauer, A. Chavaillaz, and D Wastell. On the effectiveness of performance-based adaptive automation. *Theoretical Issues in Ergonomics Science*, pages 1–19, 2016.

[714] L. J. Prinzel, F. C. Freeman, M. W. Scerbo, P. J. Mikulka, and A. T. Pope. A closed-loop system for examining psychophysiological measures for adaptive task allocation. *International Journal of Aviation Psychology*, 10(4):393–410, 2000.

[715] L. Yekhshatyan and J. D. Lee. Changes in the correlation between eye and steering movements indicate driver distraction. *IEEE Transactions on Intelligent Transportation Systems*, 14(1):136–145, 2013.

[716] S. Y. Hu and G. T. Zheng. Driver drowsiness detection with eyelid related parameters by Support Vector Machine. *Expert Systems with Applications*, 36(4):7651–7658, 2009.

[717] L. Sherry and P. Polson. Shared models of flight management system vertical guidance. *The International Journal of Aviation Psychology*, 9(2):139–153, 1999.

[718] A. McClumpha and M. James. Understanding automated aircraft. In M. Mouloua and R. Parasuraman, editors, *Human Performance in Automated Systems: Recent research and trends*, pages 314–319. Erlbaum, Hillsdale; NJ, 1994.

[719] M. Bekier, B. R. C. Molesworth, and A. Williamson. Tipping point: The narrow path between automation acceptance and rejection in air traffic management. *Safety Science*, 50(2):259–265, 2012.

[720] M. Kurth, C. Schleyer, and D. Feuser. Smart factory and education: An integrated automation concept. *EEE 11th Conference on Industrial Electronics and Applications (ICIEA)*, pages 1057–1061, 2016.

[721] M. Bekier and B. R. C. Molesworth. Altering user' acceptance of automation through prior automation exposure. *Ergonomcis*, pages 1–9, 2016.

[722] W. Karwowski, H J Warnecke, M Hueser, and G. Salvendy. Human factors in manufacturing. In G Salvendy, editor, *The Handbook of Human Factors and Ergonomics*. Wiley, New York, 2nd edition, 1997.

[723] P. M. Sanderson. The human planning and scheduling role in advanced manufacturing systems: An emerging human factors domain. *Human Factors*, 31(6):635–666, 1989.

[724] N. Moray. Human factors in process control. In G Salvendy, editor, *The Handbook of Human Factors and Ergonomics*. Wiley, New York, 2nd edition, 1997.

[725] P. P. Read. *Ablaze: The story of the heroes and victims of Chernobyl*. Random House, New York, 1993.

[726] C. M. Burns. Navigation strategies with ecological displays. *International Journal of Human-Computer Studies*, 52(1):111–129, 2000.

[727] C. D. Wickens. Automation in air traffic control: The human performance issues. In *Third Human Factors in Automation Conference*, Norfolk , VA, 1998. Dominion University.

[728] A. Gunasekaran. Agile manufacturing: A framework for research and development. *International Journal of Production Economics*, 62(1-2):87–105, 1999.

[729] D. D. Woods, J Tittle, M Feil, and A Roesler. Envisioning human-robot coordination in future operations. *IEEE Transactions On Systems, Man, and Cybernetics - Part C: Applications and Reviews*, 34(2):210–218, 2004.

Index

60221766R00243

Made in the USA
Lexington, KY
31 January 2017